IET MATERIALS, CIRCUITS AND DEVICES SERIES 49

Advanced Technologies for Next Generation Integrated Circuits

Other volumes in this series

Volume 2	**Analogue IC Design: The current-mode approach** C. Toumazou, F.J. Lidgey and D.G. Haigh (Editors)
Volume 3	**Analogue–Digital ASICs: Circuit techniques, design tools and applications** R.S. Soin, F. Maloberti and J. France (Editors)
Volume 4	**Algorithmic and Knowledge-Based CAD for VLSI** G.E. Taylor and G. Russell (Editors)
Volume 5	**Switched Currents: An analogue technique for digital technology** C. Toumazou, J.B.C. Hughes and N.C. Battersby (Editors)
Volume 6	**High-Frequency Circuit Engineering** F. Nibler *et al.*
Volume 8	**Low-Power High-Frequency Microelectronics: A unified approach** G. Machado (Editor)
Volume 9	**VLSI Testing: Digital and mixed analogue/digital techniques** S.L. Hurst
Volume 10	**Distributed Feedback Semiconductor Lasers** J.E. Carroll, J.E.A. Whiteaway and R.G.S. Plumb
Volume 11	**Selected Topics in Advanced Solid State and Fibre Optic Sensors** S.M. Vaezi-Nejad (Editor)
Volume 12	**Strained Silicon Heterostructures: Materials and devices** C.K. Maiti, N.B. Chakrabarti and S.K. Ray
Volume 13	**RFIC and MMIC Design and Technology** I.D. Robertson and S. Lucyzyn (Editors)
Volume 14	**Design of High Frequency Integrated Analogue Filters** Y. Sun (Editor)
Volume 15	**Foundations of Digital Signal Processing: Theory, algorithms and hardware design** P. Gaydecki
Volume 16	**Wireless Communications Circuits and Systems** Y. Sun (Editor)
Volume 17	**The Switching Function: Analysis of power electronic circuits** C. Marouchos
Volume 18	**System on Chip: Next generation electronics** B. Al-Hashimi (Editor)
Volume 19	**Test and Diagnosis of Analogue, Mixed-Signal and RF Integrated Circuits: The system on chip approach** Y. Sun (Editor)
Volume 20	**Low Power and Low Voltage Circuit Design with the FGMOS Transistor** E. Rodriguez-Villegas
Volume 21	**Technology Computer Aided Design for Si, SiGe and GaAs Integrated Circuits** C.K. Maiti and G.A. Armstrong
Volume 22	**Nanotechnologies** M. Wautelet *et al.*
Volume 23	**Understandable Electric Circuits** M. Wang
Volume 24	**Fundamentals of Electromagnetic Levitation: Engineering sustainability through efficiency** A.J. Sangster
Volume 25	**Optical MEMS for Chemical Analysis and Biomedicine** H. Jiang (Editor)
Volume 26	**High Speed Data Converters** A.M.A. Ali
Volume 27	**Nano-Scaled Semiconductor Devices** E.A. Gutiérrez-D (Editor)
Volume 28	**Security and Privacy for Big Data, Cloud Computing and Applications** L. Wang, W. Ren, K.R. Choo and F. Xhafa (Editors)
Volume 29	**Nano-CMOS and Post-CMOS Electronics: Devices and modelling** Saraju P. Mohanty and Ashok Srivastava
Volume 30	**Nano-CMOS and Post-CMOS Electronics: Circuits and design** Saraju P. Mohanty and Ashok Srivastava
Volume 32	**Oscillator Circuits: Frontiers in design, analysis and applications** Y. Nishio (Editor)
Volume 33	**High Frequency MOSFET Gate Drivers** Z. Zhang and Y. Liu
Volume 34	**RF and Microwave Module Level Design and Integration** M. Almalkawi
Volume 35	**Design of Terahertz CMOS Integrated Circuits for High-Speed Wireless Communication** M. Fujishima and S. Amakawa
Volume 38	**System Design with Memristor Technologies** L. Guckert and E.E. Swartzlander Jr.
Volume 39	**Functionality-Enhanced Devices: An alternative to Moore's law** P.-E. Gaillardon (Editor)
Volume 40	**Digitally Enhanced Mixed Signal Systems** C. Jabbour, P. Desgreys and D. Dallett (Editors)

Volume 43 **Negative Group Delay Devices: From concepts to applications** B. Ravelo (Editor)
Volume 45 **Characterisation and Control of Defects in Semiconductors** F. Tuomisto (Editor)
Volume 47 **Understandable Electric Circuits: Key concepts, 2nd Edition** M. Wang
Volume 51 **Modelling Methodologies in Analogue Integrated Circuit Design** G. Dundar and M.B. Yelten (Editors)
Volume 53 **VLSI Architectures for Future Video Coding** M. Martina (Editor)
Volume 54 **Advances in High-Power Fiber and Diode Laser Engineering** Ivan Divliansky (Editor)
Volume 55 **Hardware Architectures for Deep Learning** M. Daneshtalab and M. Modarressi
Volume 58 **Magnetorheological Materials and Their Applications** S. Choi and W. Li (Editors)
Volume 60 **IP Core Protection and Hardware-Assisted Security for Consumer Electronics** A. Sengupta and Saraju P. Mohanty
Volume 64 **Phase-Locked Frequency generation and Clocking: Architectures and circuits for modem wireless and wireline systems** W. Rhee (Editor)
Volume 67 **Frontiers in Securing IP Cores: Forensic detective control and obfuscation techniques** A Sengupta
Volume 68 **High Quality Liquid Crystal Displays and Smart Devices: Vol. 1 and Vol. 2** S.Ishihara, S. Kobayashi and Y. Ukai (Editors)
Volume 69 **Fibre Bragg Gratings in Harsh and Space Environments: Principles and applications** B. Aïssa, E.I. Haddad, R.V. Kruzelecky, W.R. Jamroz
Volume 70 **Self-Healing Materials: From fundamental concepts to advanced space and electronics applications, 2nd Edition** B. Aïssa, E.I. Haddad, R.V. Kruzelecky, W.R. Jamroz
Volume 71 **Radio Frequency and Microwave Power Amplifiers: Vol. 1 and Vol. 2** A. Grebennikov (Editor)
Volume 73 **VLSI and Post-CMOS Electronics Volume 1: VLSI and Post-CMOS Electronics and Volume 2: Materials, devices and interconnects** R. Dhiman and R. Chandel (Editors)

Advanced Technologies for Next Generation Integrated Circuits

Edited by
Ashok Srivastava and Saraju P. Mohanty

The Institution of Engineering and Technology

Published by The Institution of Engineering and Technology, London, United Kingdom

The Institution of Engineering and Technology is registered as a Charity in England & Wales (no. 211014) and Scotland (no. SC038698).

First published 2020

The Institution of Engineering and Technology
Michael Faraday House
Six Hills Way, Stevenage
Herts, SG1 2AY, United Kingdom

www.theiet.org

British Library Cataloguing in Publication Data
A catalogue record for this product is available from the British Library

ISBN 978-1-78561-664-8 (hardback)
ISBN 978-1-78561-665-5 (PDF)

Typeset in India by MPS Limited

Contents

Chapter 1

Graphene and other than graphene materials technology and beyond

Ashok Srivastava[1]

1.1 Introduction—graphene and graphene nanoribbon

Three-dimensional (3D) graphite and diamond have been known for centuries. In the published literature, *graphene* has been widely studied for more than 60 years as a basic building block for graphite materials. Pencil uses graphite for writing. The material silicon of the diamond crystallographic structure has micro-miniaturized electronics following the well-known Moore's law, where components double on a silicon chip nearly every 18 months. The current technology is rapidly moving in a direction where physics, biology and chemistry meet, i.e., nano-technology. For more than ten years, one-dimensional (1D) carbon nanotube (CNT), which is a graphene rolled in tubular form, and zero-dimensional (0D) fullerenes have been the subject of intensive research. Since CNT is in tubular form, research also continued to look for two dimensional (2D) flat materials for electronics.

Figure 1.1 shows crystal structures of different allotropes of carbon. Graphene is a monolayer of carbon atoms packed into a dense hexagonal honeycomb crystal structure, as shown in Figure 1.1(a), which can be separated and viewed as an individual atomic plane extracted from graphite as shown in Figure 1.1(b) or as an unrolled single wall CNT shown in Figure 1.1(c) or as a giant flat fullerene molecule as shown in Figure 1.1(d). Single layer of graphite or graphene was presumed not to exist in free stable form until 2004 when Novoselov *et al.* [1] experimentally first isolated single-layer graphene by micromechanical cleavage technique, peeling off repeatedly from graphite crystal using adhesive scotch tape, and reported their seminal work on the field effect study of such atomically thin carbon film.

The historical background of graphene goes back to Brodie [2] in 1859, who discovered the lamellar structure of thermally reduced graphene oxide, a multilayer carbon oxide material often used as an analogy to graphene. Kohlschutter and Haenni [3], in 1919, studied the properties of graphene oxide papers, a composite material with graphene skeleton. Three decades later, Reuss and Vogt [4] in 1948 reported the first transmission emission microscopy of a few layers graphite dry

[1]Division of Electrical and Computer Engineering, Louisiana State University, Baton Rouge, LA, USA

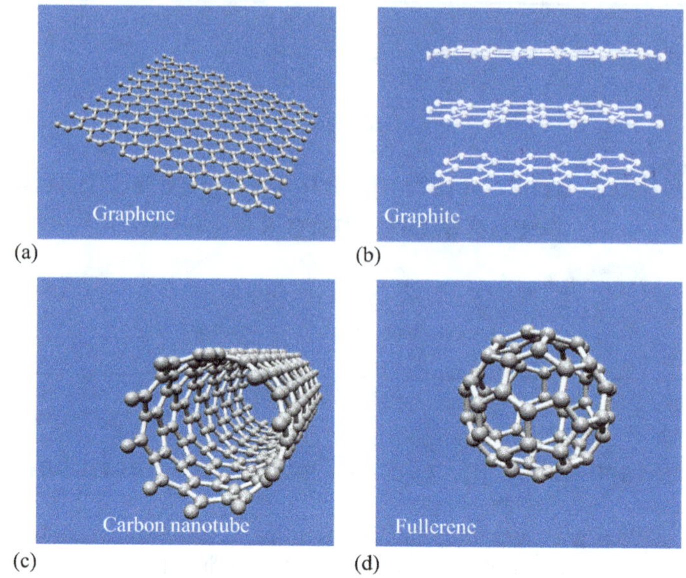

Figure 1.1 Different allotropes of carbon in different dimensions (a) two-dimensional (2D) atomically thick graphene, (b) three-dimensional (3D) graphite, (c) one-dimensional (1D) CNT, and (d) zero-dimensional (0D) fullerene

residue which is structurally a multilayer graphene. This remained the best observation of graphene for several decades. The theoretical groundwork of graphene also goes back to Wallace [5], who in 1947 first described the zone structure, number of free electrons and conductivity of a single hexagonal layer of graphite.

Between the late 1970s and early 1990s, major attention was focused on fullerenes (buckyballs) and CNTs which were discovered in 1985 [6] and 1991 [7], respectively. However, some key features of currently known graphene were reported during that period. Semenoff [8] found in 1984 that the wave functions of graphene are similar to the solutions of relativistic Dirac equation. Finally in 1987, Mouras *et al.* [9] coined the term "graphene" for single crystalline 2D carbon allotrope, before which graphene was commonly termed as "thin graphite lamellae". Surprisingly, even before the experimental observation of two different types of edge states, zigzag and armchair in graphene nanoribbon (GNR), a nanometer dimensional form of infinite graphene sheet, Nakada *et al.* [10] in 1996 extensively and accurately predicted their edge states with corresponding energy band structure.

From 2004 to 2008, research on graphene spurred tremendously considering graphene as an exciting condensed matter physics problem. Novoselov *et al.* [11] found that the electron transport in graphene is governed by relativistic Dirac equation where the charge carriers resemble Dirac fermions, relativistic particles with zero rest mass (massless particle) with an effective speed in the range of light.

Moreover, Katsnelson *et al.* [12] reported that, by using electrostatic barriers in single- and bilayer graphene, the massless Dirac fermions in graphene demonstrates Klein tunneling which is the unhindered penetration of relativistic particles through a wide potential barrier [12]. The quantized quantum Hall conductance, which is generally observed at low temperature and strong magnetic field, was also observed in graphene at room temperature [13]. Bolotin *et al.* [14] found that the low temperature carrier mobility is three times that of the best semiconductor. Thermal conductivity of graphene is also reported to be at least twice as large as that of copper for similar geometry [15]. The electron mobility in suspended graphene is found as 200,000 cm^2/V-s, which is 143 times greater than that of Si (1,400 cm^2/V-s at 300 K) [16,17].

It is the experimental discovery of 2D single-layer atomic thick *graphene* which has put it in forefront of current advanced technologies and as a substitute for silicon-based electronics. *Graphene* has demonstrated exceptional electronic properties such as the current density 2–3 order of magnitude higher than that of the copper interconnect used in current silicon technologies. It is immune to electromigration due to strong carbon-carbon bonds. It is easier to fabricate devices due to its planar nature. The process of making *graphene* layer is compatible with the standard photolithographic process used in semiconductor processing. It can be structured as a metallic or insulator. Since its discovery, researchers have been dreaming to make *graphene* a semiconductor material—a key requirement for making semiconductor chips. Feng Wang *et al.* at the UC Berkeley and Lawrence Berkeley National Laboratory (US Department of Energy, Lawrence Berkeley National Laboratory News Center: News Release, July 10, 2009) have succeeded in creating a tunable energy bandgap, though small, but a major breakthrough in realization of *semiconductor graphene* for making transistors, switches, lasers, and several types of solid state devices. Figure 1.2 explains the formation of tunable energy bandgap in *graphene*. Though graphene created excitement in the field of electronics and numerous applications, the problems started showing up. Lack of bandgap became a serious barrier for digital electronics and opening of bandgap became very problematic than initially thought. However, narrow strips of graphene demonstrated the necessary bandgap needed for semiconductor electronics

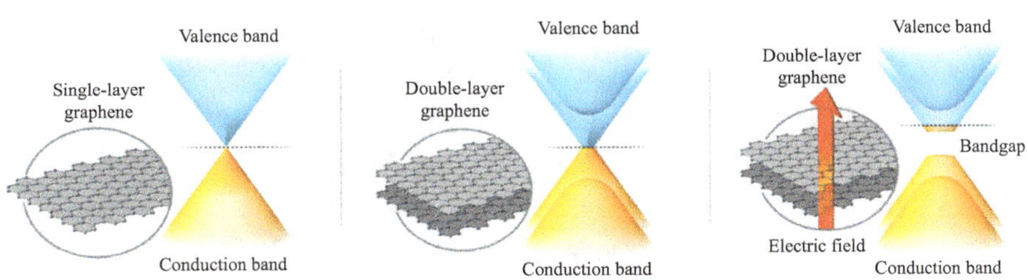

Figure 1.2 Left and middle—no bandgap, right—generation of bandgap after application of perpendicular electrical field to the layers of graphene [US DoE News Release, July 10, 2009]

which became known as GNRs. Experimentally feasibility of *GNRs* has also been demonstrated for making transistors and interconnects. Some of the recently reported devices are room-temperature ballistic transport field-effect transistors (FETs), single-electron transistors, spin transistors, and solar batteries. Researchers at MIT have already demonstrated a *graphene chip* which could reach 1,000 GHz (MIT Technical Talk, April 1, 2009).

According to Prof. Novoselov, one of the inventors of graphene "Being able to control the resistivity, optical transmittance and a material's work function would all be important for photonic devices like solar cells and liquid crystal displays, for example, and altering mechanical properties and surface potential is at the heart of designing composite materials. Chemical modification of graphene—with graphene as its first example—uncovers a whole new dimension of research. The capabilities are practically endless."

The synthesis and growth are significantly different from traditional bulk three-dimensional materials because graphene is a two-dimensional atomically thin material. While the current process technology for complementary metal-oxide semiconductor (CMOS)-integrated circuit is mature, graphene process technology is still under development, and extensive research has been carried out in this direction. Moreover, the challenges associated with obtaining large-area single crystal graphene and bilayer graphene are also present. In this chapter, synthesis of graphene and its growth mechanism are presented followed by electronic structure and properties.

1.2 Synthesis of graphene

Different methods are used for the synthesis and deposition of graphene. Figure 1.3 summarizes some of the methods used for graphene synthesis. One of the popular

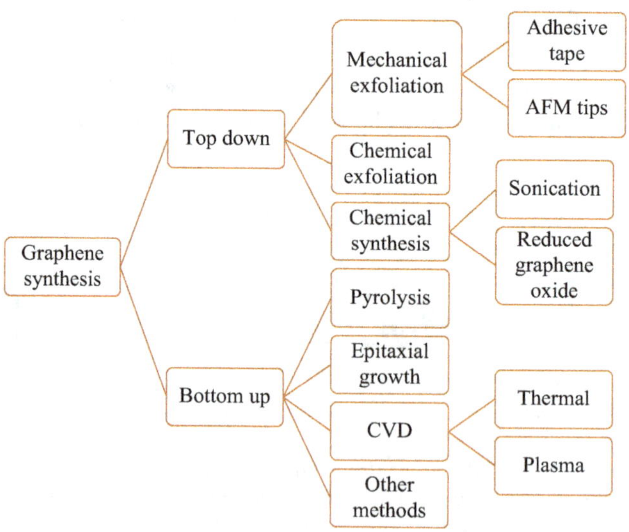

Figure 1.3 Synthesis methods for graphene

methods is the mechanical exfoliation from highly oriented pyrolytic graphite (HOPG) crystal. The other is through high temperature thermal chemical vapor deposition (CVD). Compared to nonscalable mechanical exfoliation, CVD method provides high-quality scalable production of atomically thin graphene. Using an adhesive scotch tape to repeatedly peel off layer by layer is the first technique adopted by Novoselov *et al.* [1]. However, large-area graphene fabrication using mechanical cleaving is a serious challenge which limits the feasibility of this process for industrialization. Hernandez *et al.* [18] reported the exfoliation of pure graphite in *N*-methyl-pyrrolidone by a simple sonication process. The reported exfoliated graphene films showed high-quality synthesis at yields of ~1%. Hazra *et al.* [19] in 2011 demonstrated plasma-assisted etching of graphite to form multilayered graphene and monolayer graphene. Direct graphene synthesis using electrochemical methods was reported by Liu *et al.* [20]. The method is environment-friendly and leads to the production of a colloidal suspension of imidazolium ion-functionalized graphene sheets by the direct electrochemical treatment of graphite. In 2006, Somani *et al.* [21] first attempted for CVD grown graphene on Ni using camphor (terpenoid, a white transparent solid of chemical formula $C_{10}H_{16}O$) as the precursor material. However, using TEM, they found that the planar few-layer graphene consists of ~35 layers of stacked single graphene sheets with an interlayer distance of 0.34 nm. Using methane (CH_4), Li *et al.* [22,23] studied the growth of large scale (1 cm^2) single-layer graphene on Ni and Cu substrates which is so far the widely used method employed for obtaining CVD graphene. Further, they developed a graphene transfer method by solution etching of Cu and then transferring of the floated graphene onto any substrate. Bae *et al.* [24] in 2010 produced a 30-inch scaled graphene sheet using roll-to-roll production on a Cu substrate and transferred by wet chemical etching of Cu.

A typical CVD process for deposition of graphene consists of four steps: (a) adsorption and catalytic decomposition of precursor gas, (b) diffusion and dissolution of decomposed carbon species on the surface and metal bulk, (c) dissolved carbon atoms segregation onto metal surface, and (d) surface nucleation and growth of graphene [25]. However, in case of metals having poor carbon affinity such as copper, the decomposition of carbon precursor is followed by the direct formation of graphene on copper where dissolution and subsequent segregation of carbon atoms are prohibited. The low solubility of the carbon in copper also makes the growth process predominantly self-limiting to single-layer graphene [22]. The most common carbon precursor for graphene growth is methane (CH_4), which has a strong C–H bond (440 $kJmol^{-1}$). For this strong C–H bond in methane, its thermal decomposition occurs at very high temperature (>1200 °C). However, such a high temperature is not easily obtained in typical thermal CVD setup. In order to reduce the decomposition temperature of methane, different transition metal catalysts (e.g. Fe, Co, Ni, Cu) are widely used and the growth of graphene on such metals can be obtained at low temperatures (<900 °C).

During the annealing step, the catalyst surface is covered with molecular hydrogen which can be referred to as dissociative chemisorption of H_2 on the metal surface [25]. Compared to Ni, Cu shows higher hydrogen solubility. This process is followed by the catalytic decomposition of the carbon precursors on the metal surface.

At this stage, the competitive process between the dissociative chemisorption of H_2 and physical adsorption and dehydrogenization of CH_4 on catalyst surface occurs. With suitable choice of thermodynamic parameters, the chemical potential of surface carbon atoms is maintained lower than the carbon in gas phases which further helps to form stable graphitic rings and grow into large graphitic structures up to graphene formation [25]. Once such nucleation of graphene structure is stable on the metal surface, the growth mechanism is followed by the attachment of carbon species onto graphene edges. The quality, uniformity, and surface coverage on metal substrate depend on suitable choice of high temperature, pressure, and exposure time. As the growth time increases, the individual graphene domains progressively increase in size and coalesce into a continuous layer. Nevertheless, after the growth and formation of a continuous layer, further exposure to carbon precursor does not lead to deposition of multilayered graphene due to the self-limiting process as described earlier in case of copper substrate. It is important to note that the graphene growth on copper is surface-related and does not occur due to out-diffusion from bulk. Using the isotope labeling, Li *et al.* [23] demonstrated that the Raman modes of ^{12}C and ^{13}C isotopes differ in energy which provided a substantial understanding of the gradual increment of the graphene layer growth laterally on copper surface providing critical structural information of graphene growth.

Figure 1.4(a) shows floating graphene film on Cu etchant $Fe(NO_3)_3$ after Cu has been fully etched. Prior to that, CVD graphene was grown on Cu foil of 25 μm thick. Figure 1.4(b) shows the floating graphene transferred on SiO_2 substrate. The optical

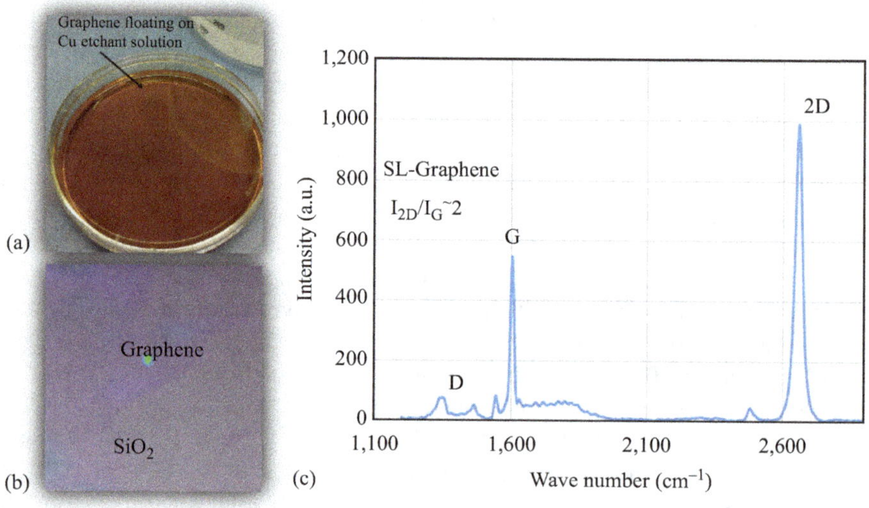

Figure 1.4 (a) CVD grown graphene floating on Cu etchant after Cu has been fully etched, (b) transferred on SiO_2, and (c) Raman spectroscopy of single-layer graphene after transferred on SiO_2

contrast confirms single-layer graphene compared to SiO_2. Figure 1.4(c) shows Raman spectroscopy of graphene transferred on SiO_2 substrate. A 632 nm laser is used for Raman spectroscopy. The graphene on Cu has been deposited using the Benchtop nanoCVD-8G System (Moorfield Nanotechnology Ltd., U.K.). A small D peak and a dominant 2D peak compared to the G peak confirm the growth of single-layer graphene on Cu foil.

The CVD reactor for the growth and synthesis of graphene on metal substrate is cold wall resistive heater type system, as shown in Figure 1.5(a) and (b). The gas flow process and standard recipe following the work of Bointon *et al.* [26] are depicted in Figure 1.5(c) and (d), respectively. Graphene grown through CVD system is typically polycrystalline in nature where a lot of graphene seeds nucleates and coalesce. Hence, the growth of graphene film on metallic substrate is twofold: (a) the nucleation and (b) growth [25].

Moreover, due to self-limiting catalytic decomposition of carbon molecules in metal substrates through diffusion and adsorption, controlled growth of bilayer graphene is challenging. Therefore, as the process technology for graphene continues, the growth of large-area single crystal as opposed to polycrystalline graphene and controlled synthesis single/poly crystalline bilayer graphene is necessary.

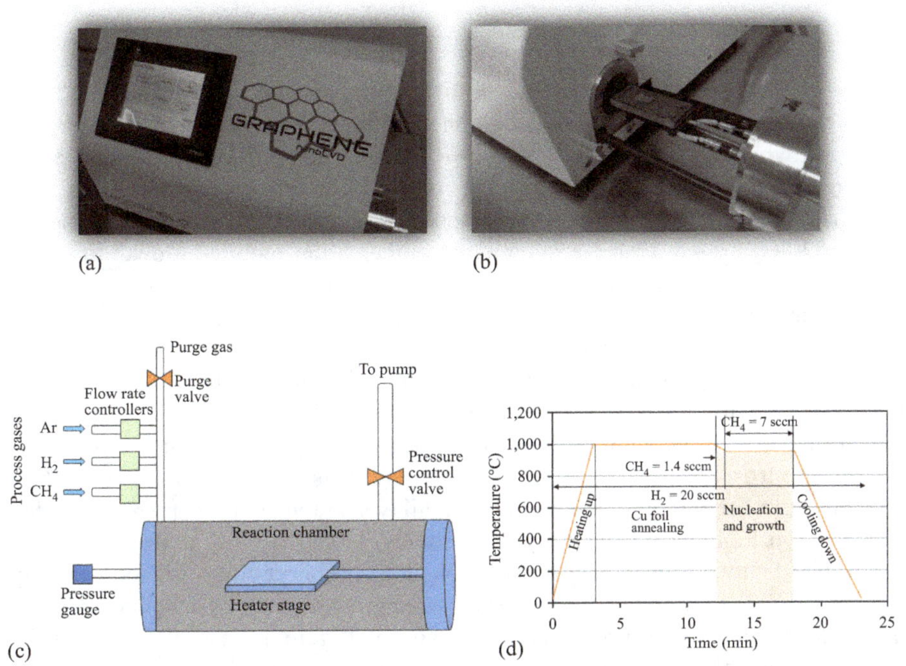

Figure 1.5 Cold wall resistive heater type CVD system by Moorefield Nanotechnology, U.K. (a) CVD assembly, (b) heater and chamber assembly, (c) schematic of gas flow process, and (d) standard growth condition for single-layer graphene

Recently, Hao *et al.* [27] have demonstrated that by controlling the oxygen on copper substrate centimeter scale graphene single crystal can be obtained repeatedly. Traditionally the size of single crystals in a polycrystalline graphene sheet is few micrometers only. Using cold wall CVD system, Misekis *et al.* [28] have grown millimeter scale graphene sheet on copper foil for 30–60 min compared to traditional hot wall CVD system which requires 3–7 h of growth time. Nevertheless, the growth of more than centimeter scale graphene single crystal is still challenging. Many attempts to grow bilayer graphene on copper have been carried out, however, majority of these studies have resulted in small domain of bilayer graphene with a large variation in the domain size [29]. Since an electric field tunable bandgap can be obtained in a bilayer graphene, it is essential to produce uniform and large domain single crystal bilayer graphene sheet. Hao *et al.* [29] recently have shown that an oxygen activated CVD process can produce as large as half-millimeter size Bernal A-B stacked bilayer graphene singe crystal on copper. Mu *et al.* [30] have shown that by controlling the partial pressure of hydrogen during the nucleation stage, bilayer graphene can be grown on copper foil. As part of growth studies of single-layer and bilayer graphene, control of chamber pressure during growth period has been modified for obtaining bilayer graphene on copper foil. Detail of the process variability effect is provided in the following section.

1.2.1 Growth of multilayer graphene film on copper

Chen *et al.* [31] proposed that switching hydrogen pressure between high and low would result in the growth of bilayer graphene. Similar results were attained by Lu *et al.* [32] whereby simply controlling the hydrogen pressure bilayer graphene has been grown. Following the work in [31] and [32], the chamber pressure has been modified for obtaining multilayer graphene using the cold wall resistive heater nanoCVD-8G System. However, compared to the earlier reported growth time, the process adopted here not only requires less time but also becomes economical. Figure 1.6(a) shows the optical image of a copper foil processed under similar growth as described in the work of Bointon *et al.* [26] for a chamber pressure of 20 Torr during the growth period. With the carbon precursor $CH_4 = 10\%$, $H_2 = 5\%$ and Ar $= 85\%$ for 120 s and a chamber pressure of 20 Torr at 1000 °C, both the bilayer and multilayer graphene have been observed along with single layer. The Raman spectroscopy performed at different areas as observed in Figure 1.6(a) confirms the observation of bilayer and multilayer graphene on copper foil which are shown in Figure 1.6(b) and (c), respectively. Note that with a growth time of only 120 s, the total processing time for such graphene sheet on copper foil was only 20 min, which is shorter than the earlier reported growth time in the work of Bointon *et al.* [26]. From the optically contrast image of Figure 1.6(a), difference in the number of layers of graphene can be easily understood as well. Compared to the lighter area, the darker area represents more graphene layers. The Raman peaks studied in the comparatively less dark area and shown in Figure 1.6(b) reveal that graphene is bilayer (BL-graphene) with an extensive level of defects or hydrogenated edges. A strong D peak compared to both G and 2D peak is a characteristic

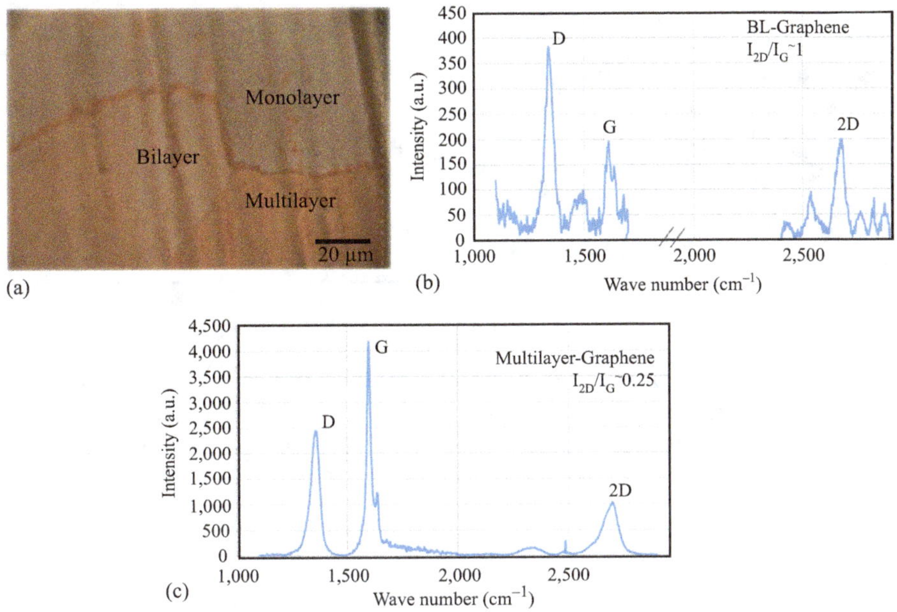

Figure 1.6 Raman spectroscopy of graphene grown on copper foil at 20 Torr pressure, (a) optical image showing three different regions, (b) Raman peaks for bilayer graphene area, and (c) Raman peaks for multilayer graphene area

feature of a graphene film with defects or halogen terminated edges. An I_{2D}/I_G ratio near 1 also reveals that the area is bilayer [33].

Further, Raman analysis of the darker region confirms that the graphene is multilayer as shown in Figure 1.6(c). With an I_{2D}/I_G ratio of nearly 0.25 confirms that the graphene in the region is more than 10 layers and similar to graphitic carbon [33]. The D peak for this region is low which informs comparatively less defects or hydrogen-terminated edges compared to the Raman spectra of Figure 1.6(b).

In order to analyze the effect of growth or exposure time on the number of graphene layers in a similar growth condition, the copper foil was exposed for 300 s instead of only 120 s. Figure 1.7(a) shows a sample area of the grown graphene for such growth condition. It has been found that compared to uniform planar graphene sheet, graphene growth for such long period of time results in not only multilayer graphene but also a graphitic carbon with an extensive level of hydrogen-terminated edges. For this reason, the Raman peaks obtained for such region of hydrogenated graphitic carbon reveal a strong D peak and poor I_{2D}/I_G ratio which are shown in Figure 1.7(b). It is to be noted that compared to the 0.25 ratio of I_{2D}/I_G, the sample exposed for 300 s provides only an I_{2D}/I_G ratio of 0.19. Therefore, based on the results obtained through Figures 1.6 and 1.7, an optimized growth period is required for the large-area bilayer graphene synthesis. Nevertheless, further process

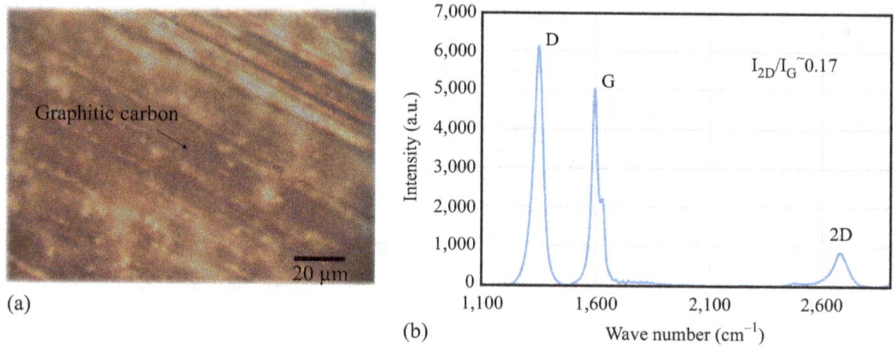

Figure 1.7 *(a) Optical image of graphitic carbon grown on copper after an exposure time of 300 s at a chamber pressure of 20 Torr and (b) Raman spectroscopy of the dark area marked with an arrow*

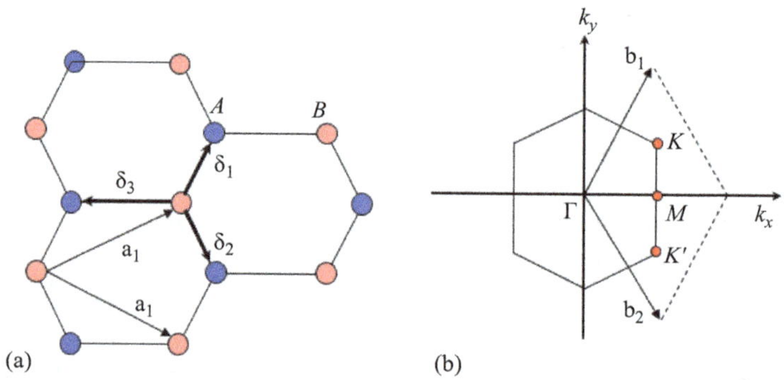

Figure 1.8 *(a) Hexagonal lattice structure of graphene consisting of two atoms A and B in a unit cell.* ***a₁*** *and* ***a₂*** *show direction of the lattice vectors in the primitive unit cell and (b) reciprocal lattice vectors* ***b₁*** *and* ***b₂*** *in the first BZ*

variation of CH_4 and H_2 concentrations, growth temperature, and chamber pressure can provide difference in graphene quality.

1.3 Electronic structure of graphene

In this section, a brief description of graphene electronic structure has been discussed. The carbon atoms in the graphene plane form strong σ-covalent bonds with three neighboring carbon atoms by in-plane sp^2 hybridization. The fourth bond is in the form of a π-bond in z-direction [34]. Electrons from this bond can move freely in the delocalized π-electronic system referred to as the π-band and π^*-band. The hexagonal lattice can be drawn as shown in Figure 1.8(a) and can be seen as a triangular lattice with a basis of two atoms per unit cell. The lattice can be written as follows [34]:

$$\mathbf{a}_1 = \frac{a}{2}\left(3, \sqrt{3}\right), \quad \mathbf{a}_2 = \frac{a}{2}\left(3, \ -\sqrt{3}\right) \tag{1.1}$$

where $a \approx 1.42$ Å is the carbon–carbon distance. The reciprocal-lattice vectors are given by,

$$\mathbf{b}_1 = \frac{2\pi}{3a}\left(1, \sqrt{3}\right), \quad \mathbf{b}_2 = \frac{2\pi}{3a}\left(1, -\sqrt{3}\right) \tag{1.2}$$

The two points K and K′ are at the corners of the Brillouin zone (BZ). They are referred to as Dirac points. The positions of these two points in a momentum space are defined as follows:

$$K = \left(\frac{2\pi}{3a}, \frac{2\pi}{3\sqrt{3}a}\right), \quad K' = \left(\frac{2\pi}{3a}, -\frac{2\pi}{3\sqrt{3}a}\right) \tag{1.3}$$

The three nearest-neighbor vectors in real space are given by,

$$\delta_1 = \frac{a}{2}(1, \sqrt{3}), \quad \delta_2 = \frac{a}{2}(1, -\sqrt{3}), \quad \delta_3 = -a(1, 0) \tag{1.4}$$

Using the expressions of lattice vector and reciprocal lattice constants, the nearest-neighbor tight binding Hamiltonian of graphene results in the following linear dispersion [34]:

$$E(\vec{k}) \ = \pm t\sqrt{\ 1 + 4\cos\frac{\sqrt{3}k_x a}{2}\ \cos\frac{k_y a}{2} + \ 4\cos^2\frac{k_y a}{2}} \tag{1.5}$$

Here, t is the nearest-neighbor hopping energy (hopping between different sublattices). The plus sign applies to the upper (π^*) and the minus sign the lower (π) band corresponding to the conduction and valence bands, respectively. It is clear from (1.5) that the spectrum is symmetric around zero energy where both the conduction and valence bands touch each other.

Figure 1.9 shows the full-band structure of graphene first BZ [34]. It can be seen that the energy dispersion around the band edges of graphene is linear. Plots are shown for the electron energy dispersion for π and π-bands in the first BZ as contour plots at equidistant energies and as pseudo-3D representations for the 2D structures. The demonstrated linear dispersion shows that the conduction and the valence bands touch each other at the charge neutrality point or known as the Dirac point shown by the arrow in Figure 1.9 at the symmetric K and K′ points which are plotted in Wolfram computational dynamic player tool. This shows that the band-gap in graphene to be zero or specifically graphene is referred to as a zero bandgap semiconductor or a semimetal.

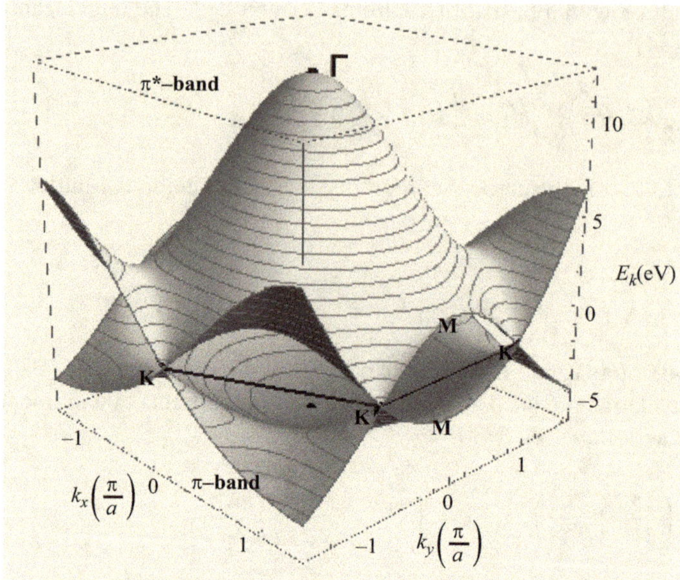

Figure 1.9 Dispersion relation of graphene first BZ shown in reciprocal lattice space (k-space) with both x- and y-axis normalized with π/a. K and K′ are the symmetric points

1.4 Bandgap engineering of graphene

Graphene is a zero bandgap semiconductor or a semi-metal. This results in transistors made of graphene difficult to turn off. In order to obtain appropriate switching behavior using graphene-based transistors, a significant bandgap is required which leads to the study of bandgap engineering of graphene. In this section, some methods for obtaining a bandgap in graphene are discussed. Figure 1.10 shows a summary of different ways of obtaining bandgap in graphene.

Castro *et al.* [34] reported that if an electric field is applied to a bilayer graphene vertically, then this opens a bandgap, making graphene as a field tunable semiconductor. Both the theoretical and experimental considerations have shown that for a field of few 10^4 kV/cm could open a bandgap of 250 meV. Recently, two unconventional methods have been reported namely: (1) graphene growth on MgO [35] and (2) by irradiation of graphene with an ion beam [36]. Being atomically thick, the interaction of graphene with underneath substrate plays a critical role in graphene electronic properties. Giovannetti *et al.* [37] reported in 2007 the ab-initio density functional theory (DFT)-based electronic structure calculation of graphene on hex-boron nitride substrate, resulting in a bandgap 53meV at the Dirac point. Recently, Nevius *et al.* [38] reported the growth of semiconducting graphene on highly ordered SiC substrate along the <0001> direction of the SiC hexagonal crystal pack (HCP). They measured the so far recorded the highest bandgap of single-layer graphene of 0.55 eV using angle-resolved photoemission spectroscopy (ARPES), which is a direct method to measure the energy bandgaps in materials.

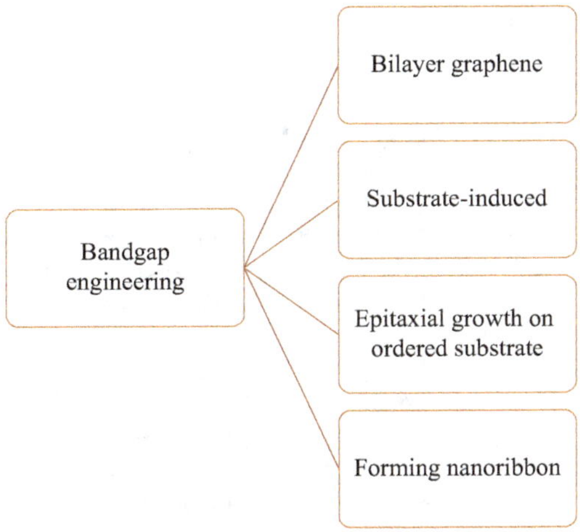

Figure 1.10 Energy bandgap engineering methods used for graphene

The most commonly known method for opening a bandgap in graphene is to confine the infinite graphene sheet into a narrow ribbon where length is much greater than width. Due to the quantum confinement of the electrons into a nanoribbon, a measurable finite bandgap can be opened [39,40]. For using graphene in transistor-level operation, it is necessary to have a finite bandgap of the material and GNR helps in this regard significantly.

1.4.1 Energy bandgaps of GNR

In GNR, the bandgap is directly proportional to the inverse of the width [39]. It has been predicted that GNR with width scaled down to 2 nm should provide a gap in excess of 1 eV [39]. It is important to note that the origin of bandgap is still under debate. Apart from considering the lateral confinement as the origin of bandgap, it has been suggested that another notable effect such as Coulomb blockade is responsible for the formation of such bandgap [40]. Han *et al.* [41] experimentally demonstrated lithographically patterned GNR with width dependent bandgap. One of the most effective methods for obtaining GNR is to unzip a single wall CNT with bottom-up chemical approach [42]. Compared to lithographically patterned GNR, this method provides smooth defect-free GNR [43].

Energy bandgap in GNR is also dependent on the edge types along which the transport occurs. Figure 1.11 shows a top view of a GNR with two types of edges, i.e. armchair and zigzag edges. Localized edge states at the Fermi level are observed in zigzag edge nanoribbon whereas such edge states are absent in armchair edge nanoribbons. These localized states are important as these infer to localized wave functions at the GNR edges and contribute to antibonding properties of GNR and electronic structure [10]. For the GNR shown in Figure 1.11, the nanoribbon width varies along the Y-direction and length along the Z-direction. The variable p is an integer. The numbering of atoms $(1, 2, 3 \ldots p)$ along the GNR width is also shown in Figure 1.11. The notation of chirality used for GNR is

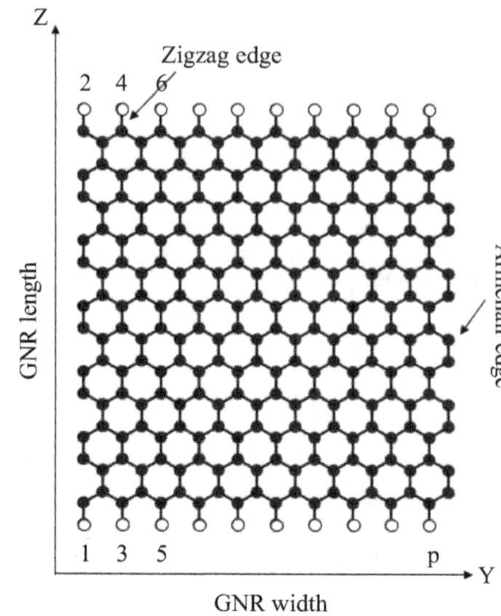

Figure 1.11 GNR, where p is an integer denoting the pth atom along the width

expressed as (p, 0), where p is the number of carbon atoms on each ring of unrolled nanotube. Generally, "p" is defined in terms of any of the configurations from $3N$, $3N + 1$ or $3N + 2$ along the GNR width. It should be noted that p is the total number of atoms considering both sides of the nanoribbons whereas N is an integer. Therefore, in a (4,0) armchair GNR, $p = 4$ with $3N + 1$ configuration considering $N = 1$. Whereas in a (5,0) armchair GNR, $p = 5$ with a $3N + 2$ configuration considering $N = 1$. For (6,0) armchair GNR, $p = 6$ with $3N$ configuration considering $N = 2$ [44].

Energy bandgap of GNR, both armchair and zigzag, differs depending on the method of calculation. Electronic structure of GNR is modeled traditionally by the simple tight binding (TB) approximation based on π-bonded p_z -orbital electrons or usually studied by Dirac equation of massless particle considering the effective speed of light (~10^6 m/s). Such assumptions lead to conclude armchair GNR to be either metallic or semiconducting. Results obtained by TB approximation considering nearest-neighbor hopping integral of 2.7 eV show that armchair GNR is metallic for $p = 3N + 2$ and semiconducting for both $p = 3N$ and $p = 3N + 1$ configurations [45]. Basically, the hierarchy of energy bandgap is maintained as $\Delta_{3N+1} > \Delta_{3N} > \Delta_{3N+2}(= 0 \text{ eV})$, Δ being the energy gap, where N is an integer. Figure 1.12 shows the width-dependent bandgap, calculated using nearest-neighbor tight binding Hamiltonian considering p_z orbital encoded in "CNT bands", available in the open-source simulation framework Nanohub [46]. In Figure 1.12, both (4,0) and (6,0) are semiconducting. Zero bandgap is observed for (5,0) GNR which is a $3N + 2$ configuration for $N = 1$.

However, the first principle calculation using self-consistent pseudopotential method by local (spin) density approximation (L(S)DA) shows that there are no

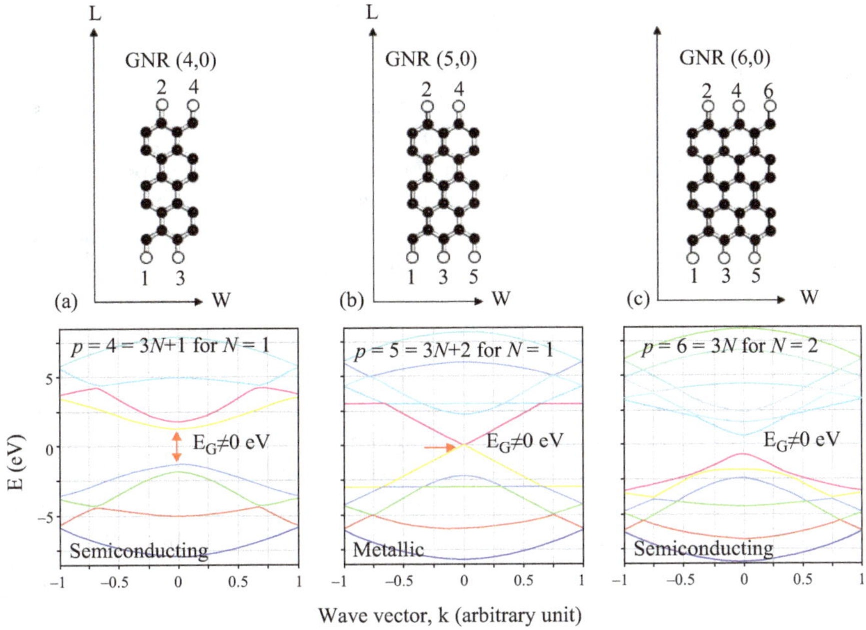

Figure 1.12 Width-dependent bandgap of GNR with increase in the number of atoms. (a) Energy band diagram for (4,0) GNR which is a $3N + 1$ configuration for $N = 1$ and semiconducting, (b) energy band diagram for (5,0) GNR which is a $3N + 2$ configuration for $N = 1$ and metallic, (c) energy band diagram for (6,0) GNR which is a $3N$ configuration for $N = 2$ and semiconducting. L denotes the length and W denotes the width of GNR. The numbers shown for chirality of GNR are depicted along the width of GNR

metallic GNR [45]. The energy gap as a function of width is now grouped in a family of energy gaps and maintains the hierarchy of $\Delta_{3N+1} > \Delta_{3N} > \Delta_{3N+2}$ ($\neq 0$ eV). Such an energy gap originates from the quantum confinement and crucial role of edge states and changes with a-GNR width. Moreover, first principle many electrons Green's function approach within the GW approximation provides quasi-particle energy gap with additional self-energy correction for both armchair and zigzag GNRs. It should be noted that GW refers to the single particle Green's function "G" and the screened coulomb interaction "W".

Recently, Kim *et al.* [47] have shown that proper consideration of higher energy levels in addition to p_z-orbitals in TB scheme gives more accurate description of the GNR band structure. It is shown that within the TB method $3N + 2$ GNRs are not really metallic if higher energy levels such as "d" orbitals are included. This is in agreement with the electronic structure obtained from rigorous first principle-based calculations.

The nearest-neighbor tight binding Hamiltonian-based calculation predicts that irrespective of nanoribbon width, zigzag edge type GNRs are metallic which is in contrast to the bandgap obtained by first principle calculation using self-consistent pseudopotential method by (L(S)DA) [45]. Based on the calculation of Son *et al.* [45], zigzag GNRs show gaps because of a staggered sublattice potential on the hexagonal lattice due to edge magnetization. Recently, the experimental work of Ruffieux *et al.* [48] have also reported that there are finite energy bandgaps in zigzag GNRs which match with the first principle-based calculation of zigzag GNRs. Therefore, the predictions based on tight binding approximation are no more valid.

1.5 GNR-based transistors, circuits, and interconnects

Zhang *et al.* [49] reported a basic structure of a p-i-n n-type armchair GNR (a-GNR) tunnel FET of 20 nm channel length and 4.9 nm channel width. Fahad *et al.* [50] presented an extensive study of single gate a-GNR TFET shown in Figure 1.13. Figure 1.13(a) shows vertical cross-section of p-type a-GNR TFET with 1 nm SiO_2 top gate dielectric. Channel length is 20 nm with 5 nm of source

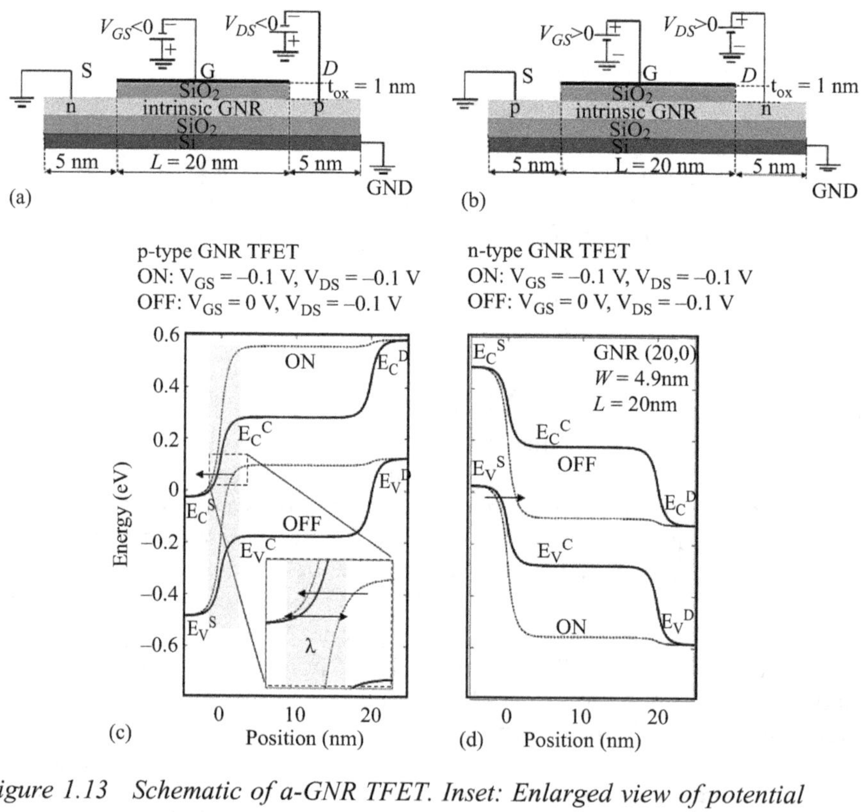

Figure 1.13 Schematic of a-GNR TFET. Inset: Enlarged view of potential variation

and drain extension making the total length of GNR 30 nm. Figure 1.13(b) shows an n-type a-GNR TFET. Figure 1.13(c) shows energy band diagram of n-i-p GNR TFET (p-type GNR TFET where both V_{GS} and V_{DS} are '−' ve). Figure 1.13(d) shows the energy band diagram of p-i-n GNR TFET (n-type GNR TFET where both V_{GS} and V_{DS} are '+' ve). It should be noted that in both Figure 1.13(c) and (d), solid line is for OFF state whereas dashed line is for ON state. OFF state is defined as $|V_{DS}| = 0.1$ V and $|V_{GS}| = 0$ V and ON state is defined as $|V_{DS}| = 0.1$ V and $|V_{GS}| = 0.1$ V. Semiconducting a-GNR (20, 0) has a bandgap of 0.289 eV for its corresponding 4.9 nm width. In CMOS technology, the interconnect material is copper and aluminum, which are different from silicon semiconductor used. Contrary to silicon CMOS technology, graphene-based integrated circuits can use the same material for both complementary transistors and interconnects.

Kang *et al.* [51] proposed all-graphene circuits as shown in Figure 1.14(a)–(g), where transistors and interconnects are fabricated from a single sheet of graphene. The figure shows a series of two all-graphene inverters. Tunable bandgap of graphene can be adjusted for GNR interconnects by pattering it with larger width and different orientation. Metallic and semiconducting GNRs are formed by changing GNR width and chirality, such that zigzag edge GNRs can be used as metallic source and drain regions and GNR interconnects while armchair GNRs are used for the semiconducting channel [52]. The fabrication process with atomic precision is required to implement zigzag-edged and armchair-edged GNRs for interconnects and transistors with smooth edges in order to maintain the metallic and semiconducting behaviors of GNR and prevent the reduction in mean free path by edge scattering.

The 3D hybrid structure of CNTs connected perpendicularly to graphene layers has been synthesized by CVD process [53,54] and theoretically investigated by ab initio calculation [55] as shown in Figure 1.15. It can be seen that one-dimensional carbon nanotube can be used as a via (vertical structure contacting two horizontal graphene layers) in this structure.

1.6 Doping of graphene

Graphene can be doped either by chemical doping or by electrostatic doping [56]. In electrostatic doping, a positive and negative gate voltage generates n- and p-type graphene, respectively [57]. Moreover, ion doping in graphene sheets can reach electron and hole density around $10^{14}/cm^2$ [58]. Traditionally boron (B) and nitrogen (N) are treated as natural candidates for doping graphene due to the same atomic size as in carbon. Wang *et al.* [58] observed experimentally n-type doping of GNR through electrochemical reaction with NH_3. Such a doping forms C–N bonds at GNR edges. Though the method provides high ON/OFF current ratio of $\sim 10^5$, mobility degrades in n-type GNR FET compared to in pristine GNR FET. One problem associated with it is that N (nitrogen)-doped graphene (NG) can be both n- and p-type based on the bonding nature of N atoms [59–61]. Recently, it has been studied experimentally that chemically functionalized array of GNR with 4-nitrobenzenediazonium (4-NBD) and diethylenetriamine (DETA) molecules can provide doping of GNR arrays to p- and n-type, respectively [62]. In both cases,

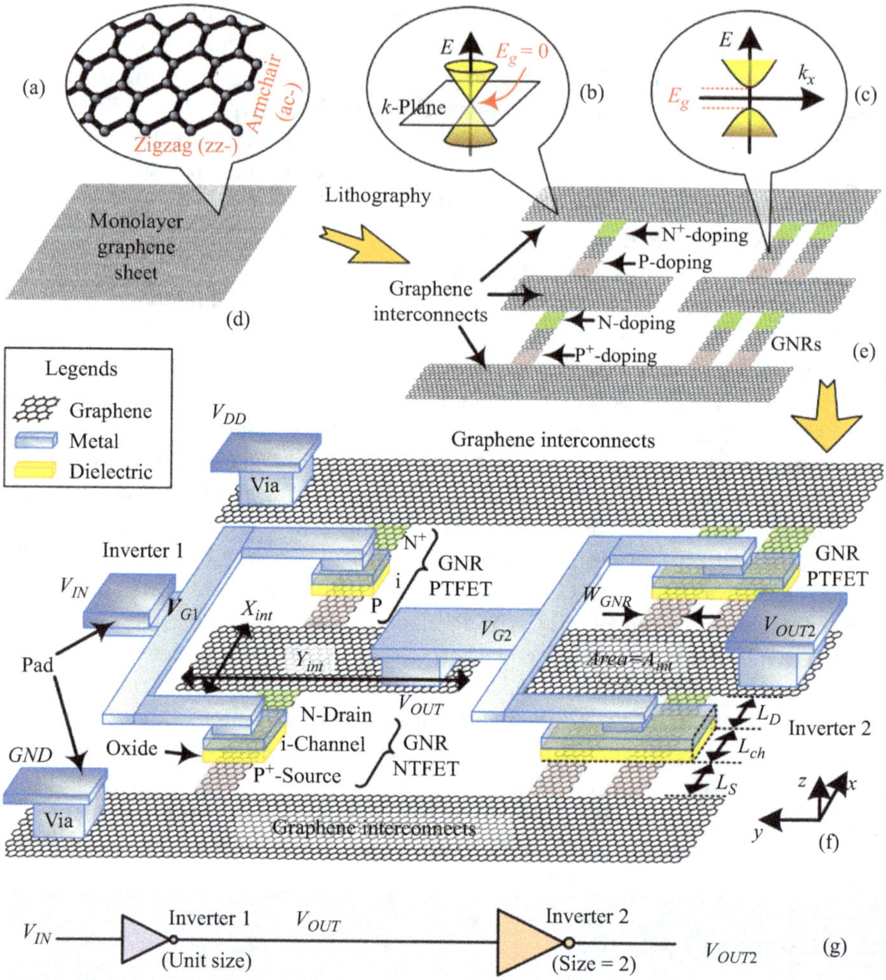

Figure 1.14 (a)–(g) Design and fabrication of all-graphene integrated circuits. (Reprinted with permission from [51], Copyright 2018, AIP Publishing)

due to the presence of a large quantity of edges, higher doping effect is observed in GNRs than that in pristine graphene sheets.

1.7 Other than graphene materials and beyond

In the current CMOS technology, transistor channel lengths are down to from 45 nm to 10 nm. By 2020, CMOS technologies are projected to reach line density of 10^{10} devices/cm^2, switching speed of 12 THz, circuit speed of 61 GHz and switching energy of 3×10^{-18} J. These are the figures which any new replacement

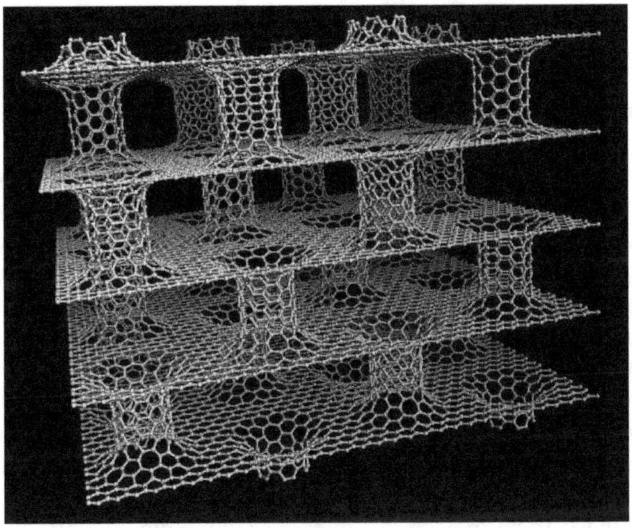

Figure 1.15 3D hybrid nanostructure of CNT and graphene. (Reprinted with permission from [55], Copyright 2018, American Chemical Society)

technology for silicon has to compete with. The main reason why graphene FETs cannot replace silicon transistors is because of the fact that channels cannot be switched-off. Transistors will leak current in the off-state. Typical current on/off ratio in digital CMOS devices is 10^4 to 10^7 whereas reported on/off current ratio in wide channel graphene FET is ~100 at room temperature (wide channel FET from SiC epitaxial graphene, $L = 10\ \mu\text{m}$, $W = 1.5\ \mu\text{m}$) with maximum field effect mobility of 7,600 cm^2 /V-s.

However, in most of the analog applications strict off-switching is not very crucial. When circuits are powered, the transistors are biased in linear region. It is the dynamic power consumption that dominates over the static power dissipation. THz frequencies are possible with 20 nm gate lengths in graphene RF transistors. One of the most severe limitations, already a limiting factor today, is power consumption—or in other words heat generated by the operation of the device. Here, graphene holds promise.

Discovery of two-dimensional atomic layer graphene in 2004 was perceived as a possible replacement of silicon but it lost promise as an integrated circuit material due to being semi-metal in electronic conduction. On the other hand, graphene is a very useful material for optical applications such as solar cells, LEDs, touch screens, photodetectors, etc. Monolayer graphene is almost transparent and when combined with its excellent electrical conductivity, the natural applications relate to transparent conductive films (TCF). TCFs are used as electrodes in solar cells, in displays and touch screens, etc.

In search of materials other than graphene, layered transition-metal dichalcogenide (TMD) type of materials denoted by MX_2, where M is a transition metal from group IV–VII, and X is a chalcogen such as S, Se, Te have shown great promise for electronics, photonics, energy harvesting, and biosensors. TMDs, one-atom thick, are superior to graphene in many ways. These have bandgap which is very important to design transistors as switches and good absorbers of circularly polarized light so they can be used as detectors. The most widely researched material is molybdenum disulfide (MoS_2) and single-layer MoS_2-based FETs are reported [63]. Bulk MoS_2 is semiconducting with an indirect bandgap of 1.2 eV. The single-layer MoS_2 is a direct bandgap semiconductor with a bandgap of 1.8 eV. The reported mobility in TMDs is too low to be used for semiconductor electronics and attempts are being made to improve carrier mobility of TMD-based devices for electronic applications. MoS_2-based transistors with hafnium oxide (HfO_2) gate dielectric have reported mobility ~200 $cm^2v^{-1}s^{-1}$ and 10^8 on/off current ratio with ultra-low power operation. Recently, Srivastava and Fahad [64] have reported a novel transistor based on combining horizontal current flow between source and drain with vertical interlayer tunneling. A schematic of MoS_2 junctionless tunneling FET considering MoS_2/*h*BN/MoS_2 is shown in Figure 1.16. The dashed line AA′ refers to the vertical direction of interlayer tunneling and BB′ refers to lateral direction of source-drain ballistic carrier transport. Compared to recently reported device structures in [65] and [66], the present device structure gives subthreshold slope close to 60 mV/decade and demonstrates upper GHz operation with relatively comparable on/off current ratio.

Other new class of emerging two-dimensional materials denoted by Xenes (silicene, germanene, and stanene) [67] based transistors remained yet to catch up

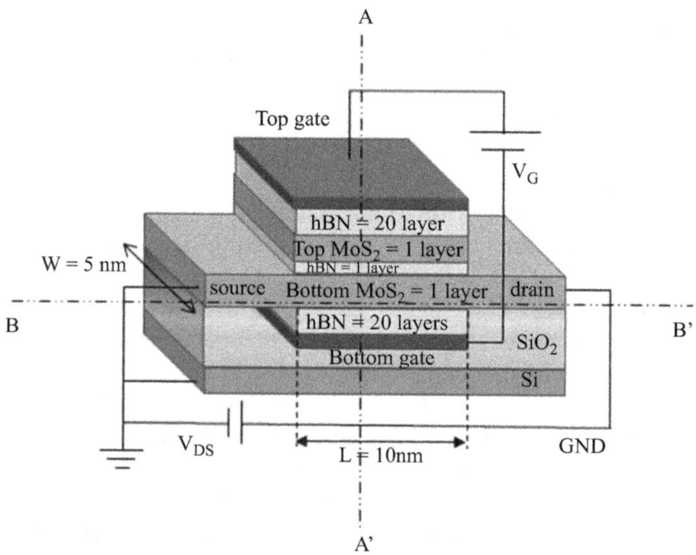

Figure 1.16 Schematic of MoS_2 junctionless tunneling FET considering MoS_2/hBN/MoS_2

theoretically and experimentally. Recent studies on some of these materials and other materials such as arsenene- and antimonene-based transistors have shown promise for the future [68]. Among some of these materials, silicene is currently the focus of research because of its closeness to graphene hexagonal structure and is predicted to be stable and compatible with the CMOS technology [69].

Similar to graphene, silicene is also a Dirac material with zero bandgap at the Dirac point. Any FET requires near 0.4 eV of bandgap which puts a critical challenge for silicene in device applications. Ni *et al.* [70] have shown that an opening of a bandgap in silicene can be obtained without degrading their electronic characteristics by metal adsorption which also led to a computational study of silicene p-i-n tunneling transistor. An on/off current ratio of 10^3 with the subthreshold slope of 77mV/decade has been reported based on the first principle calculations [70].

1.8 Conclusion

Graphene with its unique electronic properties is highly suitable for numerous electronic applications. Among different growth techniques, CVD is most promising due to its low cost and large area. However, growth of large-area single crystal graphene is still challenging. Owing to its zero bandgap property, graphene is not yet suitable for digital applications. However, finite bandgap can be obtained in the form of GNR which demonstrates width and edge-type dependent energy bandgap. GNR TFET can be a viable option for low power high-performance integrated circuit design. By utilizing the zero band properties of graphene, the promise of graphene interlayer tunnel transistor can also be explored. Other than graphene 2D materials such as layered transition-metal dichalcogenide (TMD) and Xenes have emerged and shown great promise for electronics, photonics, energy harvesting, and biosensors.

References

[1] Novoselov, K. S., Geim, A. K., Morozov, S. V. *et al.* 'Electric field effect in atomically thin carbon films'. *Science*. 2004; **306**: 666–69

[2] Brodie, B. C., 'On the atomic weight of graphite'. *Philosophical Trans. of the Royal Society of London*. 1859; **149**: 249–59

[3] Kohlschütte, V. and Haenni, P. 'Zur kenntnis des graphitischen kohlenstoffs und der graphitsäure'. *Zeitschrift für anorganische und allgemeine Chemie*. 1919; **105**: 121–44

[4] Ruess, G. and Vogt, F. 'Höchstlamellarer kohlenstoff aus graphitoxyhydroxyd'. *Monatshefte für Chemie und verwandte Teile anderer Wissenschaften*.1948; **78**: 222–42

[5] Wallace, P. R. 'The band theory of graphite' *Phys. Rev.* 1947; **71**: 622–34

[6] Kroto, H. W., Heath, J. R., O' Brien, S.C., Curl, R. F., and Smalley, R. E. 'C60: Buckminsterfullerene'. *Nature*. 1985; **318**: 162–63

[7] Iijima, S. 'Helical microtubules of graphitic carbon'. *Nature*. 1991; **354**: 6–58

[8] Semenoff, G. W. 'Condensed-matter simulation of a three-dimensional anomaly'. *Phys. Rev. Lett.* 1984; **53**: 2449–52

[9] Mouras, S., Hamm, A., Djurado, D. and Cousseins, J. C. 'Synthesis of first stage graphite intercalation compounds with fluorides'. *Revue de chimie minérale.* 1987; **25**: 572–82, 1987

[10] Nakada, K., Fujita, M., Dresselhaus, G. and Dresselhaus, M. S. 'Edge state in graphene ribbons: Nanometer size effect and edge shape dependence'. *Phys. Rev. B.* 1996; **54**: pp. 17954–961

[11] Novoselov, K. S., Geim, A. K., Morozov, S. V., *et al.* 'Two-dimensional gas of massless Dirac fermions in graphene'. *Nature.* 2005; **438**: 197–200

[12] Katsnelson, M. I., Novoselov, K. S. and Geim, A. K. 'Chiral tunnelling and the Klein paradox in graphene'. *Nat Phys.* 2006. **2**: 620–25

[13] Novoselov, K. S., Jiang, Z., Zhang, Y., *et al.* 'Room-temperature quantum hall effect in graphene'. *Science.* Mar 9, 2007; **315**: 1379

[14] Bolotin, K. I., Sikes, K. J., Hone, J., Stormer, H. L. and Kim, P. 'Temperature-dependent transport in suspended graphene'. *Phys. Rev. Lett.* 2008; **101**: 096802, 2008

[15] Prasher, R. 'Graphene Spreads the Heat'. *Science.* 2010; **328**:185–186, 2010

[16] Meyer, J. C., Geim, A. K., Katsnelson, M. I., Novoselov, K. S., Booth, T. J. and Roth, S. 'The structure of suspended graphene sheets' *Nature.* 2007; **446**: 60–63

[17] Bolotin, K. I., Sikes, K. J., Jiang, Z., *et al.* 'Ultrahigh electron mobility in suspended graphene'. *Solid State Comm.* 2008: **146**: 351–55

[18] Hernandez, Y., Nicolosi, V., Lotya, M., *et al.* 'High-yield production of graphene by liquid-phase exfoliation of graphite'. *Nat Nano.* 2008; **3**: 563–68

[19] Hazra, K. S., Rafiee, J., Rafiee, M. A., *et al.* 'Thinning of multilayer graphene to monolayer graphene in a plasma environment'. *Nanotechnology.* 2011; **22**: 025704

[20] Liu, N., Luo, F., Wu, H., Liu, Y., Zhang, C. and Chen, J. 'One-step ionic-liquid-assisted electrochemical synthesis of ionic-liquid-functionalized graphene sheets directly from graphite'. *Advanced Func. Mat.* 2008; **18**: 1518–25

[21] Somani, P. R., Somani, S. P. and Umeno, M. 'Planer nano-graphenes from camphor by CVD'. *Chemical Phys. Lett.* 2006; **430**: 56–59

[22] Li, X., Cai, W., An, J., *et al.* 'Large-area synthesis of high-quality and uniform graphene films on copper foils' *Science.* 2009; **324**: 1312–14, 2009

[23] Li, X., Cai, W., Colombo, L. and Ruoff, R. S. 'Evolution of graphene growth on Ni and Cu by carbon isotope labeling'. *Nano Letters.* 2009; **9**: 4268–72

[24] Bae, S., Kim, H., Lee, Y., *et al.* "Roll-to-roll production of 30-inch graphene films for transparent electrodes' *Nat Nano.* 2010; **5**: 574–78

[25] Yan, K., Fu, L., Peng, H. and Liu, Z. 'Designed CVD growth of graphene via process engineering' *Accounts of Chemical Research.* 2013; **46**: 2263–74

[26] Bointon, T. H., Barnes, M. D., Russo, S. and Craciun, M. F. 'High quality monolayer graphene synthesized by resistive heating cold wall chemical vapor deposition'. *Advanced Materials.* 2015; **27**: 4200–06

[27] Hao, Y., Bharathi, M. S., Wang, L., *et al.* "The role of surface oxygen in the growth of large single-crystal graphene on copper. *Science.* 2013; **342**: 720
[28] Miseikis, V., Convertino, D., Mishra, N., *et al.* "Rapid CVD growth of millimeter-sized single crystal graphene using a cold-wall reactor'. *2D Materials.* 2015; **2**: 014006
[29] Hao, Y., Wang, L., Liu, Y., *et al.* 'Oxygen-activated growth and bandgap tunability of large single-crystal bilayer graphene'. *Nat Nano.* 2016; **11**: 426–31
[30] Mu, W., Fu, Y., Sun, S., Edwards, M., Ye, L., Jeppson, K. and Liu, J. 'Controllable and fast synthesis of bilayer graphene by chemical vapor deposition on copper foil using a cold wall reactor' *Chemical Engineering Journal.* 2016; **304**: 106–14
[31] Chen, W., Cui, P., Zhu, W., Kaxiras, E., Gao, Y. and Z. Zhang, Z. Atomistic mechanisms for bilayer growth of graphene on metal substrates'. *Phys. Rev. B.* 2015; **91**: 045408
[32] Liu, Q., Gong, Y., Wilt, J. S., Sakidja, R. and Wu, J. Synchronous growth of AB-stacked bilayer graphene on Cu by simply controlling hydrogen pressure in CVD process'. *Carbon.* 2015; **93** (11): 199–206
[33] Ferrari, A.C., Meyer, J. C., Scardaci, V., *et al.* 'Raman spectrum of graphene and graphene layers'. *Physical Review Letters.* 2006; **97**: 187401
[34] Castro, E. V., Novoselov, K. S., Morozov, S. V., *et al.* 'Biased bilayer graphene: semiconductor with a gap tunable by the electric field effect'. *Phys. Rev. Lett..* 2007; **99**: 216802
[35] Kelber, J. A., Zhou, M., Gaddam, S., Pasquale F. L., Kong, L. M., and Dowben, P. A. 'Direct graphene growth on oxides: interfacial interactions and band gap formation'. *ECS Trans.* 2012; **45**: 49–61
[36] Nakaharai, S., Iijima, T., Ogawa, S., *et al.* 'Electrostatically-reversible polarity of dual-gated graphene transistors with He ion irradiated channel: toward reconfigurable CMOS applications'. *Tech. Digest IEEE Int. Elect. Dev. Meeting*, 2012; 4.2.1–4.2.4
[37] Giovannetti, G., Khomyakov, P. A., Brocks, G., Kelly, P. J. and van den Brink, J. 'Substrate-induced band gap in graphene on hexagonal boron nitride: Ab initio density functional calculations'. *Phys. Rev. B.* 2007; **76**: 073103
[38] Nevius, M. S., Conrad, M., Wang, F., *et al.* 'Semiconducting graphene from highly ordered substrate interactions'. *Phys. Rev. Lett.* 2015; **115**: 136802
[39] Yang, L., Park, C.-H., Son, Y. W., Cohen, M. L., and Louie, S. G. 'Quasiparticle energies and band gaps in graphene nanoribbons'. *Phys. Rev. Lett.* 2007; **99**: 186801
[40] Sols, F., Guinea, F., and Neto, A. H. C. 'Coulomb blockade in graphene nanoribbons'. *Phys. Rev. Lett.* 2007; **99**: 166803
[41] Han, M. Y., Özyilmaz, B., Zhang, Y. and Kim, P. 'Energy band-gap engineering of graphene nanoribbons'. *Phys. Rev. Lett..* 2007; **98**: 206805
[42] Xie, L., Wang, H., Jin, C., *et al.* 'Graphene nanoribbons from unzipped carbon nanotubes: atomic structures, raman Spectroscopy, and electrical properties'. *J. of the American Chemical Society.* 2011; **133**: 10394–97

[43] Kosynkin, D.V., Higginbotham, A. L., Sinitskii, A., *et al.* 'Longitudinal unzipping of carbon nanotubes to form graphene nanoribbons'. *Nature.* 2009; **458**: 872–76

[44] Fahad, Md .S. *Physical modeling of graphene nanoribbon field-effect transistor using non-equilibrium Green function approach for integrated circuit design.* Baton Rouge: Louisiana State University. 2016; Ph.D. (Electrical Engineering) Dissertation

[45] Y. W. Son, Y. W., M. L. Cohen, M. L. and S. G. Louie, S.G. 'Energy gaps in graphene nanoribbons'. *Phys. Rev. Lett.*. 2006; **97**: 216803

[46] Yang, L., Anantram, M. P., Han, J. and Lu, J. P. 'Band-gap change of carbon nanotubes: effect of small uniaxial and torsional strain'. *Phys. Rev. B.* 1999; **60**(19): 13874–78. Available from http://www.nanohub.org/cntbands [Accessed 15 May 2018]

[47] Kim, S., Luisier, M., Boykin, T. B., and Klimeck, G. 'Computational study of heterojunction graphene nanoribbon tunneling transistors with p-d orbital tight-binding method'. *Appl. Phys. Lett.* 2014; **104**: 243113

[48] Ruffieux, P., Wang, S., Yang, B., *et al.* 'On-surface synthesis of graphene nanoribbons with zigzag edge topology'. *Nature.* 2016; **531**, 489–92

[49] Zhang, Q., Fang, T., Xing, H., Seabaugh, A. and Jena. D. 'Graphene nanoribbon tunnel transistors'. *IEEE Elect. Dev. Lett.* 2008; **29**(12): 1344–46

[50] Fahad, Md. S., Srivastava, A., Sharma, A. K., Mayberry, C. 'Analytical current transport modeling of graphene nanoribbon tunnel field-effect transistors for digital circuit design'. *IEEE Trans. on Nanotechnology.* 2016; **15** (1): 39–50

[51] Kang, J., Sarkar, D., Khatami, Y., Banerjee, K. 'Proposal for all-graphene monolithic logic circuits'. *Applied Physics Letters*, 2013; **103**: 083113

[52] Y. M. Banadaki Y. M., Srivastava, A. 'Investigation of the width-dependent static characteristics of graphene nanoribbon field effect transistors using non-parabolic quantum-based model'. *Solid-State Electronics*, 2015; **111**: 80–90

[53] Wang, X. Sun, G., and Chen, P. 'Three-dimensional porous architectures of carbon nanotubes and graphene sheets for energy applications' *Frontiers in Energy Research.* 2014 Aug 14; **2**: 1–33

[54] Kondo, D. Sato, S. and Awano, Y. 'Self-organization of novel carbon composite structure: graphene multi-layers combined perpendicularly with aligned carbon nanotubes' *Applied Physics Express.* 2008; **1**: 074003

[55] Dimitrakakis, G. K., Tylianakis, E. and Froudakis, G. E. 'Pillared graphene: a new 3-D network nanostructure for enhanced hydrogen storage' *Nano letters.* 2008; **8**: 3166–70

[56] Lv R., and Terrones, M. 'Towards new graphene materials: doped graphene sheets and nanoribbons'. *Materials Letters.* 2012; **78**: 209–18

[57] Chiu, H. Y., Perebeinos, V., Lin, Y.-M., and Avouris, P. 'Controllable p-n Junction formation in monolayer graphene using electrostatic substrate engineering'. *Nano Letters.* 2010; **10**: 4634–39

[58] Wang, X., Li, X., Zhang, L., *et al.* 'N-Doping of graphene through electrothermal reactions with ammonia'. *Science.* 2009; **324**: 768–71

[59] Fujimoto Y. and Saito, S. 'Formation, stabilities, and electronic properties of nitrogen defects in graphene'. *Phys. Rev. B.* 2011; **84**: 245446

[60] Schiros, T., Nordlund, D., Pálová, L., *et al.* 'Connecting dopant bond type with electronic structure in N-doped graphene' *Nano Letters.* 2012; **12**: 4025–31

[61] Hou, Z., Wang, X., Ikeda, T., Terakura, K., Oshima, M., and Kakimoto, M. 'Electronic structure of N-doped graphene with native point defects'. *Phys. Rev. B.* 2013; **87**: 165401

[62] Solis-Fernandez, P., Bissett, M. A., Tsuji, M., and Ago, H. 'Tunable doping of graphene nanoribbon arrays by chemical functionalization.' *Nanoscale.* 2015; **7**: 3572–80

[63] Radisavljevic B., Radenovic A., Brivio J., Giacometti V. and Kis A., "Single-layer MoS2 transistors," *Nature Nanotechnology*, vol. **6**, pp. 147–50, March 2011

[64] Srivastava, A., and Fahad, Md. S. 'Vertical MoS2/hBN/MoS2 interlayer tunneling field-effect transistor'. *Solid-State Electronics*. 2016; **126** (12): 96–103

[65] Fiori, G., Bruzzone, S., and Iannaccone, G. 'Very large current modulation in vertical heterostructure graphene/hBN transistors.' *IEEE Trans. on Elect. Dev.* 2013; **60**: 268–73

[66] Ghobadi N., and Pourfath, M. 'A comparative study of tunneling FETs based on graphene and GNR heterostructures'. *IEEE Trans. on Elect. Dev.* 2014; **61**: 186–92

[67] Molle, A. 'Xenes: a new two-dimensional materials platform for nanoelectronics'. *ECS Transactions.* 2016; **75**(5): 163–73

[68] Pizzi, G., Gibertini, M., Dib, E., Marzari, N., Iannaccone, G., and Fiori, F. 'Performance of arsenene and antimonene double-gate MOSFETs from first principles'. *Nature Communications.* 2016; **7:12585/DOI:10.1038/ncomms12585**: 1–9

[69] Lew Yan Voon, L. C., Zhu, J., and Schwingenschlögl, U., 'Silicene: recent theoretical advances'. *Applied Physics Reviews.* 2016; **3**: 040802-1–040802-13

[70] Ni, Z., Zhong, H., Jiang, X., *et al.* 'Tunable band gap and doping type in silicene by surface adsorption: towards tunneling transistors'. *Nanoscale.* 2014 Jul 7; **6**(13): 7609–18

Chapter 2

Emerging graphene-compatible biomaterials

Hindumathi R. Dhanasekaran[1], *Jagannatham Madiga*[2], *Chandra P. Sharma*[3] *and Prathap Haridoss*[2]

2.1 Introduction

Present-day challenges in modern healthcare systems include invasive procedures for the diagnosis of diseases and treatment, time-consuming lab tests, centralized medical facilities which are not easily accessible to all people, and requirement of expertise in medical prognosis. All these make modern healthcare costlier. These factors also make early detection of onset of diseases and monitoring chronic conditions difficult. Advances in bio-electronic devices make future healthcare easier. Integrated circuits play a major role in all the three areas of biomedical applications – diagnostic, monitoring and therapeutic. Biosensors and imaging techniques help in diagnostics by sensing the change in different vital body signals and by scanning the internal organs, respectively. Continuous monitoring of these vital signals, electrochemically analysing the gas and chemical levels (e.g. glucose sensing) and monitoring controlled delivery of therapeutic molecules are possible only because of the advances in functional materials and integrated circuits.

Biomedical materials have come a long way, from inert supportive materials to bioactive and responsive implants. Biocompatible and bioabsorbable materials have made life with implants better and manageable. Portable devices which could continuously monitor the condition of the patients by measuring vital signals and biosensing, wirelessly transmitting the data to medical practitioner who could give timely feedback and initiate therapy from remote location are nearing commercialization [1]. The advances in nano materials and nano characterization techniques have made this feasible.

Sensor technologies are based on either electrochemical, optical or acoustic wave sensing. In addition, magnetic nanoparticles could be introduced in vivo and used for

[1]Department of Biotechnology, Indian Institute of Technology Madras, Chennai, India
[2]Department of Metallurgical and Materials Engineering, Indian Institute of Technology Madras, Chennai, India
[3]Biomedical Technology Wing, Sree Chitra Tirunal Institute for Medical Sciences and Technology, Thiruvananthapuram, India

magnetic resonance imaging and sensing the path of drug delivery vehicles [2]. The building blocks for bioelectronic systems include sensors for sensing physiological signals, amplifier for the amplification of received signals, data encoder for converting the signal into data, power source to power the components, control unit and data storage or wireless transmitter. Miniaturized energy storage devices with high power, high energy density and optical transparency are critical for portable and wearable sensors. Micro and nano electromechanical systems (MEMs and NEMs) are miniaturized, lightweight and ultra-sensitive devices and these are recent addition to the biomedical systems. But miniaturized implantable devices face challenges due to limited battery capacity and few energy sources [3].

While organic materials are naturally stretchable, inorganic materials could be physically stretched and reconfigured while retaining their intended properties [4]. Inorganic nanoparticles are also attractive over organic nanoparticles for imaging and biomedical applications as they are highly inert and stable with good mechanical, optical and magnetic properties. They could also be easily functionalized and surface-modified, which improves their functionality and processability for different fabrication methods [5]. As further miniaturization of modern silicon-based microelectronic semiconductor devices is difficult, carbon-based nanomaterials have emerged as next generation electronic materials. These nanocarbons have very high carrier mobility and mechanical flexibility. Graphene, which is two-dimensional (2D) planar arrangement of carbon atoms, has particularly gained interest and widely studied. The strong sigma bonds between sp^2-hybridized carbon atoms renders graphene very high Young's modulus (in the range of 1 TPa) and exceptionally high 2D failure strength. Combined with their piezo-resistive properties, graphene nanomechanical resonators can sense ultra-low forces, charges and single atomic masses and hence they are promising candidates for nano sensors [6]. They can also be fabricated into highly efficient and cost-effective devices for energy harvesting and real-time imaging.

2.1.1 Carbon nanomaterials

Carbon is one of the most available elements in the Earth's crust. Till 1985, only two crystalline carbon forms such as diamond and graphite were known. This section describes detailed information on various carbon nanomaterials such as carbon nanotubes (CNTs), carbon nanohorns (CNHs) and graphene.

After the discoveries of fullerene in 1985 [7] and CNTs in 1991 [8], two other forms, CNHs in 1999 [9] and graphene in 2004 [10] were included into the new carbon allotropes family. Graphene is a 2D material and based on the number of layers, it can be divided as single, double and multi-layered graphene sheets. Individual CNH is short tubules (diameter 2–3 nm and length 30–50 nm) of wrapped graphene sheets with irregular horn-shaped or conical structures. Carbon nanotubes vary with graphene in the structure and morphology. The structure of CNTs is tubular and they are wrapped from planar graphene sheets. Based on the number of walls present in CNTs during their production, they can be single-, double- and multi-walled in nature. Various synthesis methods are available to produce CNTs and prominent among them are arc-discharge, chemical vapour deposition, laser ablation and pyrolysis. The number of walls in the produced CNTs differ based on the process parameters and the catalyst used during their production.

Graphite has a layered structure, with each layer containing hexagonally bonded carbon atoms. Each single layer is called graphene. The hybridization of carbon atoms in a graphene sheet is sp^2. The covalent bonds between the carbon atoms in the sheet are planar with the C–C bond length 1.42 Å and bond angle 120°. There is also π bonding between the carbon atoms in the sheet. This delocalized π bonding is responsible for the electrical conductivity. In comparison, the hybridization of carbon atoms in diamond is sp^3, and the bonds are arranged in a tetrahedral geometry, with a bond angle of 109.5°. The lack of free electrons makes the diamond an insulator (band gap of 5.5 eV) while the network structure helps in fast conduction of lattice vibrations (phonons) responsible for the extremely high thermal conductivity. The C–C distance in diamond is 1.54 Å. As evidenced from the C–C distance in these two allotropes of carbon, the bonding in graphene is stronger than that in diamond. The strength of graphene is not evident from the measurement of mechanical properties of graphite, since graphite has a layered structure (Bernal Stacking). The bonding between the layers is weak because of van der Waals forces and therefore enables the individual sheets of strongly bonded carbon atoms to slide with respect to each other.

Two-dimensional graphene is one of the crystalline allotropes of carbon. Carbon atoms in the graphene are densely packed in a hexagonal pattern with regular sp^2 bonds. Graphene is a basic structure for any sp^2-bonded carbon materials including graphite and it can be described as a one-atom thick layer of graphite hexagonal sheet. High purity graphene has several advantages such as its strength, low density and nearly transparent. Graphene also has high thermal and electrical conductivities due to the phonon scattering and free electrons availability, respectively. Graphene with unique physical, mechanical and electrical properties [11], is widely being used in several applications such as metal-nanoparticles support [12], gas storage [13], electrochemical energy storage [14], capacitors [15], etc.

Various types of graphene are available which differ in their three-dimensional structure and properties. Graphene nanoribbons, also called nanostrips in the zigzag orientation at low temperatures, show spin-polarized metallic edge currents, which have potential applications in the field of spintronics. In the 'armchair' orientation, the edges of the nanostrips behave like semiconductors. Using papermaking techniques on dispersed, oxidized and chemically processed graphite in water, the monolayer flakes form a single sheet of graphene and create strong bonds. These sheets, called graphene oxide paper, have a measured tensile modulus of 32 GPa. Graphene oxide flakes in polymers display enhanced photo-conducting properties. Graphene-based membranes are impermeable to all gases and liquids (vacuum-tight). However, water evaporates through them as quickly as if the membrane was not present. In 2013, a three-dimensional honeycomb of hexagonally arranged carbon was termed 3D graphene.

Bilayer graphene displays anomalous quantum Hall effect, tunable band gap and potential for excitonic condensation – making them promising candidates for optoelectronic and nano electronic applications. Bilayer graphene typically can be found either in twisted configuration where the two layers are rotated relative to each other, or graphitic Bernal stacked configuration where half the atoms in one layer lie atop half the atoms in the other. Stacking order and orientation govern the optical and electronic properties of bilayer graphene. One way to synthesize bilayer graphene is via chemical vapour deposition which can produce large bilayer regions that almost exclusively conform to Bernal stack geometry.

Though various conventional methods are available to synthesis graphene, the following section mostly focuses on the advanced and recent synthesis methods and describes the properties of synthesized and processed graphene.

2.2 Graphene synthesis and properties

Properties of graphene depend mainly on its method of preparation. Commonly graphene is prepared from graphite by mechanical exfoliation [16], liquid-phase exfoliation [17], chemical vapour deposition [18] and chemical reduction [19]. Synthesis of graphene from CNT or CNHs by aforementioned methods is tedious, because it involves cleavage of high energy multiple C–C bonds. However, graphene nanoribbons have already been achieved by the unzipping of CNTs using chemical [20–22], plasma etching [23], electrochemical [24], laser irradiation [25] methods and chemical conversion of CNH [26]. In particular, oxidized graphene nanoribbons obtained from chemical methods [22] were further reduced to enhance its electronic properties [20]. The first method to prepare graphene is probably the reduction of graphite oxide. In 1962, Boehm synthesized monolayer flakes of graphene from the reduction of graphite oxide. In their study, graphite oxide is rapidly heated and then exploited to obtain highly dispersed carbon powder with only a few graphene flakes. It is found from the graphite oxide reduction that the quality of the synthesized material is lower compared to other methods. This is due to the incomplete removal of the functional groups and also introduces structural defects on over-oxidation. Some researchers enhanced the protocol of oxidation to achieve high yield graphene films by removal of most of the functional groups. The charge carrier mobility exceeded 1,000 cm/Vs. The addition of a graphite oxide film to a DVD and burning it in DVD writer produced a thin graphene film having high electrical conductivity (1738 S/m) and specific surface area (1520 m^2/g).

Recently, Sahu *et al.* [26] converted CNHs into graphene. CNHs were heated up to 100 °C and H_2O_2 was added drop-wise onto the CNHs, and was left to react for 10 min. Then, the suspension was washed and dried. Eight milligrams of graphene was produced by this process. It was noted that the formation of graphene sheets was very slow between 50 and 80 °C and no reaction was observed below 50 °C. The transparent single-layer graphene sheets had folded edges. The authors have also mentioned that as prepared graphene was not oxidized unlike that obtained by Devi *et al.* [27]. Graphene oxide prepared by Hummers method shows a characteristic peak at 230 nm corresponding to $\pi \rightarrow \pi^*$ and a hump around 300 nm corresponding to $n \rightarrow \pi^*$ transition. The conversion of the CNH could be through a radical mechanism. Similar reactions of H_2O_2 on activated carbon were also reported in the literature [28]. In general, single-layer graphene can be identified by selected area electron diffraction (SAED) analysis [29,30] and Raman spectrum [31,32]. Sahu *et al.* also measured the electrical properties of CNHs and the as-prepared graphene sheets and claimed that the conductivity value of synthesized graphene sheets showed higher magnitude than that of CNHs. This could be due to the change of amorphous nature of CNHs to crystalline nature of graphene after H_2O_2 treatment [26]. The electrical properties obtained are better than that of the reported graphene nanoribbons prepared from CNT by the unzipping method [20] as well as graphene nanosheets prepared from graphite by solution method [17].

Currently, researchers are successful in producing graphene sheets using arc discharge method [33], which is mostly utilized for the synthesis of variety of CNTs including single-walled and multi-walled in nature. Figure 2.1 shows the TEM image of multi-layered graphene produced by arc discharge method. The high-purity multi-layer graphene sheets with 100–200 nm size and large surface area are identified. High-resolution TEM images (Figure 2.1 (b–d)) show that the arc discharge synthesized multi-layered graphene sheets are with two to four number of layers.

The graphene materials have potential applications including durable display screens, electric circuits and various medical devices. Varieties of graphene materials can be synthesized commercially and have been scaled up by several companies to sell graphene in large quantities. The membranes of graphene films have larger surface area due to their nano size and hence these films can be utilized in water purification as it allows only water to pass through and restrict all other liquids and gases. Further development and commercialization of such type of

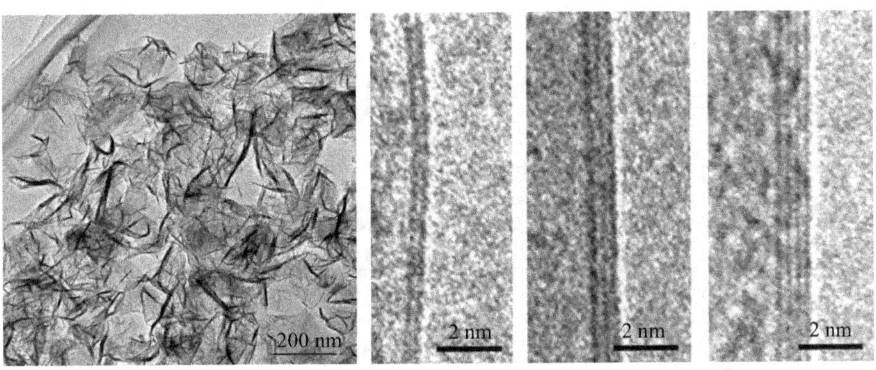

(a) (b) (c) (d)

Figure 2.1 (a) TEM image of multi-layered graphene sheets produced by DC arc-discharge. (b-d) HRTEM images showing the edge of multi-layered graphene sheets consisting of two (b), three (c) and four (d) layers. [Reprinted from [33] with permission from Elsevier] graphene membranes could be useful in biofuel production and the beverage industry and could enhance the economy of the industries.

Graphene is thermodynamically unstable when its size is less than about 20 nm, as 6,000 atoms are necessary for the least stable structure. For the molecules larger than 24,000 atoms, it becomes the most stable fullerene. The thermal conductivity of the graphene at room temperature (4.8×10^3 to 5.3×10^3 W·m^{-1}·K^{-1}) made by a non-contact optical technique is very high compared to the thermal conductivity measurements of the CNTs and diamond. The thermal conductivity of carbon material depends on the ratio of its isotopes (i.e. the ratio of 12C to 13C). Most of the graphene materials have pure 12C and hence give higher thermal conductivity values. The highest thermal conductivity of the graphene makes it to be used as additive in coolants. Graphene is known to be mechanically the strongest with stiffness (elastic modulus) of 1 TPa. Moreover, the density of graphene is very low, which makes graphene lightweight compared to other carbon materials. The graphene with its nano characteristics and higher specific surface area is useful in many biomedical and related applications. The high electrical conductivity and high optical transparency of the graphene make it a suitable material for transparent conducting electrodes, touch screens, solar cells, liquid crystal displays, organic photovoltaic cells and organic light-emitting diodes. The excellent mechanical properties and high flexibility of the graphene are advantageous over traditional materials which are very brittle in nature. Table 2.1 shows the typical characteristics of graphene.

High carrier mobility and low noise are useful properties of graphene for integrated circuits, in which it can be used as the channel in a field-effect transistor (FET). It is very difficult to produce single-layer graphene directly on suitable substrate. Though researchers face challenges to fabricate such single sheet of graphene, so far, the smallest transistor made from the graphene is one atom thick and ten atoms wide. The fabricated graphene-based transistors also can operate at GHz frequencies. Using graphene, both n-type and p-type transistors can be created, and it is possible to demonstrate a functional graphene-based integrated circuit with a complementary inverter with such type of one n-type and one p-type graphene transistors.

Table 2.1 Characteristics of graphene material

Density	1.1 g/cc
Number of layers	Single, double or multiple
Specific surface area	400 m^2/g
Average flake thickness	4 nm
Melting point	3,000 °C
Electrical resistivity	10^{-6} Ω cm
Thermal conductivity	5,000 W·m^{-1}·K^{-1}
Young's modulus	1 TPa

2.3 Functionalization of graphene

This section describes the functionalization of graphene and its surface enhancement. Functionalization of graphene enhances the surface characteristics and thereby its properties. There are several functionalization methods which include covalent and non-covalent bonding, and functionalization of graphene surface with nanoparticles, deposition of quantum dots and substitutional doping. The individual functionalization method has its own advantages to enhance the surface properties of synthesized graphene.

The functionalization of synthesized graphene-based materials has been developed to improve the mechanical, electrical and thermal properties of the graphene to be used in several applications. Majority of the functionalization methods help to improve the dispersibility of graphene after functional attachments. The dispersion of graphene in solvents play a crucial role in the formation of nanocomposites with graphene and enhances the mechanical properties of composites. The band gap of graphene can be altered using chemical doping for nanoelectronic devices [34]. Most of the functionalization methods include the covalent addition of free ions to C–C bonds and sp^2 carbon atoms of graphene. Graphene can be produced in many ways and graphite is an inexpensive raw material to produce bulk amount of graphene. The synthesized pure and pristine graphene sheets cannot be dissolved in polar solvents as they are hydrophobic in nature. Hence, the non-covalent functionalization is essential to dissolve graphene in common solvents especially in organic solvents. This enables the graphene to be utilized in several advanced applications by avoiding the $\pi-\pi$ stacking between graphene sheets. Similar to that in CNTs, the electronic network of the graphene sheets is not disturbed by non-covalent functionalization by π-interactions [35,36]. Graphene has sigma and π-bonds. The free electrons available in the π-bonds are responsible for the electrical conductivity of graphite and graphite-based derivatives including graphene. Electrical properties of the pristine and functionalized graphene structures have been investigated extensively in the literature. Non-covalent functionalization of intermolecular interactions involving π-systems plays a crucial role in stabilizing the proteins, complexes, organic molecules and nanomaterials [37–40]. The interaction of π-systems with the functionalized groups drastically changes the electronic structure of the nanosystems and hence plays a major role to fabricate the electronic nanodevices and also the design of nanomaterial systems. Several research studies have been performed to identify the π-interaction with functional groups such as nonpolar gas$-\pi$, H$-\pi$, $\pi-\pi$, cation$-\pi$ and anion$-\pi$ interactions [41,42]. Extensive investigations have been done on the energetic and geometrical importance of π-interactions. The strength of the π-interactions is determined by the combined effect of attractive and repulsive forces. A pristine graphene has large active surface area, in comparison with CNT, CNHs, amorphous carbon and other carbon nano structures. Moreover, pristine graphene has high conductivity and it is highly pure since the synthesis of graphene is free of metallic catalyst or carbon impurities [43]. The other forms of graphene such as graphene oxide and partially reduced graphene oxide materials also exhibit these advantages except electrical conductivity. The deposition of nanoparticles can be performed on the synthesized graphene using various methods. The most commonly used methods are electroless and electrolytic deposition to decorate the carbon materials. Pre-coating methods such as sensitization and activation are necessary before electroless plating to achieve better coatings. In addition, chemical methods are also used to perform the deposition of nanoparticles on graphene. To achieve the uniform and fine coating of the metal nanoparticles on graphene, several factors and parameters need to be considered such as temperature, time, pH of the solution, concentration of the metal-reducing agent and concentration of the metal source. Recently, silver nanoparticles were decorated onto as-prepared graphene using silver nitrate ($AgNO_3$) in aqueous solution. The high electron density of graphene is enough to reduce $AgNO_3$ to Ag (0.8 V vs. NHE). Uniform dispersion of silver nanoparticles were obtained on the surface of graphene sheet with the mean size of 6 ± 2 nm. Graphene sheets obtained from CNHs show better electrical properties due to strong electron density [26].

Quantum dots are nanostructures and they exhibit exciting optical and electronic properties. However, many challenges need to be overcome for their effective use. In solar cells, the utilization of quantum dots suffers from accumulation of charge carriers in the device [44]. Graphene oxide nanoplatelets decorated with quantum dots overcome these challenges by acting as nanowires which promote direct and efficient charge transfer to quantum dots. Thus, the efficiency of the solar cells is increased. Substitutional doping of graphene involves the replacement of carbon atoms from the hexagonal honeycomb lattice of graphene by N_2 or B atoms. The doped graphene sheets show n- or p-type behaviour depending on the electrophilic character of the atoms that substitute the carbon atoms. The doping can be controlled by various parameters and by controlling the degree of doping, electrical properties can be monitored, and hence the applications of the modified graphene can be explored in nanoelectronics devices.

Wrinkled structures could be formed from graphene nanosheets on surface modification, which could also reduce the aggregation tendency of graphene nanosheets [45].

2.4 Graphene-based nanocomposites

Generally, composites can be defined as a heterogeneous mixture of two or more materials with unique combination of properties, which depend on the characteristics of individual components, size and shape distribution of reinforcement and also on the interaction between the two at their interface. Composites can be classified based on the matrix as polymer, ceramic and metal matrix; the type of the reinforcement, which includes chemical nature (oxides, carbides, nitrides), shape (continuous fibres, short fibres and particles) and orientation (oriented or non oriented); or based on processing route of the composites (in-situ, ex-situ, liquid metal-based, solid state processed, etc.). Polymer matrix, often called a resin, is used in several commercially produced composites. Depending upon the starting raw ingredients, various polymer materials are available in the market. Proper selection of reinforcement is mandatory to obtain better load transfer from the matrix to reinforcement and thereby to enhance the mechanical properties of the composites. Several literature studies are reported on the preparation and properties of graphene/GNP-reinforced epoxy composites [46–58]. It is found that with the addition of graphene-based nanoparticle, the electrical properties of the composites are significantly improved. Guadagno *et al.* have found that 1 wt. % GNP-reinforced epoxy is nearly two times stronger compared to pure epoxy. However, the addition of higher amount of GNP results in poor dispersion and thereby decreases the strength of the composite [59]. Due to the large specific surface area of GNP and high viscosity of polymers, dispersion of carbon nano-fillers is challenging. To disperse the nano-fillers uniformly in polymer matrices, several dispersion techniques have been used [60]. The addition of more than 1 wt. % GNP in composites leads to agglomeration of reinforcements and thereby decrease in shear strength [61].

The agglomeration of carbon reinforcement also leads to the deterioration of electrical properties due to the disturbance of the electron flow in the composites. Stankovich *et al.* [20] measured the polystyrene composites reinforced with graphene and found that the electrical conductivity of the composites increases with increase in the amount of reinforcement. The increase in conductivity is due to homogeneous distribution of graphene in the composite, as the homogeneity results in formation of continuous network of conductive filler paths through the insulating matrix. Filling of conductive materials in elastic conductive composites could be designed in five different ways: implanting conductive fillers in elastomers, filling microchannels in composite with metals, filling elastomers in conductive networks, blending conductive fillers and elastomeric materials and synthesizing metal fillers within elastomers [62].

With metal matrix, though the mechanical properties are enhanced, electrical properties will deteriorate with the addition of carbon reinforcement. This decrement in the electrical conductivity is because the metal is electrically conductive, and the addition of carbon reinforcement leads to the obstacle of the electrical flow in the metal matrix [63].

2.5 Advances in diagnostic sensors

Sensors are devices that detect the input from an object and send the information to processors. In general, sensor will have two major components – the sensing component and the transducer. Based on the function, typically, transducer is of several types such as voltammetric, amperometric, conductometric, spectrophotometric, etc. In case of nanosensors, nanomaterial acts as the sensing device. The reduced size and better performance in terms of sensitivity of the nanosensors are major advantages over conventional sensors. Wide variety of nanomaterials can be fabricated into sensors, depending on their sensitivity to different stimuli such as gases, difference in electrical potential, fluid flow, optical, magnetic, pressure, thermal, mechanical strain, etc.

Sensors that identify the biological components like cells, tissues, proteins, microorganisms, nucleic acids, biomimetic polymers and antibodies are known as biosensors. Biosensor is an analytical device which associates with the electronics and signal processors. Biosensors detect the biological components with physio-chemical transducer. The biosensors are used to detect and monitor vital body signals of patients, premature infants, children, elderly, athletes, psychiatric patients, and people in remote regions where medical services are not easily accessible. Hence, they are significantly effective in prevention, timely diagnosis, treatment and control of diseases. Vital body signals include heart rate, blood pressure, body temperature and respiration rate. Other important physiological signals that need to be monitored are electrocardiogram (ECG), muscle current or electromyogram and brain electric field using electroencephalogram. Biosensors are also used to detect various biomolecules and disease-causing pathogens.

Surface plasmon resonance (SPR) is the resonant oscillation of electrons at the interface between the positive and negative permittivity of the material stimulated by an incident light. SPR-based biosensors play a role in interacting with biomolecules and chemical detection. They allow label-free and real-time detection with high sensitivities. Surface plasmons are longitudinal charge density wave illuminated along the interface of metal/dielectric film, when the incident light polarizes with the traverse magnetic direction. This is a function of refractive indices of the metal, dielectric and the analyte [64]. Adding graphene layer to SPR-based biosensors increases their sensitivity, as the optical property of each graphene layer could change the SPR angle by 40% in prism and 45% in planar waveguide-based SPR biosensors [65].

Microelectrode array (MEA) is an arrangement of multiple electrodes targeting several sites in parallel for extracellular recording of neural and cardiac signals. They are also used for the stimulation of biological tissues and cells whose response is to be studied. The number and arrangement of electrodes in a MEA decides its spatial resolution. Increased spatial resolution results in the increased sensitivity of the sensor. MEAs with multichannel systems are identified to have the highest spatial resolution. Very small electrodes with less impedance and low noise level are achieved by introducing nanomaterials as electrode material. Transistors are semiconductor devices with at least three terminals, and they are used to amplify the signals and power. A current or voltage applied to one pair of the terminals is controlled by another pair. The output power is higher than the input power and in the transistor. When compared to MEA, transistors are active elements and hence are more functional and tuneable.

2.5.1 Graphene-based field-effect transistors

Field effect transistors (FET) controls the electrical behaviour of the devices by using an electric field. The major terminals in the FET are source, drain and gate. The electrical conductivity between the drain and source is monitored by the voltage difference between the gate and the body of the device. Based on the charge to the gate to the body, FETs are classified as n-type (negative charge) and p-type (positive charge) FETs. They are also categorized depending on the material used and some examples are metal-oxide semiconductor (MOS) and metal-nitride-oxide semiconductor (MNO) FETs. Recently, development in fabrication techniques led to the fabrication of next-generation FETs used in bioelectronic devices. Graphene-based field-effect transistors (G-FETs) are also developed for biosensor applications. A G-FET consists of a conductive graphene-based channel through two metal contacts which act as the source and drain electrodes. Graphene-based FETs could largely reduce the size of the device, and the transconductance is linearly dependent on width/length ratio of graphene. Graphene is used as an active layer and also as an electrode in FETs and is advantageous over conventional organic materials as they are more compatible with flexible and stretchable materials [66]. Graphene-based FETs have the advantages of high sensitivity, low cost and time for preparation, and high throughput detection with low detection limits. The high specific surface area of the graphene is an advantage for bio-sensing applications. The exceptionally high electron mobility and transconductance property of graphene make it an ideal candidate for high sensitive field-effect signal transducers. The source-drain current can be modulated using a change in an external field. Similarly, changes in biological environments that need to be measured/detected could trigger or act as a switch for effecting the performance of graphene-based electronic devices [67]. Electrically responsive tissue such as brain, ear and skeletal muscle could be stimulated, and the energy can be utilized in neural prostheses in the treatment of vagal nerve, cochlear implants, retinal implants and spinal cord [68].

Single-layer graphene FET (single-layer graphene flakes on oxidized Si substrate) could successfully record electrogenic signals and have the unique capability of recording signals as both p- and n-type devices, simply by changing the water gate potential. Signals recorded with larger graphene device could represent the average of extracellular potential from different sources and give broader peak-to-peak signal width [69]. Distinguishable action potentials could be recorded with graphene FETs from ex vivo heart tissue, in vitro cardiac-like cell line and in vitro cortical neurons [70]. For applications such as prosthetic skin and minimally invasive surgery, sensors with multi-axial detection capabilities, high sensitivity and reliability are required [71]. Monolithic graphene-graphite designed as nano FET sensors have 3D sensing capabilities with superior sensitivity, structural flexibility and nanoscopic sensing resolution. Electrical detection from nanoscale electric filed modulation of the graphene channel, detection of localized chemical changes with high sensitivity could be achieved with these sensors [72].

2.5.2 Gas and chemical sensors

Detectability of wide range of chemicals such as glucose, lactate, ascorbic acid, dopamine and uric acid in blood enables the identification of related diseases. G-FET sensor designed on flexible polyester substrate could detect glucose level in the range of 3.3–10.9 mM and is useful for the diagnosis of diabetes [73]. Lactate excreted in sweat and blood are biomarker for a variety of diseases such as heart failure, liver diseases, drug toxicity, metabolic disorders and microbial contamination. Flexible graphene biosensor could detect as low as 0.08 μM lactate in a steady-state measuring time of 2 s [74].

Aptamers such as adenosine triphosphate, nicotinamide adenine dinucleotide, acetylcholine, cholesterol, benzenediol isomers, epinephrine; gases and ions can be detected using graphene nanopores [75]. Aptamers are peptide molecules that bind to a specific target molecule. Graphene–gold nanoparticle composite could serve as a stable substrate for aptamer mobilization in a microfluidic chip designed with an aptamer tagged with ferrocene as redox probe for detection of norovirus in spiked blood samples, and also helps in signal amplification [76].

2.5.3 Magnetic and electromagnetic sensors

Microelectrodes capable of recording neural signals and simulation are valuable tools for the study of the biophysical aspects of the central nervous system. The data can be used to diagnose and treat neuronal disorders. Though the traditional electrodes made of platinum, titanium, gold, glassy carbon, etc. have good impedance, they lack long-term stability in performance. Carbon-based nanomaterials have better stability and performance in terms of sensitivity of neural signals [68]. Soft electronic materials with strength and pliability are preferred as implantable sensors. High conductivity and magnetically responsive, graphene-based nanomaterials and composites are largely used for bioimaging [77]. Electromagnetic waves of terahertz range are non-invasive, non-ionizing and have unprecedented sensing ability for a wide range of biological materials [78].

2.5.4 pH and temperature sensors

pH responsive poly (4-vinyl pyridine) added to graphene can be used in the design of pH-responsive switchable biosensor that can detect the presence and quantity of enzyme glucose oxidase. Graphene oxide-glucose oxidase – poly (4-vinyl pyridine) solution was drop-cast onto glassy carbon electrodes and dried. This electrode could show appreciable difference in contact angle with different pH. At pH 6 (off state), polymer turns to shrunken state and at pH 4 (on state), polymer is swollen, as shown in Figure 2.2. It is capable of amperometric glucose sensing and pH-dependent loading of enzyme glucose oxidase [79].

Similarly, temperature-responsive switchable interface can be used to control electrochemical bioreactions such as bio-catalysis, using a zipper-like mechanism (Figure 2.3). The mechanism consists of a 2D graphene donor and a polymeric receptor which are rationally assembled. At low temperature of 20 °C, considerable shrinkage was observed in donor–receptor interface, causing restricted access to the associated enzyme cholesterol oxidase to its substrate. At higher temperature of 40 °C, the surface is made more accessible, increasing the permeability and easy diffusion of electro-active species through the electrode surface. The output is obtained as large peak currents. Thus, the response with change in temperature results in amplified electrical signals [80].

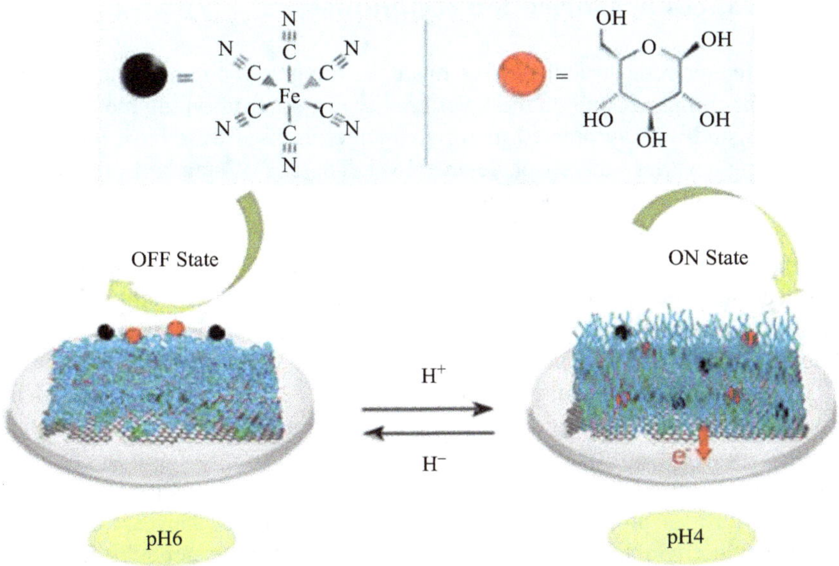

Figure 2.2 Schematic representations of a pH-encoded switchable graphene oxide interface at two different states. The tunable character of the interface was tested using redox-active ferri/ferrocyanide probe (black circle) and glucose as a substrate (red circle). Reprinted from [79], Published by The Royal Society of Chemistry

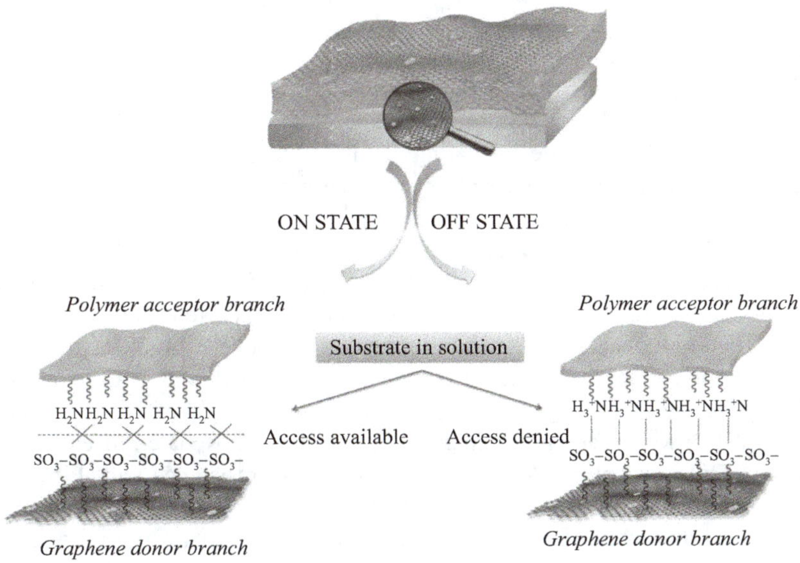

Figure 2.3 Schematic representation of on/off switchable bioelectrocatalytic graphene interface. Reprinted from [80] © John Wiley and Sons

2.6 Advances in fabrication techniques

Nanoparticle electrode structures have made the integrated circuits much smaller in size, without compromising on the functionality. Rapid prototyping methods for 3D printing assisted with automated motorized stages and software have made it easy for designing a variety of sensor devices [81]. The use of standard printed circuit board manufacturing techniques has also proved efficient in building solid contact ion selective electrodes [82]. Conductive coating on flexible materials is advantageous than single-layer sensors. Printed and flexible electronic systems could be embedded into clothing with ease, without affecting the functionality. Biosensors printing techniques include drop-casting, screen printing and inkjet printing [83]. Lithography and plasma etching are other familiar microfabrication techniques.

Graphene inks are gaining attention because of their optoelectronic, electrochemical and mechanical properties. Stable dispersion of graphene with rheological properties appropriate for printing is important to achieve maximum flexibility and functionality. Capasso *et al.* first used liquid-phase exfoliation of graphite in water/ethanol mixture for preparation of graphene flakes and then they are dispersed to produce conductive ink, which is used to print flexible polyester substatres [84].

Inorganic-based laser lift-off process could be used for dry etching and transfer of high-performance inorganic thin films onto flexible substrates for manufacturing large-area flexible inorganic devices [85]. Broad range of vertical nanostructure arrays (VNAs) could be fabricated using plasma etching of semiconductors, oxides, metals, glass and polymers. These VNAs with sharp tips enable field electron emission (FEE) devices to operate at low voltage with stable current. FEE is the basis of main electron source used in microscopy, display and vacuum electronics. VNAs are also used as DNA sensors, biomimetic structures, gas sensors and transdermal drug delivery [86].

Figure 2.4 (a) shows typical fabrication steps required to form flexible supporting film for G-FETs. A 200 nm sacrificial aluminium layer is laid in between carrier silicon wafer and the insulating flexible substrate polyimide, so that the flexible layer can be easily released after lithography and metallization processes. Graphene synthesized onto copper foil using CVD method is carefully transferred on a resist substrate as a carrier and washed in deionized water in clean room. This graphene sheet attached to the resist substrate is transferred to flexible layer onto which graphene is to be fabricated. The resist substrate is lifted off using acetone as solvent. Then graphene and electrode patterning are done using laser lithography. Finally, aluminium sacrificing layer is removed by deep etching and the fabricated flexible graphene layer (Figure 2.4(b)) is released. The field-effect measurements in Figure 2.4(c) were made in liquid gate conditions in phosphate-buffered saline at 30 mV. G-FETs on polyimide exhibit higher contact resistance and shifted Dirac point [87].

A typical assembly of SiO_2 microelectrode array is shown in Figure 2.5. The electrode probe is coated with graphene for improving biocompatibility and four different-sized graphene FETs were microfabricated onto insulating silica layer.

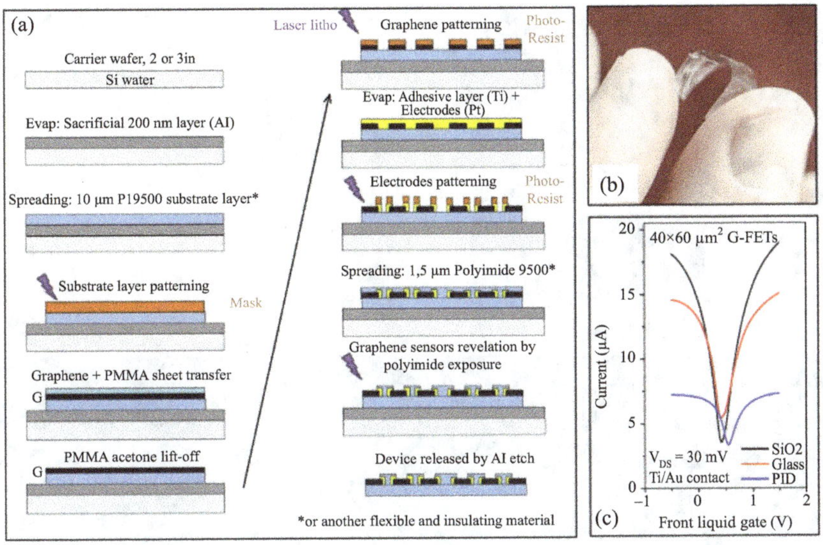

Figure 2.4 Fabrication of polyimide-flexible supporting device for G-FETs and transconductance measurements: (a) process flow to build a flexible device on polyimide or any flexible insulating substrate using bottom-up approach, (b) a flexible device released from silicon wafer and (c) field effect in 40 × 60 µm G-FETs on different substrates. Reprinted with permission from [87]

This MEA probe can be used as motor cortex implant to detect the activity of motor neurons in brain [87].

Graphene printed onto water-soluble silk film could permit the biotransfer of graphene nanosensor transducer onto tooth enamel. A parallel inductor–resistor–capacitor resonant circuit was simulated, designed using simulation tool. This circuit was fabricated as biosensor using planar coil antennae with gold inductive coil for wireless transmittance and graphene as resistive electrode. Through self-assembly of peptides on the graphene transducer, pathogenic bacteria in saliva are detected [88]. The fabrication and transfer onto tooth enamel is shown in Figure 2.6. This flexible biosensor could also be successfully transferred onto muscle tissue.

2.7 Advances in monitoring and therapy

For multifunctional nanoparticles which are used for imaging, targeting and therapy, precise control of the surface chemistry is important [89]. Graphene quantum dots have strong photoluminescence property and hence explored for bioimaging. They are also non-toxic and nano size enables them to be internalized and transported by

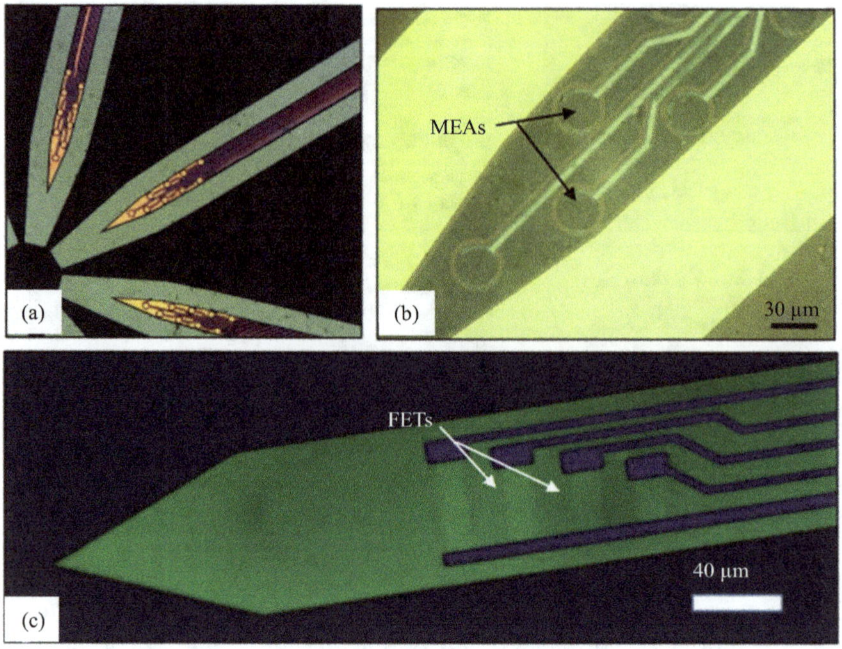

Figure 2.5 Fabrication of graphene biosensors on 3D micromachined devices: (a) optical microscope image of microfabricated MEA probes, (b) expanded view showing graphene MEAs, and (c) Optical micrograph of microfabricated probe showing four different-sized graphene transistors. Reprinted with permission from [87]

most cells. Hence their path inside human body could be easily traced from outside through bioimaging techniques. This property can be used for nanoparticle-mediated drug delivery to infected cells and monitoring the path of drugs. Sulphur doping could increase the emission of blue colour of graphene quantum dots, which is effective material for live cell imaging [90].

2.7.1 Microfluidics

Microfluidics is a network of 10 to 100 μm-size small channels which can handle liquids in nanolitre or picolitre scale. As fluids could be precisely manipulated using a micro-scale device, small-scale interaction of cells such as interaction with other cells, biomolecules, toxins, pharmaceutical compounds and nanomaterials could be studied easily, which was not possible by conventional methods [91]. Thus microfluidics has a major role in future biomedical research and analysis. Because of miniaturized devices and liquid handling, microfluidics has made possible the 'lab-on-a-disc' concept, for a wide range of medical diagnostics applications.

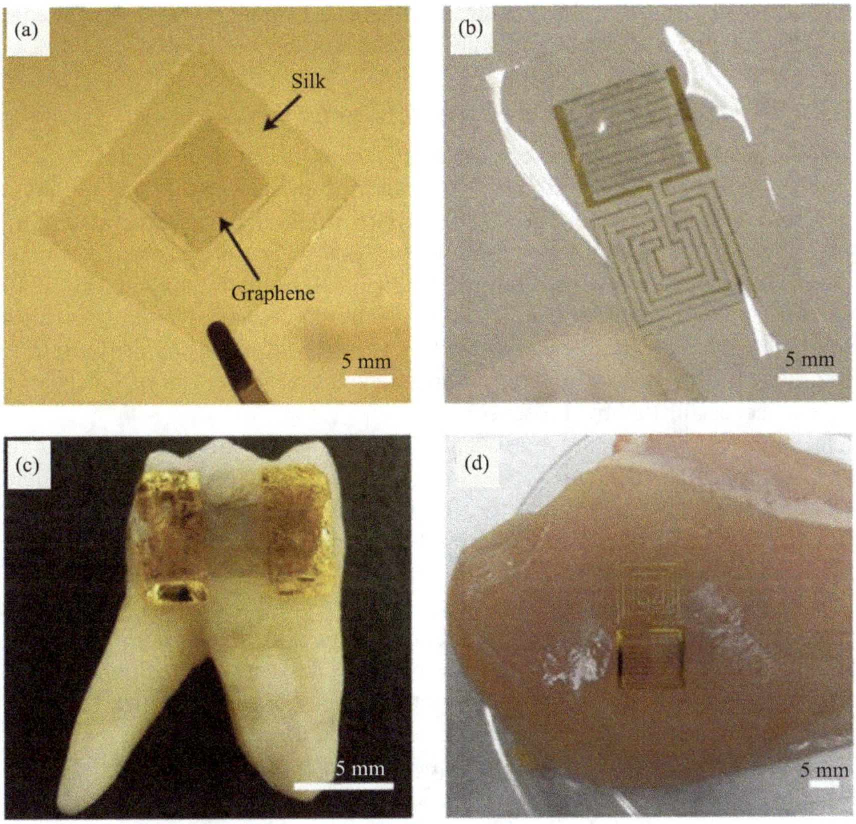

Figure 2.6 *Graphene biotransfer and characterization. (a) Graphene printed onto bioresorbable silk film. (b) Passive wireless telemetry system consisting of a planar meander line inductor and interdigitated capacitive electrodes integrated onto the graphene/silk film. (c, d) Graphene nanosensor biotransferred onto the surface of a human molar (c) and onto muscle tissue (d). Reprinted by permission from Springer Nature, [88] © 2012*

Porous graphene sponge coated with shape-memory polymer *trans*-1,4-polyisoprene (TPI) could be a slippery film with electrochemically tunable wettability [92]. This graphene sponge could rapidly recover from 85% strain compression, for 10 cycles, without shrinkage or cracks, and about 90% height was retained in the sponge even after 5,000 cycles. This kind of shape memory sponge with tuneable property could be used for liquid handling in microplates. Microplates are widely used in biotesting, where many numbers of tests involving very low quantity liquid (in terms of few μl) are used. The main challenge of microplates is that pipetting different liquids in each microplate multiple times is labour intensive and time-consuming. High throughput liquid handling for microplate can be done using this graphene/TPI hybrid film, by controlling with a circuit. This shape memory material can be reused by compressing and cleaning off the liquid.

2.7.2 Wireless, portable and wearable electronics

Because stretchable strain sensors can easily conform to the complex nature of human physique, they are preferred for monitoring human motion and to measure bio-signals. Graphene nano-flakes infused into rubber-like adhesive pad can be used as piezo-resistive strain sensor to measure heartbeat and wide range of human motion. Ultrathin but with exceptional mechanical strength and stability of graphene makes them very useful for flexible electronic devices. Flexible devices are used in wearable consumer electronics, soft robotics, medical prosthetics, electronic skin and health monitoring [93]. Graphene and transition meta dichalcogenides are most successful flexible biosensors [94].

Textiles and clothing serve as ideal material for wearable electronics, as they can accommodate different types of sensors in different locations, signal-processing units and transmitters for transmitting the data to remote location and to get the feedback, yet comfortable and utilizing minimum power [95]. Fabric made of rGO nanosheets and electrospun nylon-6 nanofibres show appreciable sensitivity to NO_2 and excellent bending stability [96]. Reduced graphene-oxide dispersion can be applied to cotton fabric using simple pad-dry technique to produce durable, washable and flexible e-textile material. This single graphene e-textile material is multifunctional as it can be used as sensor and flexible heating element powered by the graphene textile supercapacitors. The change in resistivity of the rGO-coated fabric with respect to bending; compressing and twisting can be utilized for mechanical sensing [97]. The schematic of the preparation processes for large-scale production of flexible e-textile material and how they can be applied as wearable electronic device are shown in Figure 2.7.

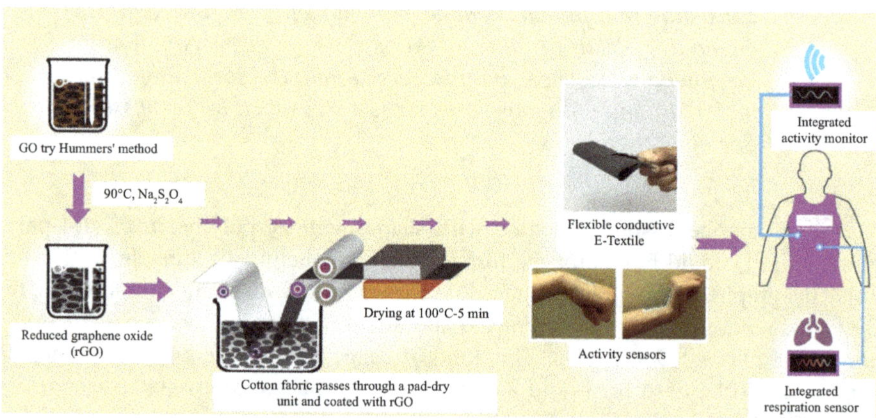

Figure 2.7 Schematic diagram of the scalable production of graphene-based wearable e-textiles. Reprinted from [97]

2.8 Bio-microelectromechanical systems (MEMS) and bio-nanoelectromechanical systems (NEMS)

MEMS are miniaturized mechanical and electromechanical devices. MEM device has various micro-electronic structure components such as micro-sensor and micro actuator. The component size in MEMS varies between 1 to 100 μm and the size of the MEM device ranges from 0.02 to 1 mm. The micro sensors and micro actuators used in MEMS are generally piezoelectric transducers, which converts measured mechanical signals into electrical. MEMS incorporated with micro-transducers improve the capabilities of micro devices used for controlled drug delivery systems [98]. Scanning tunnelling-tip microscope (STM), which is used to detect individual atoms and molecules in nanometre scale and atomic force microscope (AFM) which is used to manipulate the position of individual atoms and molecules on the surface of a substrate, are typical examples of electromechanical based devices. Surface stress biosensors such as micro cantilever, and micro membrane which act as transducers are highly sensitive, fast and economic methods for drug screening and monitoring therapeutic effect [99]. Biochemical liquid samples such as metabolites, macromolecules, proteins, nucleic acids, cells and viruses can also be analysed using bio-MEMS [100].

NEMS are the devices in which the electrical and mechanical behaviour is integrated at the nanoscale. The major difference between MEMS and NEMS is the size of the functional components of the devices. Compared to MEMS, NEMS have the high surface area to volume ratio, lower mass and large quantum effects. NEMS are generally integrated with transistors with mechanical actuators or motors. Based on the application, carbon nanomaterials can be utilized in both MEMS and NEMS. In MEMs, they can be used as molecular wires and sensors.

NEMS are fabricated in two different approaches – top-down and bottom-up methods. Top-down approaches are generally from manufacturing of MEMS structures using electron beam lithography. Bottom-up approaches are by assembly of the atoms and molecules as building blocks. Nanoparticles form a bridge between bulk materials and molecular structures. Intrinsic graphene has symmetric properties and do not show piezoelectric nature. But the piezoelectricity can be introduced in graphene by inducing defects or adding foreign atoms [101]. Also, monolayer or few layers graphene sheets on a supporting material can oscillate at its natural frequency and could serve as a nanomechanical resonator. Graphene cantilever, graphene clamped-clamped resonator and graphene drum resonator are few examples [102]. Graphene-based materials are of light weight, and as graphene has large surface area, they can bind over the entire device with lower quantity of the material with enhanced properties. Graphene-based NEMS are applicable in bioinspired technologies such as biomimetics which means mimicking the biology found in nature. These NEMS also can be used in biotechnology to enable new discoveries, for the amplification and identification of DNA structures, nanomachined STMs, biochips for the detection of hazardous biological agents and nanosystems for bio-screening applications. Various research studies are still going on in the field of biotechnological applications of NEMS devices.

2.9 Advanced power sources and control systems

Miniaturized devices should be complemented with low-power consumption strategies. Power is required for energy delivery, data acquisition, processing and transmission, and feedback systems. Leakage and control losses should also be minimized. Energy-harvesting approaches from wireless systems or ambient energy such as vibrations and thermal could eliminate the need for batteries in miniaturized devices [103]. Lightweight graphene-based smart generators have the capability to harvest electricity from chemical potential energy in response to external stimuli such as moisture, friction, liquid flow, pressure force and heat. They also conduct electricity and store the power generated, in a controllable manner [104].

Flexible thin electronics based on inorganic materials could also be used to generate power from human body by using one or more of devices such as piezoelectric generator, optoelectronic system, thermoelectric generator and triboelectric energy harvester [3]. Triboelectric generators could be formed directly on human skin using atomically thin graphene of <1 nm thickness as electrode, polydimethyl siloxane of <1.5 μm thickness as electrification layer and <0.9 μm thick polyethylene terephthalate as substrate. They have the advantage of higher contact area, enabling more sensitive self-powered touch sensors. This system could be self-powered by the electricity generated on contact with human body, and human motion can be effectively sensed and communicated as digital signals [105]. Similarly, the powerful contractions from the heart could serve as a battery substitute for implants which are near heart, such as pacemakers, defibrillators or ECG recorder [106]. Zarrabi *et al.* developed 'cross'-shaped nano aperture on gold substrate and coated it with graphene, this device could serve as nano-antenna for transmittance at mid-infrared frequency. This device could also act as energy-harvesting device [107].

2.10 Bioelectronics safety

Irradiation with 3 MeV proton resulted in decreased neuronal frequency of the human brain tissue with increase in the radiation dose, the loss is about 50% at an acute dose of 300 Gy and the electrophysical activity was completely lost at radiation dose of 800 Gy [108]. Any bioelectronic device should be studied for the electrostatic charge they dissipate to their surroundings, and ways to reduce the electrostatic discharge should be identified. A dual directional silicon-controlled rectifier device was able to protect the biomedical integrated circuits in complementary metal oxide semiconductor (CMOS) technologies [109]. Electric signal transmission at tissue–electrode interface involves capacitive stimulation because of electrode potential on tissue interface or charge transfer due to alternate reduction–oxidation. For neuronal signal conduction, charge transfer is predominantly required to stimulate neuronal conduction pathways, which also results in surrounding tissue damage. To avoid damage, insulation with nanoporous coating of inorganic materials such as Al_2O_3 and silicon are suggested by Arsiwala [110].

References

[1] Ning, C., Zhou, L., and Tan, G.: 'Fourth-generation biomedical materials' *Mater. Today*, 2016, 19, pp. 2–3.

[2] Gong, H., Peng, R., and Liu, Z.: 'Carbon nanotubes for biomedical imaging: The recent advances' *Adv. Drug Deliv. Rev.*, 2013, 65, pp. 1951–63.

[3] Lee, S.H., Jeong, C.K., Hwang, G.T., and Lee, K.J.: 'Self-powered flexible inorganic electronic system' *Nano Energy*, 2014, 14, pp. 111–25.

[4] Nassar, J.M., Rojas, J.P., Hussain, A.M., and Hussain, M.M.: 'From stretchable to reconfigurable inorganic electronics' *Extrem. Mech. Lett.*, 2016, 9, pp. 245–68.

[5] Huang, H.-C., Barua, S., Sharma, G., Dey, S.K., and Rege, K.: 'Inorganic nanoparticles for cancer imaging and therapy' *J. Control. Release*, 2011, 155, pp. 344–57.

[6] Khan, Z.H., Kermany, A.R., Ochsner, A., and Iacopi, F.: 'Mechanical and electromechanical properties of graphene and their potential application in MEMS'*J. Phys. D. Appl. Phys.*, 2017, 50, p. 053003.

[7] Kroto, H.W., Heath, J.R., O'Brien, S.C., Curl, R.F., and Smalley, R.E.: 'C60: Buckminsterfullerene' *Nature*, 1985, 318, pp. 162–63.

[8] Iijima, S.: 'Helical Microtubules of Graphitic Carbon' *Lett. To Nat.*, 1991, 354, pp. 56–58.

[9] Iijima, S., Yudasaka, M., Yamada, R., *et al.*: 'Nano-aggregates of single-walled graphitic carbon nano-horns' *Chem. Phys. Lett.*, 1999, 309, pp. 165–70.

[10] Novoselov, K.S., Geim, A.K., Morozov, S. V, *et al.*: 'Electric field effect in atomically thin carbon films' *Science*, 2004, 306, pp. 666–70.

[11] Geim, A.K.: 'Graphene: status and prospects' *Science.*, 2009, 324, pp. 1530–34.

[12] De Jong, K.P., and Geus, J.W.: 'Carbon nanofibers: catalytic synthesis and applications' *Catal. Rev. – Sci. Eng.*, 2000, 42, pp. 481–510.

[13] Gadipelli, S., and Guo, Z.X.: 'Graphene-based materials: Synthesis and gas sorption, storage and separation' *Prog. Mater. Sci.*, 2015, 69, (April), pp. 1–60.

[14] Arico, N.S., Bruce, P., Scrosati, B., Tarascon, J.-M., and Schalkwijk, W. Van: 'Nanostructured materials for advanced energy conversion and storage devices' *Nat. Mater.*, 2005, 4, pp. 366–77.

[15] Sekitani, T., and Someya, T.: 'Stretchable, large-area organic electronics' *Adv. Mater.*, 2010, 22, pp. 2228–46.

[16] Novoselov, K.S., Geim, A.K., Morozov, S. V, *et al.*: 'Electric field effect in atomically thin carbon films' *Science*, 2004, 306, pp. 666–70.

[17] Paton, K.R., Varrla, E., Backes, C., *et al.*: 'Scalable production of large quantities of defect-free few-layer graphene by shear exfoliation in liquids' *Nat. Mater.*, 2014, 13, pp. 624–30.

[18] Zhang, Y., Zhang, L., and Zhou, C.: 'Review of chemical vapor deposition of graphene and related applications' *Acc. Chem. Res.* 2013, 46 pp. 2329–39.

[19] Shen, Y., Lua, A.C.: 'A facile method for the large-scale continuous synthesis of graphene sheets using a novel catalyst' *Sci. Rep.*, 2013, 3, p. 3037.

[20] Stankovich, S., Dikin, D.A., Dommett, G.H.B., Kohlhaas, K.M., and Zimney, E.J.: 'Graphene-based composite materials' *Nature*, 2006, 442, pp. 282–86.

[21] Kang, Y.R., Li, Y.L., and Deng, M.Y.: 'Precise unzipping of flattened carbon nanotubes to regular graphene nanoribbons by acid cutting along the folded edges' *J. Mater. Chem.*, 2012, 22, pp. 16283–87.

[22] Cho, S., Kikuchi, K., and Kawasaki, A.: 'Radial followed by longitudinal unzipping of multiwalled carbon nanotubes' *Carbon N. Y.*, 2011, 49, pp. 3865–72.

[23] Jiao, L., Zhang, L., Wang, X., Diankov, G., and Dai, H.: 'Narrow graphene nanoribbons from carbon nanotubes' *Nature*, 2009, 458, pp. 877–80.

[24] Shinde, D. B., Debgupta J., Kushwaha A., Aslam, M., and Pillai, V. K.: 'Electrochemical unzipping of multi-walled carbon nanotubes for facile synthesis of high-quality graphene nanoribbons' *J. Am. Chem. Soc*, 2011, 133, pp. 4168.

[25] Kumar, P., Panchakarla, L.S., and Rao, C.N.R.: 'Laser-induced unzipping of carbon nanotubes to yield graphene nanoribbons' *Nanoscale*, 2011, 3, pp. 2127–29.

[26] Sahu, S.R., Rikka, V.R., Jagannatham, M., *et al.*: 'Synthesis of graphene sheets from single walled carbon nanohorns: Novel conversion from cone to sheet morphology' *Mater. Res. Express*, 2017, 4, p. aa5ee2.

[27] Devi, M.M., Sahu, S.R., Mukherjee, P., Sen, P., and Biswas, K.: 'Graphene: A self-reducing template for synthesis of graphene-nanoparticles hybrids' *RSC Adv.*, 2015, 5, pp. 62284–89.

[28] Rey, A., Zazo, J.A., Casas, J.A., Bahamonde, A., and Rodriguez, J.J.: 'Influence of the structural and surface characteristics of activated carbon on the catalytic decomposition of hydrogen peroxide' *Appl. Catal. A Gen.*, 2011, 402, pp. 146–55.

[29] Horiuchi, S., Gotou, T., Fujiwara, M., *et al.*: 'Carbon nanofilm with a new structure and property' *Japanese J. Appl. Physics, Part 2 Lett.*, 2003, 42, pp. 1073–76.

[30] Meyer, J. C., Geim, A. K., Katsnelson, M. I., Novoselov K S, Booth T J and Roth S.: 'The structure of suspended graphene sheets' *Nature*, 2007, 446, pp. 60–63.

[31] Malard, L.M., Pimenta, M.A., Dresselhaus, G., and Dresselhaus, M.S.: 'Raman spectroscopy in graphene' *Phys. Rep.*, 2009, 473, pp. 51–87.

[32] Pimenta, M.A., Dresselhaus, G., Dresselhaus, M.S., Cançado, L.G., Jorio, A., and Saito, R.: 'Studying disorder in graphite-based systems by Raman spectroscopy' *Phys. Chem. Chem. Phys.*, 2007, 9, pp. 1276–91.

[33] Li, N., Wang, Z., Zhao, K., Shi, Z., Gu, Z., and Xu, S.: 'Large scale synthesis of N-doped multi-layered graphene sheets by simple arc-discharge method' *Carbon N. Y.*, 2010, 48, pp. 255–59.

[34] Niyogi, S., Bekyarova, E., Itkis, M.E., *et al.*: 'Spectroscopy of covalently functionalized graphene' *Nano Lett.*, 2010, 10, pp. 4061–66.

[35] Karousis, N., Tagmatarchis, N., and Tasis, D.: 'Current progress on the chemical modification of carbon nanotubes' *Chem. Rev.*, 2010, 110, pp. 5366–97.

[36] Tasis, D., Tagmatarchis, N., Bianco, A., and Prato, M.: 'Chemistry of carbon nanotubes' *Chem. Rev.*, 2006, 106, pp. 1105–136.

[37] Meyer, E.A., Castellano, R.K., and Diederich, F.: 'Interactions with aromatic rings in chemical and biological recognition' *Angew. Chem. Int. Ed. Engl.*, 2003, 42(11), pp. 1210–50.

[38] Hong, B.H., Lee, J.Y., Lee, C.W., Kim, J.C., Bae, S.C., and Kim, K.S.: 'Self-assembled arrays of organic nanotubes with infinitely long one-dimensional H-bond chains [6]' *J. Am. Chem. Soc.*, 2001, 123, pp. 10748–49.

[39] Singh N. J., Lee H. M., Hwang I.-C., and Kim K. S.: 'Designing ionophores and molecular nanotubes based on molecular recognition' *Supramol. Chem.*, 2007, 19, pp. 321–32.

[40] Lee, J.Y., Hong, B.H., Kim, W.Y., *et al.*: 'Near-field focusing and magnification through self-assembled nanoscale spherical lenses' *Nature*, 2009, 460, pp. 498–501.

[41] Kim, K.S., Tarakeshwar, P., and Lee, J.Y.: 'Molecular clusters of π-systems: theoretical studies of structures, spectra, and origin of interaction energies' *Chem. Rev.*, 2000, 100, pp. 4145–85.

[42] Riley, K.E., Piton, M., Jurecka, P., and Hobza, P.: 'Stabilization and structure calculations for noncovalent interactions in extended molecular systems based on wave function and density functional theories' *Chem. Rev.*, 2010, 110, pp. 5023–63.

[43] Shi, Y., Kim, K.K., Reina, A., Hofmann, M., Li, L.J., and Kong, J.: 'Work function engineering of graphene electrode via chemical doping' *ACS Nano*, 2010, 4, pp. 2689–94.

[44] Haremza, J.M., Hahn, M.A., Krauss, T.D., Chen, S., and Calcines, J.: 'Attachment of single CdSe nanocrystals to individual single-walled carbon nanotubes' *Nano Lett.*, 2002, 2, pp. 1253–58.

[45] Parlak, O., Beyazit, S., Tse-Sum-Bui, B., Haupt, K., Turner, A., and Tiwari, A.: 'Programmable bioelectronics in a stimuli-encoded 3D graphene interface' *Nanoscale*, 2016, 8, pp. 9976–81.

[46] Yang, K., Gu, M., Guo, Y., Pan, X., and Mu, G.: 'Effects of carbon nanotube functionalization on the mechanical and thermal properties of epoxy composites' *Carbon*, 2009, 47, pp.1723–37.

[47] Allaoui, A., Bai, S., Cheng, H-M., and Bai, J.: 'Mechanical and electrical properties of a MWNT/epoxy composite' *Compos. Sci. Technol.*, 2002, 62, pp. 1993–98.

[48] Zhou, Y., Pervin, F., Lewis, L., and Jeelani, S.: Fabrication and characterization of carbon/epoxy composites mixed with multi-walled carbon nanotubes' *Mater. Sci. Eng. A*, 2008, 475, pp.157–65.

[49] Xu, X., Thwe, M. M., Shearwood, C., and Liao, K.: 'Mechanical properties and interfacial characteristics of carbon-nanotube-reinforced epoxy thin films' *Appl. Phys. Lett.*, 2002, 81, pp. 2833–35.

[50] Tang, L.C., Wan, Y. J., Yan, D., *et al.*.: 'The effect of graphene dispersion on the mechanical properties of graphene/epoxy composites' *Carbon* 2013, 60, pp. 16–27.

[51] Chatterjee. S., Nafezarefi, F., Tai, N., Schlagenhauf, L., Nüesch, F., and Chu, B.: 'Size and synergy effects of nanofiller hybrids including graphene nanoplatelets and carbon nanotubes in mechanical properties of epoxy composites' *Carbon*, 2012, 50, pp. 5380–86.

[52] Wang, F., Drzal, LT., Qin, Y., and Huang, Z.: 'Mechanical properties and thermal conductivity of graphene nanoplatelet/epoxy composites' *J. Mater. Sci.*, 2015, 50, pp. 1082–93.

[53] Rafiee, M. A., Rafiee, J., Wang, Z., Song, H., Yu, Z. Z., and Koratkar, N.: 'Enhanced mechanical properties of nanocomposites at low graphene content' *ACS Nano*, 2009, 3, pp. 3884–90.

[54] Qin, W., Vautard, F., Drzal, L. T., and Yu, J.: 'Mechanical and electrical properties of carbon fiber composites with incorporation of graphene nanoplatelets at the fiber–matrix interphase' *Compos. Part B: Eng.*, 2015, 69 pp. 335–41.

[55] Wehnert, F., Langer, M., Kaspar, J., and Jansen, I.: 'Design of multifunctional adhesives by the use of carbon nanoparticles' *J. Adhes. Sci. Technol.*, 2015, 29, pp. 1849–59.

[56] Fidelus, J., Wiesel, E., Gojny, F., Schulte, K., and Wagner, H.: 'Thermomechanical properties of randomly oriented carbon/epoxy nanocomposites' *Compos. Part A: Appl. Sci. Manuf.*, 2005, 36, pp. 1555–61.

[57] Szeluga, U., Kumanek, B., and Trzebicka, B.: 'Synergy in hybrid polymer/nanocarbon composites. A review' *Compos. Part A: Appl. Sci. Manuf.*, 201, 73 pp. 204–31.

[58] Lachman. N., and Wagner, H. D.: 'Correlation between interfacial molecular structure and mechanics in CNT/epoxy nano-composites' *Compos Part A: Appl. Sci. Manuf.*, 2010, 41, pp. 1093–98.

[59] Guadagno, L., Sarno, M., Vietri, U., Raimondo, M., Cirillo, C., and Ciambelli, P.: 'Graphene-based structural adhesive to enhance adhesion performance' *RSC Adv.*, 2015, 5, pp. 27874–86.

[60] Ma, P.C., Siddiqui, N.A., Marom, G., and Kim, J.K.: 'Dispersion and functionalization of carbon nanotubes for polymer-based nanocomposites: A review' *Compos. Part A Appl. Sci. Manuf.*, 2010, 41, pp. 1345–67.

[61] Jojibabu, P., Jagannatham, M., Haridoss, P., Janaki Ram, G.D., Deshpande, A.P., and Bakshi, S.R.: 'Effect of different carbon nano-fillers on rheological properties and lap shear strength of epoxy adhesive joints' *Compos. Part A Appl. Sci. Manuf.*, 2016, 82, pp. 53–64.

[62] Park, M., Park, J., and Jeong, U.: 'Design of conductive composite elastomers for stretchable electronics' *Nano Today*, 2014, 9, pp. 244–60.

[63] Rajkumar, K., and Aravindan, S.: 'Tribological behavior of microwave processed copper–nanographite composites' *Tribol. Inter.*, 2013, 57, pp. 282–96.

[64] Kim, J.A., Hwang, T., Dugasani, S.R., *et al.*: 'Graphene based fiber optic surface plasmon resonance for bio-chemical sensor applications' *Sensors Actuators, B Chem.*, 2013, 187, pp. 426–33.

[65] Saifur Rahman, M., Anower, M.S., Bashar, L. Bin, and Rikta, K.A.: 'Sensitivity analysis of graphene coated surface plasmon resonance biosensors for biosensing applications' *Sens. Bio-Sensing Res.*, 2017, 16, pp. 41–45.

[66] Sharma, B.K., and Ahn, J.H.: 'Graphene based field effect transistors: Efforts made towards flexible electronics' *Solid State Electron.*, 2013, 89, pp. 177–88.

[67] Choi, J., Wang, M.C., Cha, R.Y.S., Park, W. Il, and Nam, S.: 'Graphene bioelectronics' *Biomed. Eng. Lett.*, 2013, 3, pp. 201–08.

[68] Melita, E.D., Purcel, G., and Grumezescu, A.M.: 'Carbon nanotubes for cancer therapy and neurodegenerative diseases' *Rom. J. Morphol. Embryol.*, 2015, 56, pp. 349–56.

[69] Cohen-Karni, T., Qing, Q., Li, Q., Fang, Y., and Lieber, C.M.: 'Graphene and nanowire transistors for cellular interfaces and electrical recording' *Nano Lett.*, 2010, 10, pp. 1098–102.

[70] Kireev, D., Brambach, M., Seyock, S., *et al.*: 'Graphene transistors for interfacing with cells: Towards a deeper understanding of liquid gating and sensitivity' *Sci. Rep.*, 2017, 7, pp. 1–12.

[71] Alvares, D., Wieczorek, L., Raguse, B., Ladouceur, F., and Lovell, N.H.: 'Development of nanoparticle film-based multi-axial tactile sensors for biomedical applications' *Sensors Actuators A Phys.*, 2013, 196, pp. 38–47.

[72] Nam, T., Park, S., Lee, S., *et al.*: 'Tumor targeting chitosan nanoparticles for dual-modality optical/MR cancer imaging' *Bionconjug Chem.*, 2010, 21(4), pp. 578–82.

[73] Kwak, Y.H., Choi, D.S., Kim, Y.N., *et al.*: 'Flexible glucose sensor using CVD-grown graphene-based field effect transistor' *Biosens. Bioelectron.*, 2012, 37, pp. 82–87.

[74] Labroo, P., and Cui, Y.: 'Flexible graphene bio-nanosensor for lactate' *Biosens. Bioelectron.*, 2013, 41, pp. 852–56.

[75] Hu, Y., Li, F., Han, D., and Niu, L.: 'Biocompatible Graphene for Bioanalytical Applications' (Springer US, 2015) pp. 81–102.

[76] Chand, R., and Neethirajan, S.: 'Microfluidic platform integrated with graphene-gold nano-composite aptasensor for one-step detection of norovirus' *Biosens. Bioelectron.*, 2017, 98, pp. 47–53.

[77] Zhao, H., Ding, R., Zhao, X., *et al.*: 'Graphene-based nanomaterials for drug and/or gene delivery, bioimaging, and tissue engineering' *Drug Discov. Today*, 2017, 22, pp. 1302–17.

[78] Yang, Y., Yang, X., Tan, Y., and Yuan, Q.: 'Recent progress in flexible and wearable bio-electronics based on nanomaterials' *Nano Res.*, 2017, 10, pp. 1560–83.

[79] Parlak, O., Turner, A.P.F., and Tiwari, A.: 'pH-induced on/off-switchable graphene bioelectronics' *J. Mater. Chem. B*, 2015, 3, pp. 7434–39.

[80] Parlak, O., Turner, A.P.F., and Tiwari, A.: 'On/off-switchable zipper-like bioelectronics on a graphene interface' *Adv. Mater.*, 2014, 26, pp. 482–86.

[81] Schneidereit, D., Kraus, L., Meier, J.C., Friedrich, O., and Gilbert, D.F.: 'Step-by-step guide to building an inexpensive 3D printed motorized positioning stage for automated high-content screening microscopy' *Biosens. Bioelectron.*, 2017, 92, pp. 472–81.

[82] Anastasova, S., Kassanos, P., and Yang, G.Z.: 'Multi-parametric rigid and flexible, low-cost, disposable sensing platforms for biomedical applications' *Biosens. Bioelectron.*, 2018, 102, pp. 668–75.

[83] Cinti, S., and Arduini, F.: 'Graphene-based screen-printed electrochemical (bio)sensors and their applications: Efforts and criticisms' *Biosens. Bioelectron.*, 2017, 89, pp. 107–22.

[84] Capasso, A., Del Rio Castillo, A.E., Sun, H., Ansaldo, A., Pellegrini, V., and Bonaccorso, F.: 'Ink-jet printing of graphene for flexible electronics: An environmentally-friendly approach' *Solid State Commun.*, 2015, 224, pp. 53–63.

[85] Lee, I., Kim, Y., Bang, S., *et al.*: 'Circuit techniques for miniaturized biomedical sensors' *Proc. IEEE 2014 Cust. Integr. Circuits Conf. CICC 2014*, 2014.

[86] He, B., Yang, Y., Yuen, M.F., Chen, X.F., Lee, C.S., and Zhang, W.J.: 'Vertical nanostructure arrays by plasma etching for applications in biology, energy, and electronics' *Nano Today*, 2013, 8, pp. 265–89.

[87] Bourrier, A.: 'Graphene bioelectronics for long term neuronal interfacing in-vivo'. *Université Grenoble Alpes*, 2017.

[88] Mannoor, M.S., Tao, H., Clayton, J.D., *et al.*: 'Graphene-based wireless bacteria detection on tooth enamel' *Nat. Commun.*, 2012, 3, pp. 763–68.

[89] De Crozals, G., Bonnet, R., Farre, C., and Chaix, C.: 'Nanoparticles with multiple properties for biomedical applications: A strategic guide' *Nano Today*, 2016, 11, pp. 435–63.

[90] Jin, K., Gao, H., Lai, L., *et al.*: 'Preparation of highly fluorescent sulfur doped graphene quantum dots for live cell imaging' *J. Lumin.*, 2018, 197, pp. 147–52.

[91] Pinpim, A., Srituravanich, W., Jeamsaksiri, W., and Sailasuta, A.: 'Microfluidic devices in biomedical research – The present and future' *Thai J Vet Med Suppl.*, 2017, 47, pp. 63–64.

[92] Wang, J., Sun, L., Zou, M., *et al.*: 'Bioinspired shape-memory graphene film with tunable wettability' *Sci. Adv.*, 2017, 3, p. e1700004.

[93] Kenry, Yeo, J.C., and Lim, C.T.: 'Emerging flexible and wearable physical sensing platforms for healthcare and biomedical applications' *Microsystems Nanoeng.*, 2016, 2, p. 16043.

[94] Li, Z., and Wong, S.L.: 'Functionalization of 2D transition metal dichalcogenides for biomedical applications' *Mater. Sci. Eng. C*, 2017, 70, pp. 1095–106.

[95] Sazonov, E., and Neuman, M.R. (Eds.): 'Wearable sensors – Fundamentals, Implementation and Applications' Academic Press, (Elsevier Inc., Cambridge) 2015.

[96] Park, H.J., Kim, W.-J., Lee, H.-K., *et al.*: 'Highly flexible, mechanically stable, and sensitive NO_2 gas sensors based on reduced graphene oxide nanofibrous mesh fabric for flexible electronics' *Sensors Actuators B Chem.*, 2017, 257, pp. 846–852.

[97] Karim, N., Afroj, S., Tan, S., *et al.*: 'Scalable production of graphene-based wearable E-textiles' *ACS Nano*, 2017, 11, pp. 12266–75.

[98] Gurman, P., Miranda, O.R., Clayton, K., Rosen, Y., and Elman, N.M.: 'Clinical applications of biomedical microdevices for controlled drug delivery' *Mayo Clin. Proc.*, 2015, 90, pp. 93–108.

[99] Sang, S., Feng, Q., Tang, X., *et al.*: 'PDMS micro-membrane capacitance-type surface stress biosensors for biomedical analyses' *Microelectron. Eng.*, 2015, 134, pp. 33–37.

[100] Jivani, R.R., Lakhtaria, G.J., Patadiya, D.D., Patel, L.D., Jivani, N.P., and Jhala, B.P.: 'Biomedical microelectromechanical systems (BioMEMS): Revolution in drug delivery and analytical techniques' *Saudi Pharm. J.*, 2016, 24, pp. 1–20.

[101] Da Cunha Rodrigues, G., Zelenovskiy, P., Romanyuk, K., Luchkin, S., Kopelevich, Y., and Kholkin, A.: 'Strong piezoelectricity in single-layer graphene deposited on SiO_2 grating substrates' *Nat. Commun.*, 2015, 6, pp. 2–7.

[102] Khan, Z.H., Kermany, A.R., Ochsner, A., and Iacopi, F.: 'Mechanical and electromechanical properties of graphene and their potential application in MEMS' *J. Phys. D. Appl. Phys.*, 2017, 50, p. 053003.

[103] Chandrakasan, A.P., Verma, N., and Daly, D.C.: 'Ultralow-power electronics for biomedical applications' *Annu. Rev. Biomed. Eng.*, 2008, 10, pp. 247–74.

[104] Ye, M., Zhang, Z., Zhao, Y., and Qu, L.: 'Graphene platforms for smart energy generation and storage' *Joule*, 2018, 2, pp. 245–68.

[105] Chu, H., Jang, H., Lee, Y., Chae, Y., and Ahn, J.H.: 'Conformal, graphene-based triboelectric nanogenerator for self-powered wearable electronics' *Nano Energy*, 2016, 27, pp. 298–305.

[106] Zurbuchen, A., Haeberlin, A., Pfenniger, A., *et al.*: 'Towards batteryless cardiac implantable electronic devices' *IEEE Trans. Biomed. Circuits Syst.*, 2017, 11, pp. 78–86.

[107] Zarrabi, F.B., Naser-Moghadasi, M., Heydari, S., Maleki, M., and Arezomand, A.S.: 'Cross-slot nano-antenna with graphene coat for biosensing application' *Opt. Commun.*, 2016, 371, pp. 34–39.

[108] Whitlow, H.J., Guibert, E., Jeanneret, P., *et al.*: 'Post-focus expansion of ion beams for low fluence and large area MeV ion irradiation: Application to human brain tissue and electronics devices' *Nucl. Instruments Methods Phys. Res. Sect. B Beam Interact. Mater. Atoms*, 2017, 404, pp. 87–91.

[109] Lin, C.Y., and Chiu, Y.L.: 'Investigation on SCR-based ESD protection device for biomedical integrated circuits in a 0.18-μm CMOS process' *Microelectron. Reliab.*, 2015, 55, pp. 2229–35.

[110] Arsiwala, A., Desai, P., and Patravale, V.: 'Recent advances in micro/nanoscale biomedical implants' *J. Control. Release*, 2014, 189, pp. 25–45.

Chapter 3

Single electron devices: concept to realization

Boddepalli Santhibhushan[1], Anurag Srivastava[1], Anu[2], and Mohammad Shahid Khan[2]

Single electron transistors are widely perceived as the next generation devices, and beyond Moore's law devices due to their promising aspects of low power consumption, high switching speed, compact size, and importantly the ability to shrink to atomic scale. The single electron devices operation is based on the quantum phenomenon called "tunneling". Though the basic structure and operating principle of these devices is quite simpler, their fabrication and real-time operation is equally difficult, as it requires multiple conditions of Coulomb blockade to be satisfied for incoherent transport. This chapter provides comprehensive information about the single electron devices, starting from their benefit over other devices in the same series, associated concepts, operating principle, and the notable advancements in experimental as well as theoretical research on these devices. This chapter is expected to work as an absolute guide for any researcher interested in single electron devices.

3.1 Introduction

3.1.1 Importance of single electron devices

The advances in integrated circuit (IC) technology have revolutionized several sectors such as consumer electronics, telecommunications, computing, and space. These advancements are led by the prediction made in 1965 by Gordon Moore, the co-founder of Intel co., popularly known as Moore's law, which states that the number of transistors on a chip will double every two years. The law worked as a roadmap for IC developers and resulted in colossal enhancement in the computational capacity. The maximum transistor count in a processor has reached from few tens to few billions over time. As of 2019, AMD's Epyc Rome processor holds the tag of having the highest number of transistors at 39.54 billion in a commercially

[1]Advanced Materials Research Group, CNT Laboratory, ABV–Indian Institute of Information Technology and Management, Gwalior, India
[2]Department of Physics, Jamia Millia Islamia, Central University, New Delhi, India

available processor. However, the stagnation of Moore's law started in the early twenty-first century and led to the invention of a dual-core processor (Figure 3.1) [1,2]. The present technology node being used by Apple and Samsung is 10 nm, while Intel is expected to release the 10 nm processors by 2019. The technology node that follows 10 nm is 7 nm, followed by 5 nm. The semiconductor experts believe 5 nm node as the end of Moore's law due to uncontrolled quantum tunneling of carriers into gate.

Thus, developing next generation nanoelectronic devices with the capability to extend Moore's law is the current research topic of interest. Single electron transistor (SET) is a nanoelectronic device that promises to extend Moore's law due to the absence of scaling limits. Researchers have developed a prototype of SET with single phosphorous atom as a quantum dot [3].

3.1.2 Theory of single electron devices

SET is a three-terminal device that works on intrinsically quantum phenomenon—tunneling. The tunneling phenomenon follows the basic principle of quantum

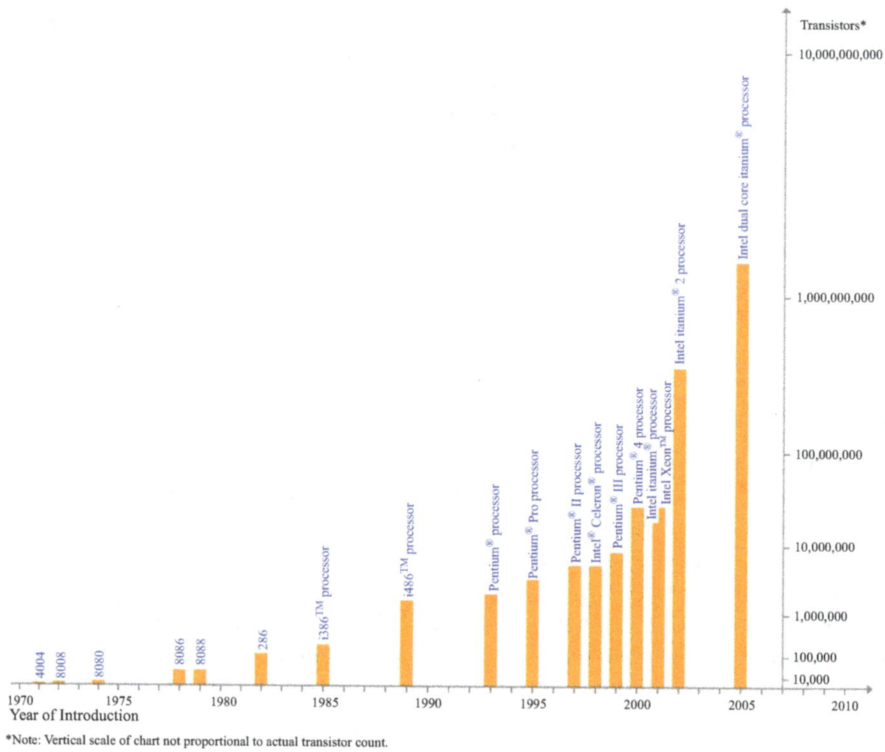

Figure 3.1 Evolution of Intel processors with time (y-axis shows the transistor count). Adapted with permission from [2]. Copyright owned by Intel Co

mechanics "all the microscopic material particles in motion can have wave characteristics whose wavelength is inversely related to the size of particle, thus an electron in motion can penetrate through a potential barrier of sufficiently small size due to the wave-particle duality". The SET contains an island, which is the heart of the device. Two electrodes—source and drain to provide the bias, and a third terminal—gate to control the flow of carriers through the island (Figure 3.2 (a)). The source and drain electrodes are separated from the island by thin insulating tunnel barriers, while a thick insulation layer separates gate from island so as to avoid unwanted electron tunneling into gate. The schematic of SET, equivalent circuit, typical transfer, and output characteristics are shown in Figure 3.2. The island is also called "quantum dot" as everything (charge, energy of electrons, energy levels, current) on the island is quantized. Here, the word "quantized" refers to the sizing to certain amount.

The working principle of SET involves one at a time tunneling of electrons from the source to quantum dot and thereby to drain, provided the mechanism is controlled by gate. In general, SETs can be operated in two regimes, coherent transport (CT) and Coulomb blockade (CB) or incoherent transport or sequential tunneling regimes. In CT regime, the quantum dot is strongly coupled between the electrodes such that the tunneled electron onto quantum dot cannot localize itself

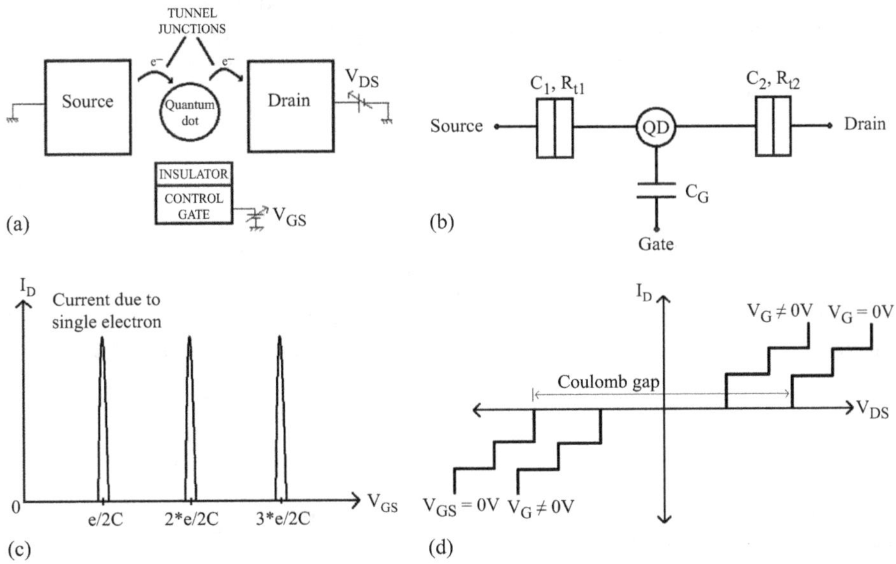

Figure 3.2 *(a) Schematic of SET, (b) equivalent circuit of SET (C and R_t indicate capacitance and tunnel resistance, respectively), (c) typical transfer characteristics drawn between gate voltage (V_{GS}) and drain current (I_D), and (d) typical output characteristics drawn between source-drain bias (V_{DS}) and drain current (I_D)*

and stays only for a short time, thereby retaining its original quantum information of source. In CB regime, the quantum dot is weakly coupled between the electrodes, such that the electron stays for sufficiently long time on the quantum dot and localizes itself so as to lose its original quantum information of source.

CB is responsible for the controlled one after another tunneling of electrons in SET. Due to CB effect, SET does not follow the Ohm's law and the resulting current–voltage (output characteristics) relation looks like a staircase. CB can be defined as the opposition offered by the existing charge of quantum dot towards the incoming charge. The term "Coulomb" is used since the blockade is a result of Coulomb interaction of charges. If CB is not maintained, the operation of SET cannot be controlled and no more acts like a SET. The following three conditions should always be satisfied to maintain the Coulomb blockade [4,5].

I. Thermal energy addition of source and Island should be less than the charging energy.
II. The uncertainty energy should be less than the charging energy.
III. Bias voltage (V_{DS}) should be less than the elementary charge (e) divided by self-capacitance of the quantum dot (C_{dot}).

Condition I

Thermal energy addition of source and island should be less than the charging energy. Charging energy (E_C) is defined as the energy required for adding one elementary charge to the quantum dot, and is expressed as a function of self-capacitance of the quantum dot as given in (3.1). When the quantum dot accommodates an excess electron under the influence of applied potentials, then its energy rises to a value equal to the charging energy.

$$E_C = \frac{e^2}{2C_{dot}} \tag{3.1}$$

It is necessary to design the SET in such a way that the quantum dot has a charging energy higher than the thermal energy of source and island combined, so as to avoid unwanted electron tunneling due to the thermal energy supplied by external and internal temperatures. The condition can be expressed mathematically as,

$$K_B T < \frac{e^2}{2C_{dot}} \tag{3.2}$$

Here, K_B is Boltzmann's constant, T is temperature, e is elementary charge, and C_{dot} is self-capacitance of the quantum dot.

Thus, it can be inferred from condition I that the quantum dot needs to have a low self-capacitance (high charging energy) to maintain high thermal range of SET. Assuming spherical shaped quantum dots, the self-capacitance is given by $C_{dot} = 2\pi\varepsilon d$, where ε and d are permittivity and diameter of quantum dot, respectively. Thus, the quantum dots with low diameter (small size) are preferable as they offer low self-capacitance and thereby high thermal range.

Condition II

Since the Island is a quantum mechanical system there must be some uncertainty in the energy. This uncertainty in energy should be less than the charging energy.

$$\text{Uncertainty energy } (\Delta E) < \frac{e^2}{2C_{\text{dot}}} \tag{3.3}$$

As per the Heisenberg's uncertainty relation of energy and time,

$$\Delta E \cdot \Delta t \approx h/2$$

$$\Delta E \approx h/2\Delta t \tag{3.4}$$

Since the quantum dot stores charge, it acts like a capacitor. Thus, it discharges with a rate,

$$\Delta t = R_t C_{\text{dot}} \tag{3.5}$$

Here, R_t is tunneling resistance.

Substituting (3.5) in (3.4),

$$\Delta E \approx \frac{h}{2R_t C_{\text{dot}}} \tag{3.6}$$

Substituting (3.6) in (3.3),

$$\frac{h}{2R_t C_{\text{dot}}} < \frac{e^2}{2C_{\text{dot}}}$$

$$\frac{h}{R_t} < e^2$$

$$R_t > \frac{h}{e^2}$$

$$R_t > \frac{4.135 * 10^{-15}}{\left(1.6 * 10^{-19}\right) * e} \frac{eV \cdot s}{C}$$

$$R_t > 26\ \text{k}\Omega \tag{3.7}$$

Equation (3.7) states that the tunneling resistance offered by the tunnel junctions should be higher than 26 kΩ to prevent any unwanted tunneling.

Condition III

The bias voltage (V_{DS}) must be less than the elementary charge divided by self-capacitance of quantum dot.

$$V_{\text{DS}} < \frac{e}{C_{\text{dot}}} \tag{3.8}$$

This condition ensures weak coupling between the source and drain electrodes, so that the tunneled electron can have enough time to localize itself on the island and lose the original quantum information of source.

If $V_{DS} > e/C_{dot}$, the source and drain electrodes get strongly coupled, which may result in unwanted tunneling of electrons without applying gate voltage due to the broadening of energy levels of the quantum dot. Also, the electron may retain its quantum information of source even after entering the drain electrode.

The three aforementioned conditions ensure one after another electron tunneling through the quantum dot by maintaining Coulomb blockade.

3.1.3 Single electron transistor: principle of operation

The energy band diagram of a SET at equilibrium condition without applying the external potentials is shown in Figure 3.3. Owing to its very small size and zero-dimensional nature, the quantum dot contains discrete energy levels. As a result of equilibrium, the Fermi level (or chemical potential) of source and drain contacts is at equal position and the quantum dot energy levels below the electrode Fermi level are filled with electrons while those above the Fermi level are empty.

Unlike the metal-oxide-semiconductor field-effect transistor (MOSFET) devices (a conventional switching device used by the semiconductor industry) that involve the formation of inversion channel between source and drain contacts to facilitate the electron transmission, the SET does not require formation of any such inversion regions but the movement of lowest unoccupied molecular orbital (LUMO) of quantum dot below the Fermi level of source/drain contacts. The operation of SET is simple to understand but rather difficult to perform and maintain due to the conditions associated with the conservation of Coulomb blockade.

The operation of SET is illustrated in Figure 3.4 with the help of energy band diagrams. The step-wise explanation of the operation is as follows.

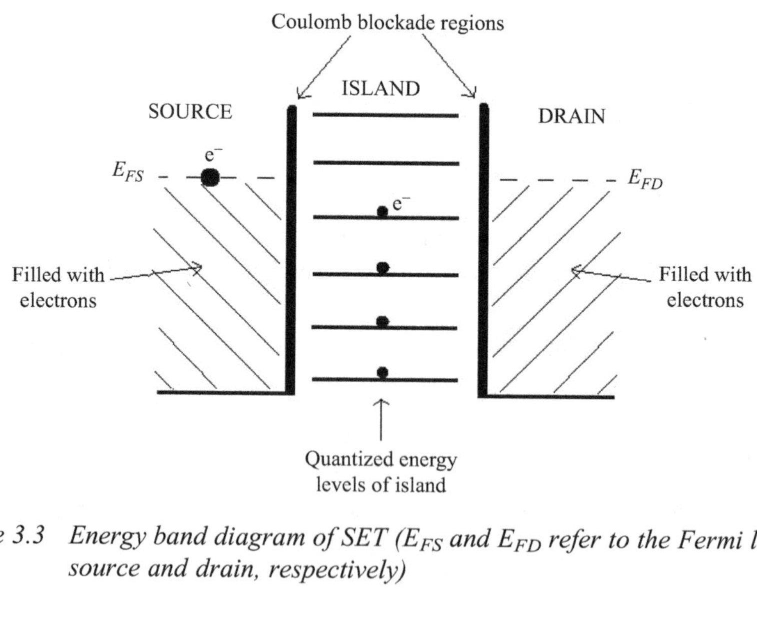

Figure 3.3 Energy band diagram of SET (E_{FS} and E_{FD} refer to the Fermi level of source and drain, respectively)

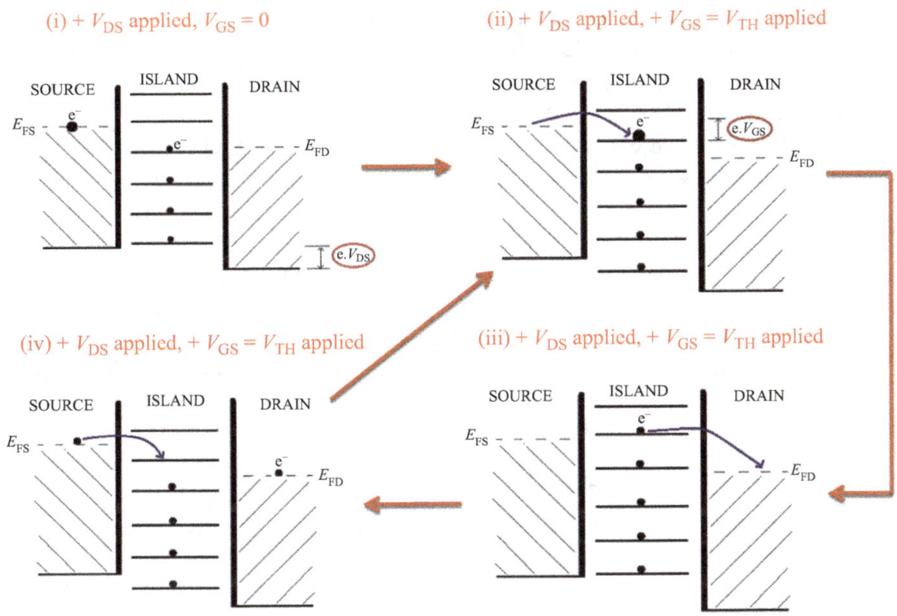

Figure 3.4 Operation of SET device explained through energy band diagrams

Step-i: The applied bias potential across the source-drain electrodes lowers the energy of drain with respect to the source. The shift in energy is proportional to $e \cdot V_{DS}$. Though this shift produces a Fermi function difference between source-drain electrodes, the electron cannot tunnel from source to drain due to the absence of unoccupied energy levels within the bias window.

Step-ii: If a positive gate potential is applied to SET, then the discrete energy levels of quantum dot shift downwards. At some gate potential equivalent to the threshold voltage of SET, an unoccupied discrete energy level can be seen within the bias window, offering a chance for a source electron to tunnel onto the quantum dot by overcoming Coulomb blockade. Since, by definition, charging energy is the energy required to add one elementary particle to the quantum dot, the shift in quantum dot energy (e.V_{GS}) must be equal to the charging energy. Thus,

$$e \cdot V_{GS} = \frac{e^2}{2C_{dot}} \tag{3.9}$$

$$V_{GS} = \frac{e}{2C_{dot}} = \text{threshold voltage}\,(V_{TH}) \tag{3.10}$$

Though we explain the operation with positive gate potentials, SET can also be operated with negative gate potentials. An applied negative gate potential shifts the discrete energy levels of quantum dot upwards. At a gate potential equivalent to the threshold voltage, an occupied discrete energy level can be seen within the bias window creating an opportunity for the quantum dot electron to tunnel onto drain.

Step-iii: As a source electron occupies a discrete energy level of the quantum dot, the energy of quantum dot raises by an amount equivalent to $e \cdot V_{GS}$. This forces the electron from the highest occupied discrete energy level of the quantum dot to tunnel onto drain.

Step-iv: As the electron tunnels from quantum dot to drain, the energy of quantum dot is again lowered by an amount equal to $e \cdot V_{GS}$ due to the applied electric field from gate. Thus, another electron from source tunnels onto the unoccupied discrete energy level of quantum dot, repeating step-ii.

3.1.4 Advantages, challenges, and applications

SET offers several **advantages** over the conventional switching devices. They are

- Ability to shrink to atomic scale due to the absence of scaling limits
- Low energy consumption
- Small size, thereby improved packing density
- High operating speed
- Simple principle of operation
- Possible co-integration with traditional CMOS circuits.

However, the following are a few **challenges** associated with the design of SET.

- Difficulty in integrating at large scale
- Complex fabrication process
- Co-tunneling problem
- Operating temperature requirement.

SETs can be used for a handful of **applications** such as the switching element of ICs, as a sensor for detection of charge, displacement, toxic gases, and DNA, single electron memory, single electron spectroscopy for the detection of microwave/IR/UV radiation, etc.

3.2 Experimental research

The idea of SET was first proposed by Dmitri Averin and Konstantin Likharev at Moscow State University in 1985 [6]. The extensive study of Josephson junctions [7] at low temperatures has laid down the foundation of the idea of single electron tunneling junctions. Two years later, the proposed SET device was fabricated by Theodore Fulton and Gerald Dolan for the first time at Bell Labs (USA) in 1987 [8].

3.2.1 First experimental observation of single electron effects [8]

Fulton and Dolan have prepared the first single electron device by depositing aluminum electrodes on an oxidized silicon wafer with 0.44 μm oxide thickness, as shown in Figure 3.5. The junctions are formed by electron-beam lithography assisted liftoff stencil usage. Thus formed junctions have offered resistance of approximately 40 kΩ. The structure shows three vertical electrodes forming junctions with a single horizontal (central) electrode. However, only two junctions are used to pump the bias current through the device, while the third electrode is used as a probe to examine the voltage in the horizontal electrode. Gate potential to the device is applied via an Au–Cr film deposited on the back of the silicon substrate.

The device is studied mainly at two different temperatures 1.7 K and 1.1 K corresponding to beyond the critical temperature of Al (1.2 K) for non-superconducting electrodes and below the critical temperature of Al for superconducting electrodes, respectively. Two junctions have been considered for the study, S—a junction with small junction area and thereby low capacitance, and L—a junction with large junction area and thereby high capacitance. At non-superconducting temperatures without applying substrate voltage, the low capacitive junctions offer a Coulomb-gap structure in the *I–V* characteristics, while the

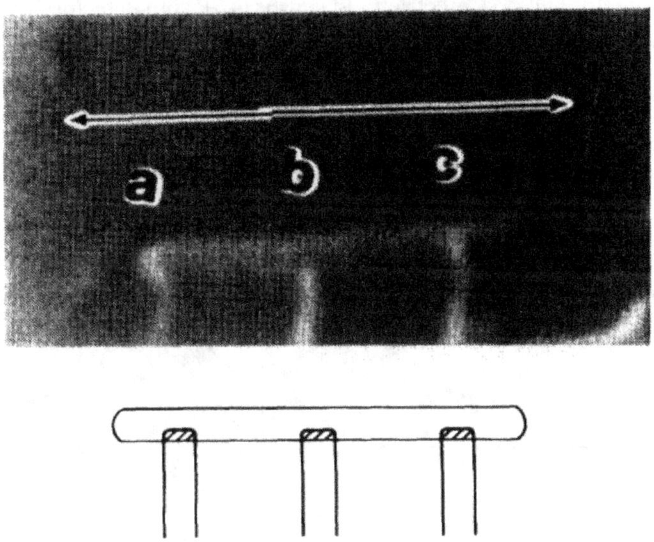

Figure 3.5 Scanning-electron micrograph of the first single electron device fabricated by Fulton and Dolan. In the image, labels a, b, and c represent the three junctions formed between the Al electrodes. Reprinted with permission from [8]. Copyright (1987) by the American Physical Society

high capacitive junctions show ohmic characteristics. Thus, smaller junctions are favorable for inducing single electron tunneling effects.

Figure 3.6(a) depicts the voltage behavior for the drive current pumped through S and L junctions with superconducting electrodes. Here, both S and L junctions show Coulomb-gap structure, with L showing the typical structure of a high capacitive junction. Curves SM and SMN depict the behavior of junction-S when a substrate voltage is applied at superconducting (1.1 K) and non-superconducting (1.7 K) temperatures, respectively. The substrate voltage is a ~0.05 Hz sawtooth wave with ~0.5 V amplitude. The substrate voltage has resulted oscillations in the voltage of horizontal electrode at both the superconducting and non-superconducting temperatures. The period of oscillations is similar for both superconducting and non-superconducting temperatures. Figure 3.6(b) gives a clear representation of *I–V* variation with the substrate voltage, where, the uniformly varied substrate voltage in increments of 1/6 has resulted in an offset of 7.5 nA between the curves. To summarize, this work verifies the single electron charging effects proposed theoretically by Averin and Likharev in [6]. The first evidence of Coulomb-gap structure and ability of substrate electric field to influence the *I–V* characteristics were presented.

In the following years, the SETs were fabricated and extensively analyzed by several researchers for various applications such as charge sensor [9,10], displacement detector [11–15], electric field sensor [16], spin detector [17], gas sensor

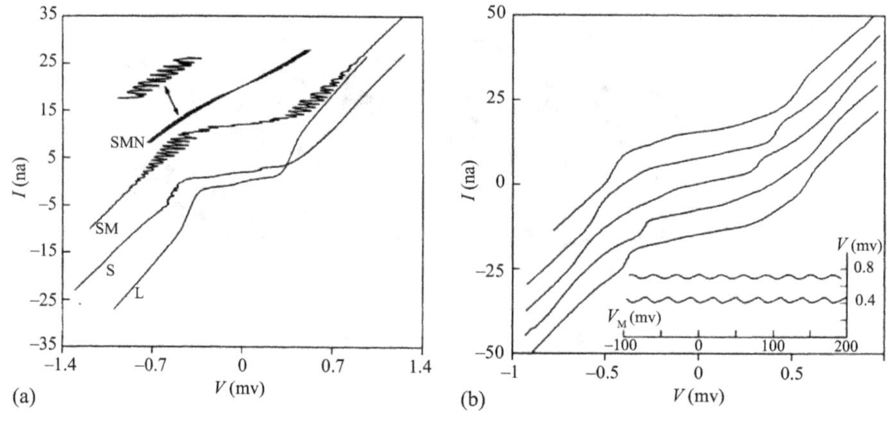

Figure 3.6 (a) I–V curves of low capacitive junction (S) and high capacitive junction (L) at superconducting temperature 1.1 K. SM and SMN curves show the behavior of junction S when substrate voltage is applied at superconducting (1.1 K) and non-superconducting (1.7 K) temperatures, respectively. (b) I–V curves of a sample at five uniformly spaced substrate voltages at superconducting temperature (1.1 K). Reprinted with permission from [8]. Copyright (1987) by the American Physical Society

[18], memory [19–24], switch [25–27]. In this section, we further highlight some key experimental developments in the SET device technology.

3.2.2 Single molecular single electron transistor

Though the early stage single electron devices were made up of metal and semiconductor crystal quantum dots, the focus was shifted steadily towards semiconductor nanostructures such as fullerenes, molecules, and nanotubes. Park *et al.* [28] have prepared the first SET made up of a single C_{60} fullerene marking a new era in single electron devices.

The device has been prepared by fabricating a pair of connected gold electrodes on an oxide layer of degenerately doped silicon wafer using e-beam lithography. A dilute toluene solution of C_{60} has been deposited onto the electrode pair. Thereafter, a gap of ~1 nm is created between the electrodes through electromigration process. Though the authors could not image the C_{60} due to its small size (~7 Å diameter), its presence between electrodes has been confirmed through improved conductance of junction in comparison to that of no C_{60} deposited. The representative device structure and the *I–V* characteristics are shown in Figure 3.7. The *I–V* curves show suppressed conductance (Coulomb-gap) near the zero bias potential and a staircase-like behavior away from the zero bias. The Coulomb-gap

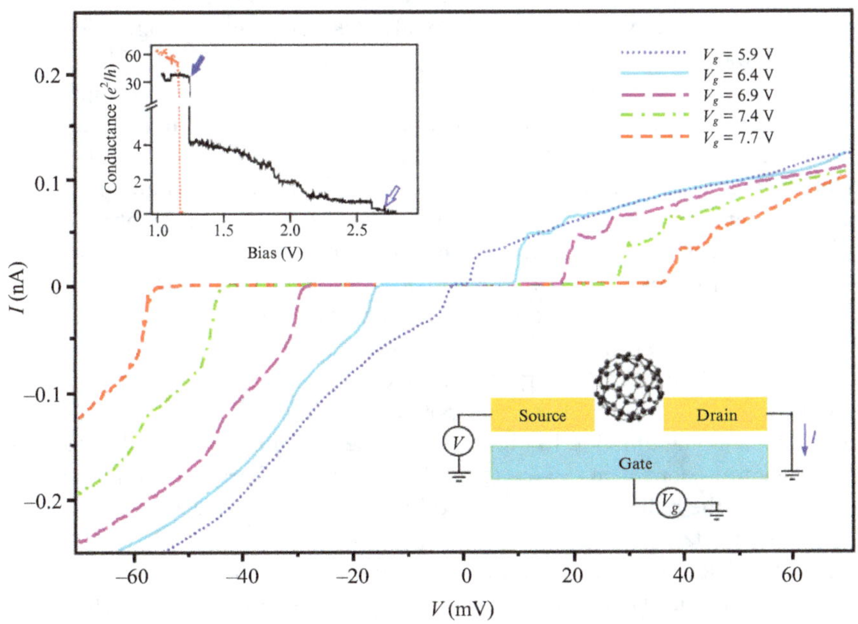

Figure 3.7 I–V characteristics of single C_{60} SET for varied gate potential at a temperature of 1.5 K. The figure also shows a schematic representation of the SET device. Reprinted with permission from [28]. Copyright (2000), Springer Nature

changes reversibly with the applied gate potential, as a higher positive gate potential stabilizes an additional electron on the C_{60} quantum dot. Another interesting aspect is the electron hopping induced nano-mechanical oscillations of C_{60} against the electrode surface. When an electron from the gold electrode hops on to the C_{60} quantum dot, the attractive interaction between the added electron and its image charge on the electrode pulls the fullerene closer to the electrode. Likewise, a similar phenomenon pulls the fullerene towards the other gold electrode when the electron from C_{60} hops onto the electrode. This process causes nano-mechanical oscillations of C_{60} with a frequency measured as ~1.2 THz.

3.2.3 *Single atom single electron transistor*

The work by Fuechsle *et al.* [3] has taken the research of single electron devices to a new high by validating the previously predicted capability of SET to scale down to atomistic limits. The authors have prepared a SET based on a single phosphorous atom quantum dot on a single silicon crystal. A combination of scanning tunneling microscopy (STM) and hydrogen-resist lithography are used to realize the device.

As shown in Figure 3.8(a), the device contains a dual-gate architecture (both gates operated at the same potential), source/drain contacts and a single phosphorous atom as quantum dot positioned precisely into a three-dimer patch of silicon between the electrodes with an accuracy of one lattice site. The electrodes and quantum dot are patterned by selectively desorbing the hydrogen resist from the surface of silicon crystal, followed by exposing to phosphine (PH_3) gas and annealing at 350 °C to create phosphorous doping. The device is operated at cryogenic temperatures so that the doped regions can only conduct, while the rest of the silicon crystal is insulated due to carrier freeze-out. Figure 3.8(c) gives schematic of the chemical reactions took place while placing a single phosphorous atom into a three-dimer patch of silicon. In Figure 3.8(b), the central bright spot is a result of an ejected silicon atom out of the dimer due to the incorporation of phosphorous atom; likewise the relative height of electrodes over the crystal surface in Figure 3.8(a) is also a result of silicon atom ejection.

The false-color plot depicted in Figure 3.9(a) confirms the presence of dopant atom as a quantum dot. A sharp needle-like peak can be seen in Figure 3.9(a) and (b), which is a result of the low-potential profile associated with the phosphorous dopant quantum dot. In Figure 3.9(c), the one electron ground state (D^0) of the phosphorous dopant in device (solid blue line) is relatively at high energy in comparison to the bulk case (gray dashed line) owing to the electrostatic screening from the electrodes.

Figure 3.10(a) shows the stability diagram of phosphorous quantum dot w.r.t the source-drain bias (V_{SD}) and gate voltage (V_G), where the charge states D^+, D^0 and D^- represent the ionized, neutral and negatively charged states of the quantum dot, respectively. The ionized state diamond (D^+) does not show an ending even for large negative gate voltages due to the fact that a single phosphorous donor cannot lose more than one electron. At zero applied bias and gate potentials, the quantum dot is in the ionized charge state, whereas the first transition $D^+ \rightarrow D^0$ occurs at a

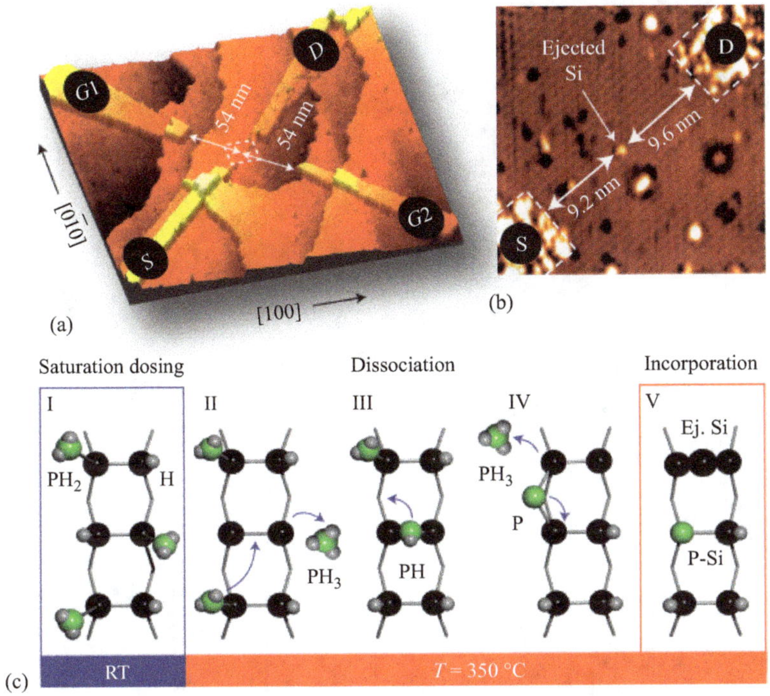

Figure 3.8 (a) STM image of the device, (b) close shot of the dotted region in (a), (c) chemical reaction depicting the incorporation of single phosphorous dopant into silicon (RT indicates Room Temperature). Reprinted with permission from [3]. Copyright (2012), Springer Nature

gate voltage of 0.45 V and the second transition $D^0 \rightarrow D^-$ occurs at a gate voltage of 0.82 V. Figure 3.10(b) shows an experimentally observed charging energy of 47 $\pm$ 3 meV. Figure 3.10(c–g) shows the tight-binding simulation results of potential profiles and orbital probability density between source-drain electrodes. Figure 3.10(c) depicts the variation of D^0 and D^- ground states position w.r.t the gate voltage, where the charge transition occurs when the state touches the electrode Fermi level. The difference in the energies of the two ground states gives the charging energy 46.5 meV, which is very close to the experimental value. In comparison to the equilibrium case depicted in Figure 3.9(c), Figure 3.10(d) and (f) shows the relative reduction in potential barrier by silicon due to the applied positive gate voltage. To summarize this work, the authors have successfully positioned a single atom dopant into a single crystal of silicon with an accuracy of one lattice point and utilized the dopant as a quantum dot for single electron tunneling. More studies of single atom transistors have followed in the subsequent years from various other researchers aiming at quantum computation application [29,30].

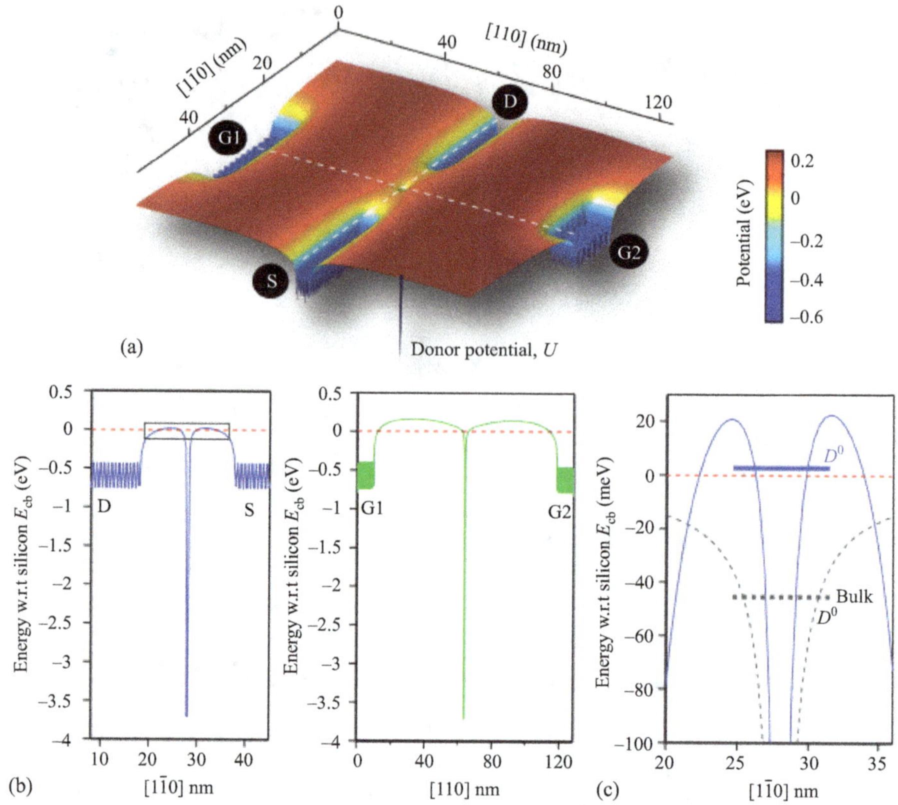

Figure 3.9 (a) False-color plot of potential distribution in the device at equilibrium, (b) Potential profile from drain to source and Gate 1 to Gate 2, (c) A close shot of the rectangularly marked portion in (b) comparing the ground state (D^0) of donor electron in the device (blue solid line) and in isolated bulk state (gray dashed line). Reprinted with permission from [3]. Copyright (2012), Springer Nature

3.3 Computational research

Though the experimental realization of single electron devices took shape in 1987, the first principle device modeling of the same started only in 2008 when Kaasbjerg *et al.* [31] introduced a semiempirical method to simulate an OPV5 molecule-based SET. Later, in 2010, Kurt Stokbro [32] has extended that framework to model molecular SETs using density functional theory (DFT). The framework by Stokbro [32] has become so popular that it prompted enormous interest in the research community in subsequent years to explore the SETs of various materials for switching [33–42] and sensing [43–53] applications using DFT.

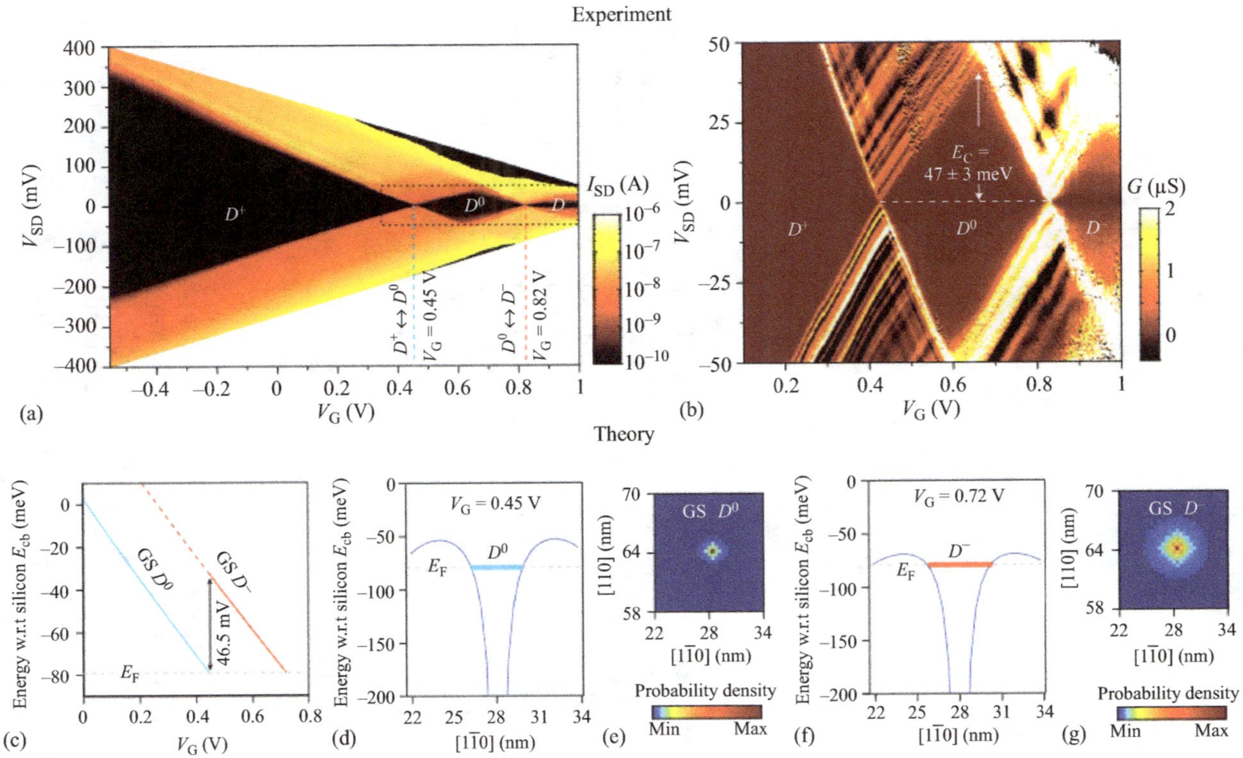

Figure 3.10 *(a) Stability diagram showing the drive current variation (log scale) w.r.t the V_{SD} and V_G, (b) differential conductance variation (linear scale) in the dotted region of (a), (c–g) tight-binding simulation results of potential profiles and orbital probability density between source and drain electrodes. Reprinted with permission from [3]. Copyright (2012), Springer Nature*

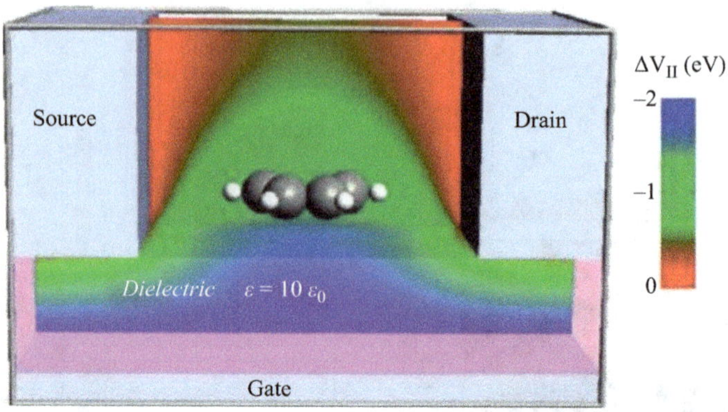

Figure 3.11 The SET model designed by Stokbro. The figure also shows the electrostatic potential induced in the device at a gate voltage of 2 V and zero source-drain bias. Reprinted with permission from [32]. Copyright (2010), American Chemical Society

Stokbro has modeled benzene and C_{60}-based molecular SETs operating in Coulomb blockade regime and estimated the charging energies [32]. Figure 3.11 shows the model of benzene SET, where the dielectric layer is assigned a dielectric constant of $10\varepsilon_0$ equivalent to Al_2O_3 and the metal slabs are assigned a work function of 5.28 eV corresponds to gold. The island molecule is positioned 1.2 Å above the dielectric layer to avoid any overlap between the dielectric layer and the compensation charge as well as screened local pseudopotential of molecular island. The total energy of the molecular quantum dot in SET device environment is estimated by combining the contributions from (3.11) and (3.12) given below. The total energy functional of the molecule is calculated as,

$$E[n] = T[n] + E^{XC}[n] + \frac{1}{2}\int \delta V^{H}(r)\delta n(r)dr + \int \sum V_i^{NA}(r)\delta n(r)dr + \frac{1}{2}\sum_{ij} U_{ij} \tag{3.11}$$

Here, T is single electron kinetic energy, E^{XC} is exchange-correlation energy, $\delta V^{H}(r)$ is difference Hartree potential, $n(r)$ is total charge density, $\delta n(r)$ is electron difference density ($\delta n(r) = n(r) - \sum_i \rho_i^{comp}(r)$), $\rho_i^{comp}(r)$ is compensation charge introduced on each atomic site i to screen the electrostatic interactions. $V_i^{NA}(r)$ is screened (neutral atom) local pseudopotential ($V_i^{NA}(r) = V_i^{loc}(r) - \int \frac{\rho_i^{comp}(r')}{|r-r'|}dr'$), $V_i^{loc}(r)$ is the local pseudopotential at site i, and U_{ij} indicates all electrostatic interactions independent of charge density.

The energy contribution of external electric field from the metallic and dielectric regions is estimated as,

$$\Delta E = \int V^{ext}(r)n(r)dr - \sum_i V^{ext}(R_i)Z_i \tag{3.12}$$

Here, $V^{ext}(r)$ is the external electrostatic potential, R_i and Z_i are the position of site i and valency of pseudopotential at site i, respectively.

The calculated charging energies of benzene and C_{60} in the isolated environment are in good agreement with the experimental reports. Moreover, the charging energies are reduced in the SET environment in comparison to the isolated case owing to the electrostatic surrounding of SET environment. The total energy variation with the applied gate potential for various charge states of benzene and C_{60} quantum dots are plotted in Figure 3.12 (a) and (b), where both the molecules show

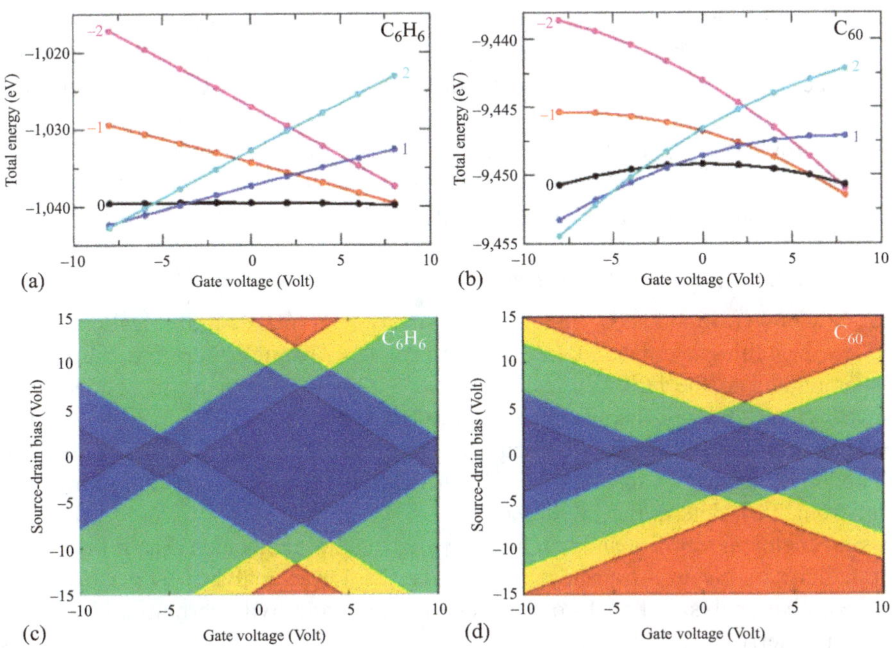

Figure 3.12 Total energy plotted as a function of gate potential for various charge states (0,1,2,-1,-2) of (a) benzene and (b) C_{60} quantum dot. Charge Stability diagram as a function of gate and source/drain potentials for (c) benzene and (d) C_{60} SET. The color code shows the number of energy levels available for conduction within the bias window (Dark blue: 0, light blue: 1, Green: 2, Yellow: 3, red: 4). Reprinted with permission from [32]. Copyright (2010), American Chemical Society

the least total energy in their neutral charge state at zero gate potential. For positive gate potentials the negative charge states are more stable and vice versa. This behavior is in agreement with the HOMO and LUMO levels following $-eV_G$, thus for positive gate potentials the LUMO level moves below the electrode Fermi level and accepts an electron, and for negative gate potentials the HOMO level moves above the Fermi level of electrode and loses an electron. The non-linear variation of total energy w.r.t gate potential in C_{60} SET is a result of charge polarization on C_{60} molecule due to the screening of gate potential by the lower atoms for the rest of the molecule.

The charge stability diagram depicted in Figure 3.12(c) and (d) is plotted from the total energies using (3.13) that represents the constraint for the current to flow in the SET device.

$$\frac{e|V|}{2} \geq \Delta E^{island}(N) + W \geq \frac{-e|V|}{2} \qquad (3.13)$$

Here, V is the source-drain bias potential, $\Delta E^{island}(N)$ is the charging energy $(\Delta E^{island}(N) = E^{island}(N+1) - E^{island}(N))$, and W is the work function of metal electrodes. The figure shows less excitation energy requirement for C_{60} SET than the benzene, meaning the C_{60} SET can be switched from OFF state (dark blue region) to ON state (light blue region) by applying relatively low gate and source-drain potentials.

Next, the various DFT-based reports based on the Stokbro's model exploring SET device for switching and sensing applications are discussed.

3.3.1 SET as switching element

The low power, high speed, and compact size of SET makes it an ideal candidate for the switching element of next generation ICs. Though there are fabrication issues related to the bulk fabrication of SET ICs in comparison to conventional MOSFET technology, the advancements in fabrication technology are expected to resolve such issues. Here, we focus on discussing the reports that studied the switching behavior of nanoscale SETs. Modeling of molecular SETs using aromatic quantum dots is the interest of Srivastava *et al.* [33–38] owing to the promising chemical stability and conductivity of aromatic molecules. The influence of aromaticity on electron transport of molecular SET was analyzed by performing a comparative analysis of SETs of Benzene (an aromatic hydrocarbon) and its non-aromatic derivative hexahydrobenzene [37]. Structure of the two molecules is shown in Figure 3.13, where benzene shows aromaticity in its carbon ring. The presence (absence) of aromaticity in benzene (hexahydrobenzene) can also be confirmed from the molecular energy spectrum and the corresponding density of states (DOS) profiles depicted in Figure 3.14. Both the HOMO and LUMO levels of benzene are composed of only p-states confirming the presence of aromaticity due to the delocalization of π-cloud, whereas the HOMO and LUMO of hexahydrobenzene are composed of both s and p states indicating the absence of π-cloud and the aromaticity. Benzene is likely to offer better electrical conduction than hexahydrobenzene owing to its relatively low HOMO-LUMO gap. The charge stability diagram analysis (Figure 3.15) predicts early transition point (or degeneracy

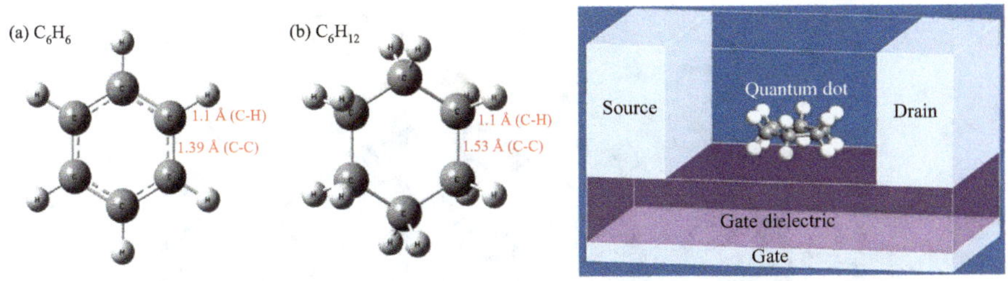

Figure 3.13 Benzene (C_6H_6) and hexahydrobenzene (C_6H_{12}) molecules, and the schematic of SET device. © [2016] IEEE. Reprinted with permission, from [37]

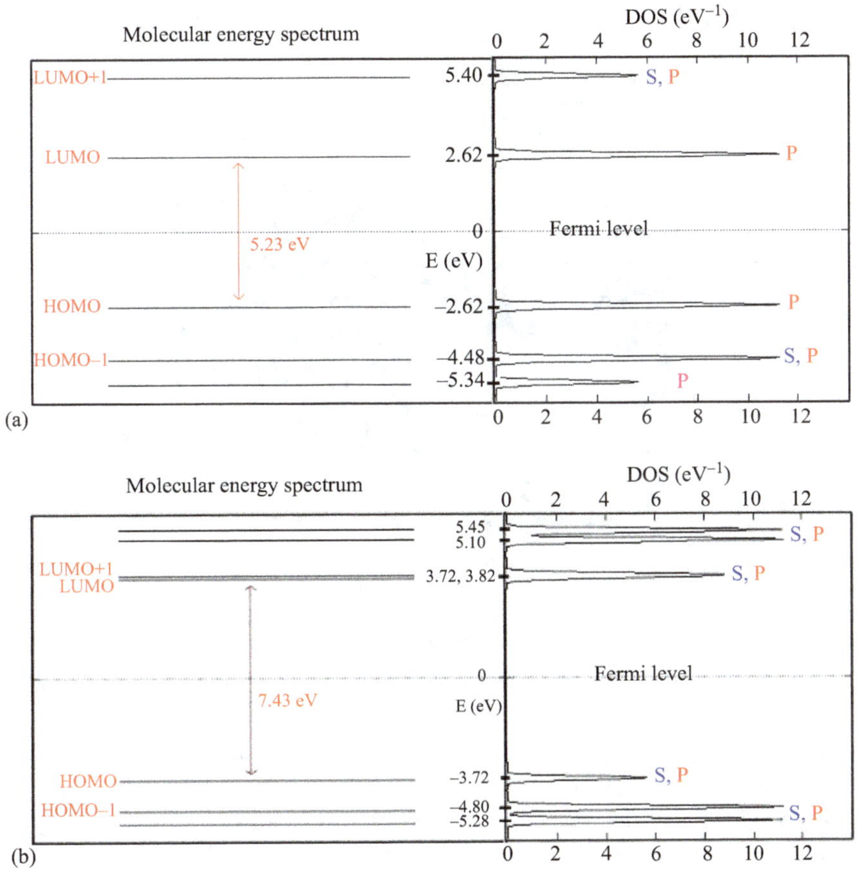

Figure 3.14 Molecular energy spectrum and the corresponding density of states (DOS) profiles for (a) benzene and (b) hexahydrobenzene. © [2016] IEEE. Reprinted with permission, from [37]

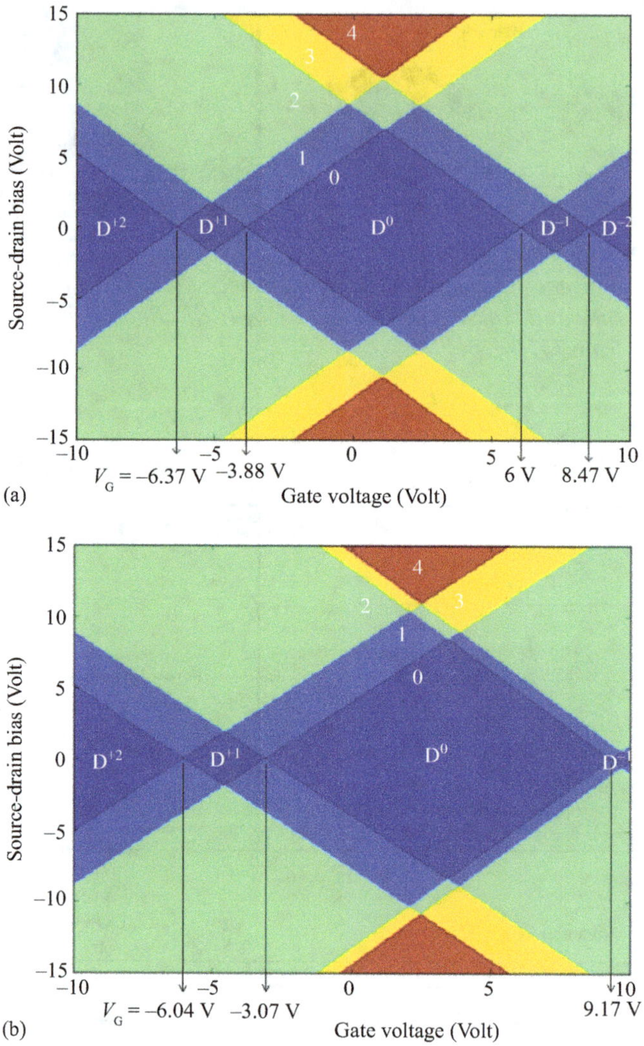

Figure 3.15 Charge stability diagram of (a) benzene and (b) hexahydrobenzene. © [2016] IEEE. Reprinted with permission, from [37]

point) for benzene than hexahydrobenzene when operated with positive gate potentials, whereas the contrary is true for negative gate potentials. In this study, the authors have further decoded the charge stability diagram by identifying and indicating the charge state of molecular quantum dot in each Coulomb blockade diamond (denoted as D^0, D^{+1}, D^{+2}, D^{-1}, D^{-2}, etc. in the dark-blue diamonds) using the total energy variation plot w.r.t gate potential for the first time.

The study of acene series aromatic quantum dots (benzene, naphthalene, anthracene, tetracene, pentacene, etc.) has revealed that the excitation energy of SET decreases with increasing number of rings in the quantum dot, which may be a result of decreasing HOMO–LUMO gap. Santhibhushan *et al.* have further reduced the excitation energy of anthracene quantum dot through boron substitution to create (10-boranylanthracene-9-yl)borane SET quantum dot [36] for low-power and high-speed switching applications. It is worth noting from the study of acene series aromatic quantum dots that the excitation energy of SET is inversely related to the size of the aromatic molecule, thus selecting an acene aromatic molecule with large number of rings not only reduces the excitation energy but also increases the device size. Hence, the selection of aromatic molecule has to be done carefully by the fabricators so as to attain a trade-off between size and excitation energy. Nasri *et al.* have successfully studied the transport properties of pentacene SET for low-power logic gate applications with various electrode materials, and reported Ti electrode as more conductive than the Au and Pt electrodes [39]. Some recent studies have explored metal organic complexes as quantum dots of SET. Anu *et al.* explored metal organic complexes of thiophene [40] and dibenzothiophene [41] as SET quantum dots for high-performance switching applications. In another work, the authors found metal dithiolenes [42] as promising materials for SET quantum dot owing to the extensive π-electron delocalization.

In another report, Anu *et al.* have proposed new metal organic complexes, namely Cr-complex of thiol-ended dibenzothiophene and W-complex of thiol-ended dibenzothiophene for high-performance switching [41]. Figure 3.16 shows the optimized structures of thiol-ended dibenzothiophene, Cr-complex of thiol-ended dibenzothiophene and W-complex of thiol-ended dibenzothiophene and the SET architecture.

The CSD and the CSD tracings presented in Figures 3.17 and 3.18 dictate the W-complex of thiol-ended dibenzothiophene SET as a promising candidate for fast switching applications in comparison to other complexes studied. The CSD tracings (Figure 3.18(a)) along the source-drain bias axis give a Coulomb staircase pattern confirming the discontinuous transport behavior. Also, the Coulomb gap is observed to follow the order: 1.744 V > 1.216 V > 1.056 V > 0.704 V for thiol-ended thiophene > thiol-ended dibenzothiophene > Cr-complex of thiol-ended dibenzothiophene > W-complex of thiol-ended dibenzothiophene, with w-complex of thiol-ended dibenzothiophene having minimum gap. Figure 3.18(b) and (c) gives the CSD tracings along the gate voltage axis. This tracing gives peaks signifying the acceptance of an electron to/from the quantum dot, while the empty regions in between the peaks signify rejection region. Here, the W-complex of thiol-ended dibenzothiophene attains the acceptance region (degeneracy point) at relatively lower gate potential than other complexes. Thus, the W-complex of thiol-ended dibenzothiophene SET quantum dot is better suitable for low-power, high-performance switching.

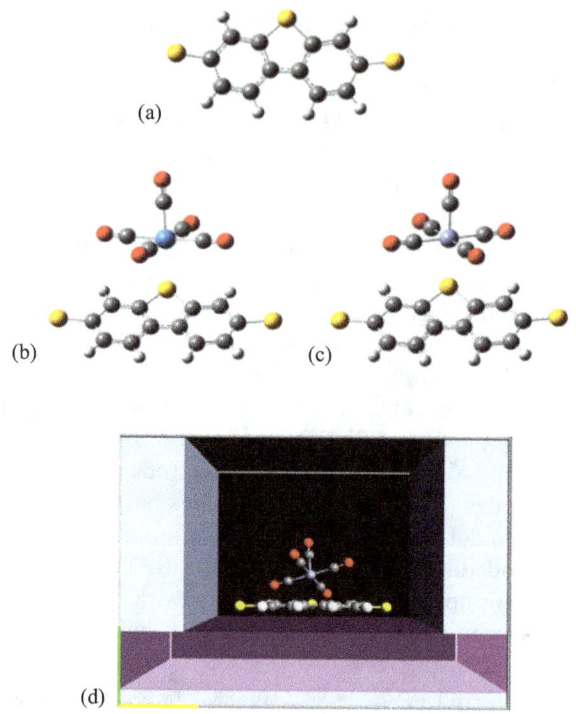

Figure 3.16 (a) Thiol-ended dibenzothiophene, (b) Cr-complex of thiol-ended dibenzothiophene, (c) W-complex of thiol-ended dibenzothiophene, and (d) the SET device architecture. Reprinted with permission from [41]. Copyright (2018), Elsevier

3.3.2 SET as sensor

SETs are emerging as a successful alternative to the conventional sensors due to their fast and unique response to the exotic species. The SET devices are been widely explored for sensing of toxic gases, DNA, drugs, charge, etc. Guo *et al.* have successfully utilized a SET nanopore to sequence DNA [43]. A nanopore is a small hole between multiple electrodes and can be constructed using various kinds of materials such as silicon, protein, metal, and graphene. The device modeled by Guo *et al.* for sensing the DNA nucleobases adenine (A), cytosine (C), guanine (G) and thymine (T) is depicted in Figure 3.19, where a sample charge stability diagram is also shown which have been used as the electronic fingerprints of detection. An interesting aspect of this sensor is that the charge stability diagram has produced very unique electronic fingerprint for each nucleobase and the fingerprints are very immune to the orientation of the nucleobase inside the nanopore (see Figure 3.20).

Later in 2014, Ray has utilized the SET nanopore to detect nicotine drug [44]. Environmental tobacco smoke (ETS) is a result of burning/consuming nicotine-containing cigarettes and long exposure to ETS can cause various chronic diseases

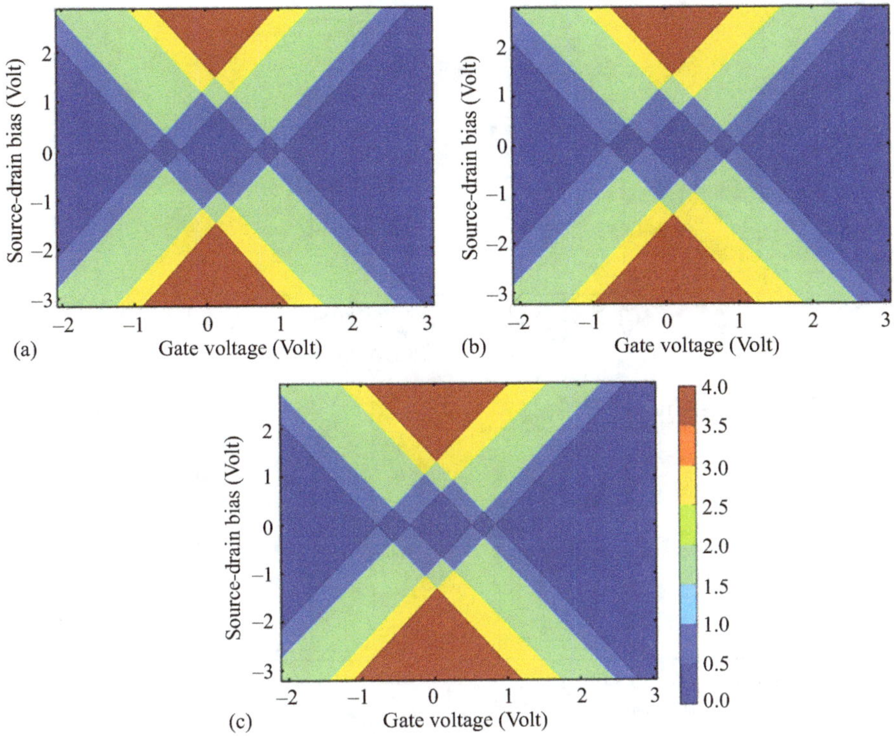

Figure 3.17 Charge stability diagrams for (a) thiol-ended dibenzothiophene, (b) Cr-complex of thiol-ended dibenzothiophene, (c) W-complex of thiol-ended dibenzothiophene. Reprinted with permission from [41]. Copyright (2018), Elsevier

like asthma, lung cancer, and heart diseases. Since, nicotine can stay significantly strong in the environment for about 2 h before metabolizing to cotine. Thus, nicotine detection is extremely important. The nanopore method proposed by the author does not require any chemical preparations as the conventional methods of nicotine detection like gas chromatography, radioimmunoassay, liquid chromatography require. Figure 3.21 shows the modeled nanopore device and the electronic fingerprint. Here also the charge stability diagram has been used as electronic fingerprint for the detection of nicotine with various possible orientations.

In another study, S. J. Ray has proposed an effective double-gated SET environment for the detection of single-atom impurities aimed at assisting the nanoscale semiconductor device fabricators with controlled impurity addition into the semiconductors [45]. Figure 3.22 shows the double-gated SET device structure and the respective electronic fingerprint. Later in 2015, S. J. Ray has proposed a very effective gate all-around structure of SET (see Figure 3.23) for humidity and toxic gas detection [46,47].

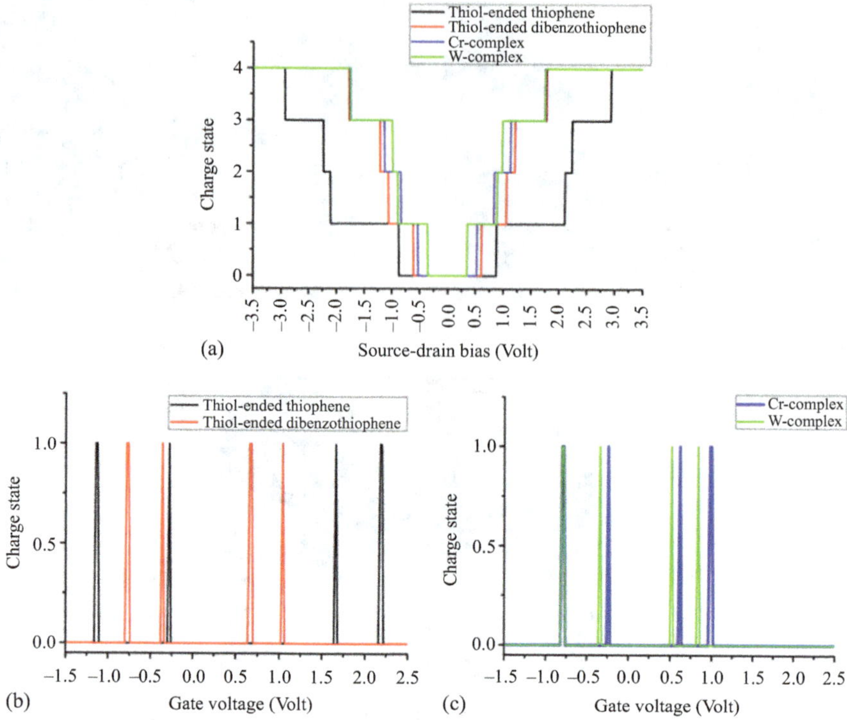

Figure 3.18 Tracings of CSDs (a) along source-drain bias for all the four molecules, (b) along the gate voltage axis for thiol-ended thiophene and thiol-ended dibenzothiophene, (c) along the gate voltage for Cr-complex of thiol-ended dibenzothiophene and W-complex of thiol-ended dibenzothiophene. Reprinted with permission from [41]. Copyright (2018), Elsevier

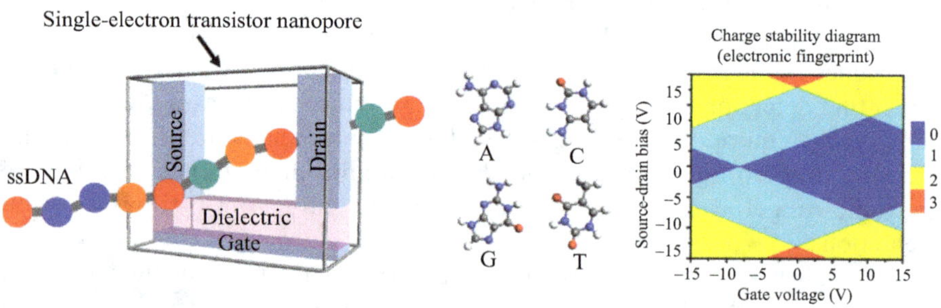

Figure 3.19 Schematic of SET nanopore device used for DNA detection, and a sample charge stability diagram as an electronic fingerprint. Reprinted with permission from [43]. Copyright (2012), American Chemical Society

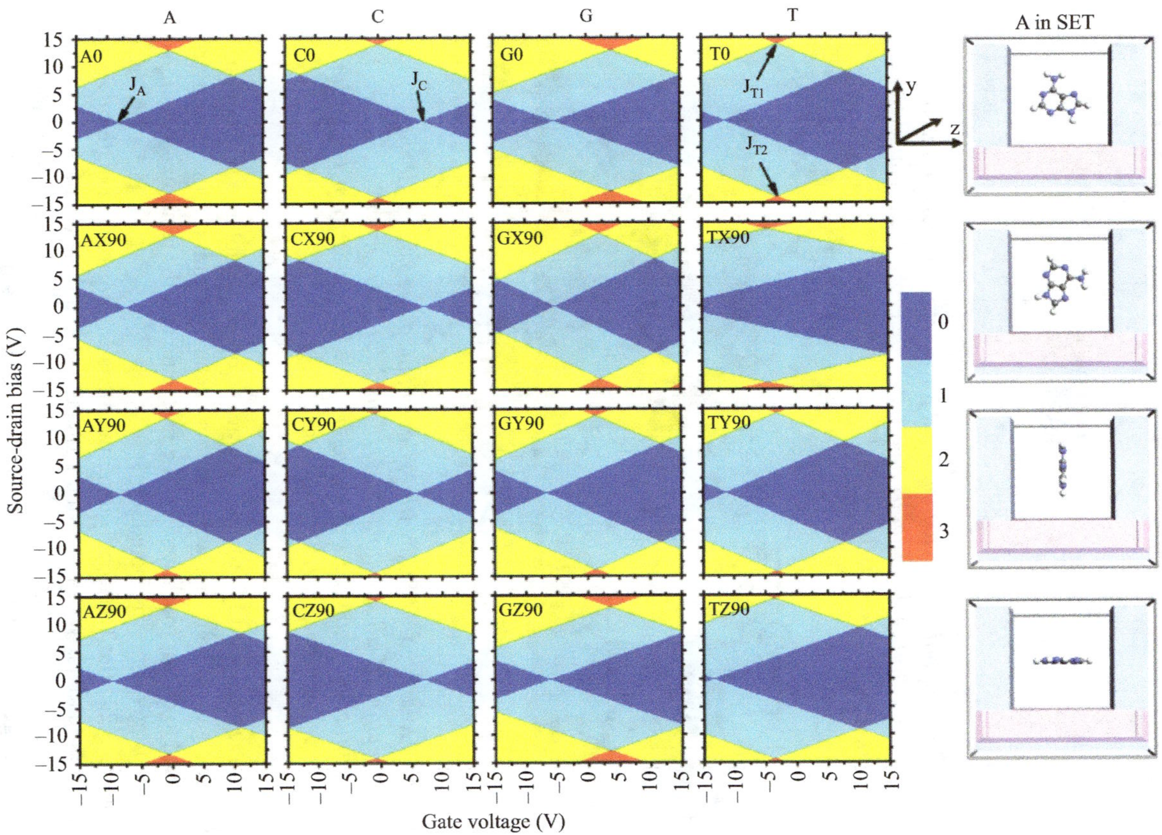

Figure 3.20 *The charge stability diagram electronic fingerprints of various DNA nucleobases adenine (A), cytosine (C), guanine (G) and thymine (T) for four different orientations of each nucleobase in the nanopore. Reprinted with permission from [43]. Copyright (2012), American Chemical Society*

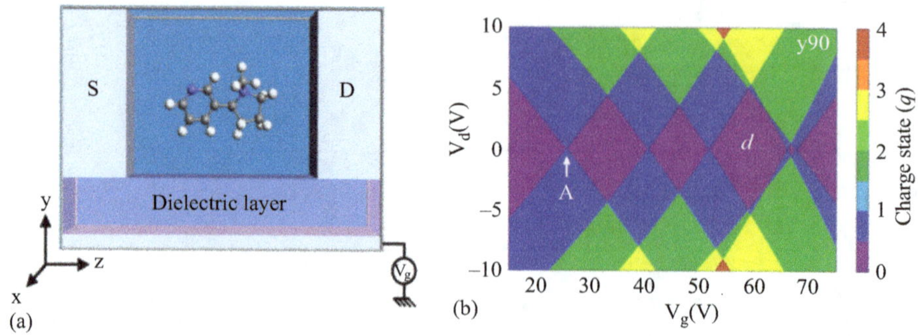

Figure 3.21 (a) SET nanopore device modeled for nicotine detection and (b) the charge stability diagram as electronic fingerprint. Reprinted from [44], with the permission of AIP Publishing

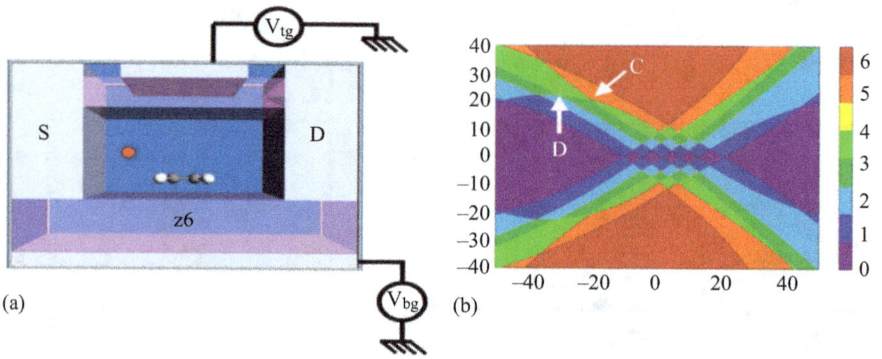

Figure 3.22 (a) Double-gated SET with 1,3-cyclobutadiene island for the detection of silicon atom impurity and (b) the respective charge stability diagram electronic fingerprint. Reprinted from [45], with the permission of AIP Publishing

Jain *et al.* have successfully utilized a tetracene quantum dot as a sensing host to detect chlorine gas [48]. In this work, the tetracene quantum dot in SET environment is exposed to an approaching chlorine molecule and the resulting variations in the electronic fingerprints are noted.

From Figure 3.24, the tetracene quantum dot is in the ionized state at zero applied external potentials. When chlorine molecule is at a distance of 4 Å, only a few degeneracy points show minute variations. At a distance of 3 Å, a definitive right shift in the degeneracy points by a voltage of 0.05 V is noted. When chlorine is close to the quantum dot (1.7 Å), a steep reduction in the excitation energy and charge state transition of zero potential Coulomb blockade diamond from $D^{+1} \rightarrow D^0$ is observed, representing the ultimate sensing ability of tetracene SET for

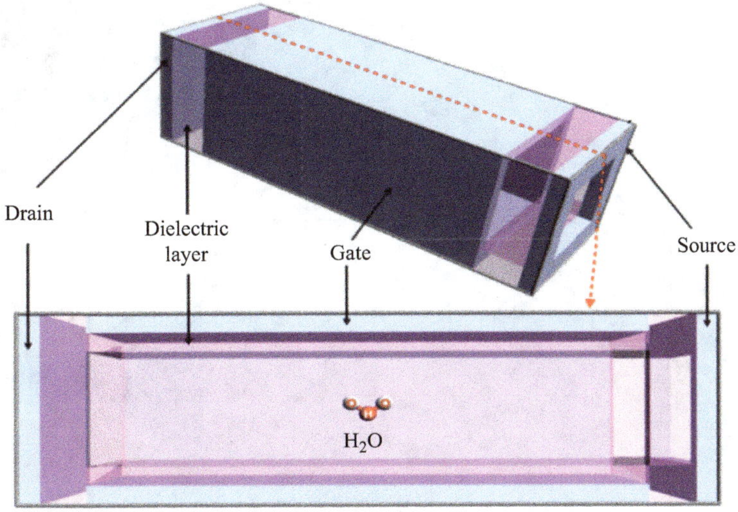

Figure 3.23 Side view and the cross-section view of the gate all-around architecture of SET designed for humidity and toxic gas detection. Reprinted from [46], with the permission of AIP Publishing

chlorine. The device is reported to possess large operational temperature range owing to the high charging energy of the quantum dot.

The significant advancements in the research and development of 2D-materials during the last two decades and their exceptional properties such as large surface area and electrical conductivity have prompted some researchers to model SET sensors using 2D-Materials. S. J. Ray has investigated graphene, MoS_2 and phosphorene monolayers as quantum dots of SET to sense CO, CO_2, NH_3, NO_2 gases [49]. As per the report, the structural and electronic properties of the monolayers remain unaffected on adsorption of toxic gases. Phosphorene offers the highest strength of physisorption for all these molecules indicating its superiority than the other two materials. It is observed that phosphorene and MoS_2 are additionally sensitive toward the N-based molecules and magnetism could be induced in the presence of a paramagnetic molecule. The sensitivity of SET has been confirmed by the charge stability diagram electronic fingerprints. Later in 2018, Sharma *et al.* have utilized Cu-doped MoS_2 sheet as SET quantum dot for sensing CO and NO gases [50]. The Cu dopant on MoS_2 acts as an active site for exotic molecule detection with enhanced adsorption energy. Figure 3.25 shows the schematic of Cu-doped MoS_2-based SET utilized for sensing CO and NO gas molecules.

Another prominent application of SET is the charge detection or electrometry, where charges on the island are critically analyzed within the SET environment. Since the discovery of the device, the quantized nature of electron transport of SET is found helpful in measuring the charges sensitively. Several experimental [51,52]

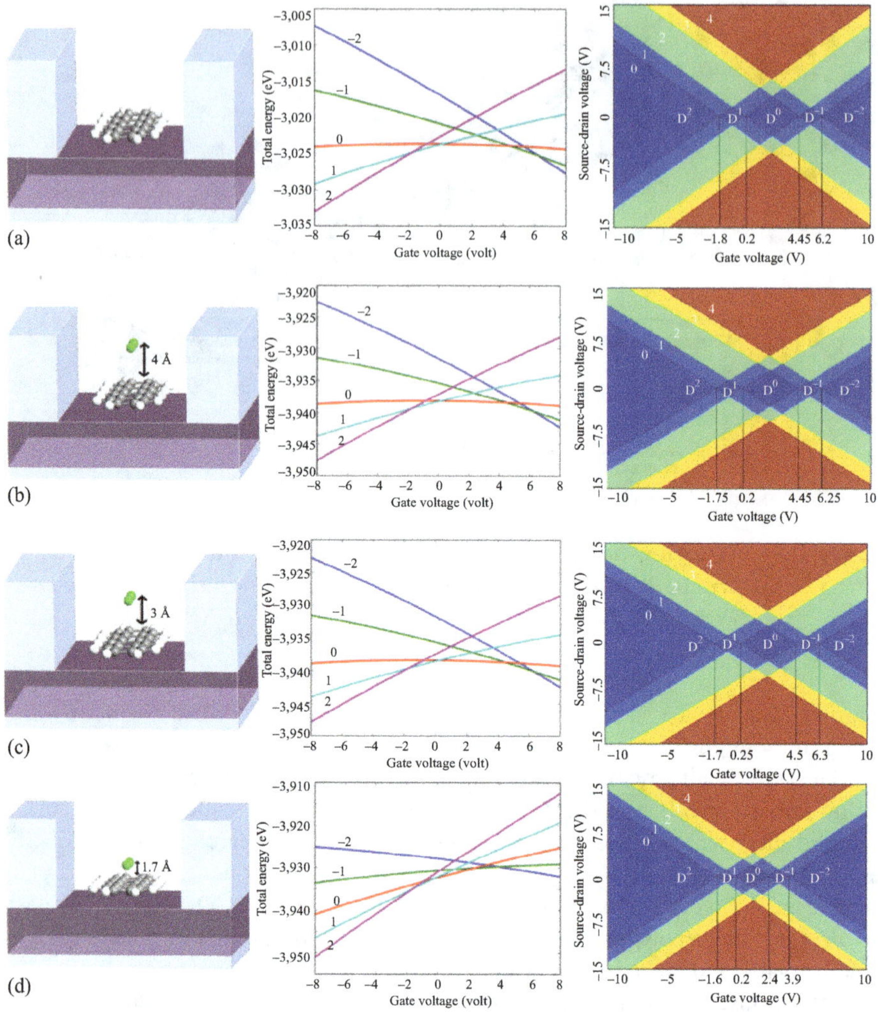

Figure 3.24 Variation in the charge stability diagram electronic fingerprints of tetracene quantum dot with approaching exotic molecule (chlorine). Reprinted with permission from [48]. Copyright (2018), Springer Nature

and theoretical [53] reports have successfully demonstrated the ability of SET for charge detection.

To summarize the whole discussion, the research of single electron devices has seen rapid advancements, since its first fabrication in 1987. Computational simulations are always regarded as a way to cut the experimental expenses as performing the simulation trials for various experimental possibilities not only save money but also time. The DFT-based computational research has taken thrust since

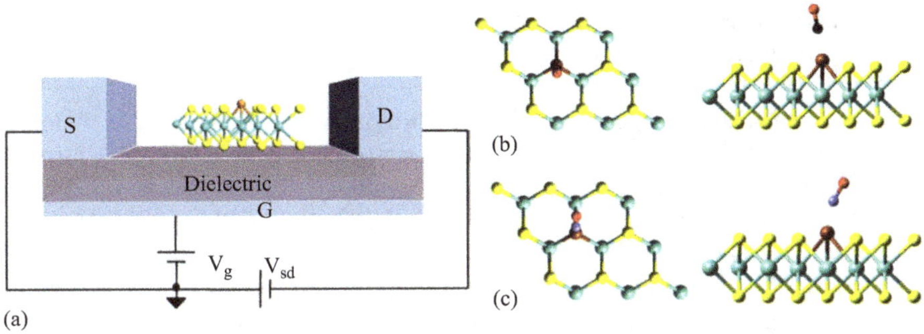

Figure 3.25 (a) Schematic of SET with Cu-doped MoS_2 quantum dot as host material. Top and side views of (b) CO adsorbed Cu-MoS_2 and (c) NO adsorbed Cu-MoS_2. © [2018] IEEE. Reprinted with permission, from [50]

the first successful modeling of SET device in 2010. So far, the device is been widely explored for various applications such as switching, sensing, electrometry, spectroscopy, and memory, both experimentally and theoretically. It is expected that this novel device will take over the conventional FET devices in near future as a switching element of ICs. Although there are few issues associated with the mass fabrication and integration of these devices in large scale at this point of time, the ever-increasing advancements in the fabrication technology may resolve such issues and pave the way for a new generation of atomic scale single electron devices with unprecedented computational capabilities.

References

[1] Kang, Sung-Mo and Yusuf, Leblebici; "CMOS Digital Integrated Circuits: Analysis and Design" 3rd Ed., McGraw Hill Pub., New York, 2003, pp. 1–4.

[2] Web reference: http://download.intel.com/pressroom/images/events/moores_law_40th/Microprocessor_Chart.jpg dated 19th Sep. 2018.

[3] Fuechsle, Martin, Jill A. Miwa, Suddhasatta Mahapatra *et al*. "A single-atom transistor." *Nature Nanotechnology* 7, no. 4 (2012): 242.

[4] Datta, Supriyo (2004) "ECE 453 Lecture 39: Coulomb Blockade," http://nanohub.org/resources/756

[5] Srivastava, Anurag, B. Santhibhushan, and Pankaj Dobwal "Performance analysis of impurity added benzene based single-electron transistor." *Applied Nanoscience* 4, no. 3 (2014): 263–69.

[6] Averin, D. V., and K. K. Likharev "Coulomb blockade of single-electron tunneling, and coherent oscillations in small tunnel junctions." *Journal of Low Temperature Physics* 62, no. 3–4 (1986): 345–73.

[7] Josephson, Brian David "Possible new effects in superconductive tunnelling." *Physics Letters* 1, no. 7 (1962): 251–53.

[8] Fulton, Theodore A. and Gerald J. Dolan "Observation of single-electron charging effects in small tunnel junctions." *Physical Review Letters* 59, no. 1 (1987): 109.

[9] Berman, David, Nikolai B. Zhitenev, Raymond C. Ashoori *et al.* "Single-electron transistor as a charge sensor for semiconductor applications." *Journal of Vacuum Science & Technology B: Microelectronics and Nanometer Structures Processing, Measurement, and Phenomena* 15, no. 6 (1997): 2844–47.

[10] Wang, Lin-Jun, Gang Cao, Tao Tu *et al.* "A graphene quantum dot with a single electron transistor as an integrated charge sensor." *Applied Physics Letters* 97, no. 26 (2010): 262113.

[11] Knobel, Robert G. and Andrew N. Cleland "Nanometre-scale displacement sensing using a single electron transistor." *Nature* 424, no. 6946 (2003): 291.

[12] Knobel, R. and A. N. Cleland "Piezoelectric displacement sensing with a single-electron transistor." *Applied Physics Letters* 81, no. 12 (2002): 2258–60.

[13] Li, Jian, J. T. Santos and M. A. Sillanpää. "High-precision displacement sensing of monolithic piezoelectric disk resonators using a single-electron transistor." *Journal of Low Temperature Physics* 191, no. 5–6 (2018): 316–29.

[14] Blencowe, Miles P. and Martin N. Wybourne "Sensitivity of a micromechanical displacement detector based on the radio-frequency single-electron transistor." *Applied Physics Letters* 77, no. 23 (2000): 3845–47.

[15] Mozyrsky, D., I. Martin and M. B. Hastings "Quantum-limited sensitivity of single-electron-transistor-based displacement detectors." *Physical Review Letters* 92, no. 1 (2004): 018303.

[16] Brenning, Henrik T. A, Sergey E. Kubatkin, Donats Erts *et al.* "A single electron transistor on an atomic force microscope probe." *Nano Letters* 6, no. 5 (2006): 937–41.

[17] González, Jhon W., F. Delgado and Joaquín Fernández-Rossier "Graphene single-electron transistor as a spin sensor for magnetic adsorbates." *Physical Review B* 87, no. 8 (2013): 085433.

[18] Karre, P. Santosh Kumar, Manoranjan Acharya, William R. Knudsen and Paul L. Bergstrom "Single electron transistor-based gas sensing with tungsten nanoparticles at room temperature." *IEEE Sensors Journal* 8, no. 6 (2008): 797–802.

[19] Inokawa, Hiroshi, Akira Fujiwara and Yasuo Takahashi. "A multiple-valued logic and memory with combined single-electron and metal-oxide-semiconductor transistors." *IEEE Transactions on Electron Devices* 50, no. 2 (2003): 462–70.

[20] Guo, Lingjie, Effendi Leobandung, and Stephen Y. Chou. "A silicon single-electron transistor memory operating at room temperature." *Science* 275, no. 5300 (1997): 649–51.

[21] Yano, Kazuo, Tomoyuki Ishii, Toshiaki Sano *et al.* "Single-electron memory for giga-to-tera bit storage." *Proceedings of the IEEE* 87, no. 4 (1999): 633–51.

[22] Matsumoto, Kazuhiko, Yoshitaka Gotoh, Tatsuro Maeda, John A. Dagata, and James S. Harris. "Room-temperature single-electron memory made by

pulse-mode atomic force microscopy nano oxidation process on atomically flat α-alumina substrate." *Applied Physics Letters* 76, no. 2 (2000): 239–41.

[23] Thelander, Claes, Henrik A. Nilsson, Linus E. Jensen, and Lars Samuelson. "Nanowire single-electron memory." *Nano Letters* 5, no. 4 (2005): 635–38.

[24] Zheng, Haisheng, Yang Zhou, and Shubhra Gangopadhyay. "Size-dependent work function and single electron memory behavior of pentacene non-volatile memory with embedded sub-nanometer platinum nanoparticles." *Journal of Applied Physics* 117, no. 2 (2015): 024504.

[25] Liu, Lu, Xueqing Li, Vijaykrishnan Narayanan, and Suman Datta. "A reconfigurable low-power BDD logic architecture using ferroelectric single-electron transistors." *IEEE Transactions on Electron Devices* 62, no. 3 (2015): 1052–57.

[26] Dutta, Bivas, J. T. Peltonen D. S. Antonenko *et al.* "Thermal conductance of a single-electron transistor." *Physical Review Letters* 119, no. 7 (2017): 077701.

[27] El Hajjam, Khalil G., Mohamed Amine Bounouar, Nicolas Baboux *et al.* "Tunnel junction engineering for optimized metallic single-electron transistor." *IEEE Transactions on Electron Devices* 62, no. 9 (2015): 2998–3003.

[28] Park, Hongkun, Jiwoong Park, Andrew KL Lim *et al.* "Nanomechanical oscillations in a single-C 60 transistor." *Nature* 407, no. 6800 (2000): 57.

[29] Tettamanzi, Giuseppe Carlo, Samuel James Hile, Matthew Gregory House *et al.* "Probing the quantum states of a single atom transistor at microwave frequencies." *ACS Nano* 11, no. 3 (2016): 2444–51.

[30] Shorokhov, V. V., D. E. Presnov, S. V. Amitonov, Yu A. Pashkin, and V. A. Krupenin. "Single-electron tunneling through an individual arsenic dopant in silicon." *Nanoscale* 9, no. 2 (2017): 613–20.

[31] Kaasbjerg, Kristen, and Karsten Flensberg. "Strong polarization-induced reduction of addition energies in single-molecule nanojunctions." *Nano Letters* 8, no. 11 (2008): 3809–14.

[32] Stokbro, Kurt. "First-principles modeling of molecular single-electron transistors." *The Journal of Physical Chemistry C* 114, no. 48 (2010): 20461–65.

[33] Srivastava, Anurag, B. Santhibhushan, and Pankaj Dobwal. "Performance analysis of impurity added benzene based single-electron transistor." *Applied Nanoscience* 4, no. 3 (2014): 263–69.

[34] Srivastava, Anurag, Boddepalli Santhibhushan, and Pankaj Dobwal. "Charge stability and conductance analysis of anthracene-based single electron transistor." *International Journal of Nanoscience* 12, no. 06 (2013): 1350045.

[35] Srivastava, Anurag, B. Santhibhushan, Vikash Sharma *et al.* "Influence of boron substitution on conductance of pyridine-and pentane-based molecular single electron transistors: First-principles analysis." *Journal of Electronic Materials* 45, no. 4 (2016): 2233–41.

[36] SanthiBhushan, Boddepalli, Mohammad Shahzad Khan, Anurag Srivastava, and Mohammad Shahid Khan. "First principle analysis of (10-Boranylanthracene-9-yl) borane-based molecular single-electron transistor for high-speed low-power electronics." *IEEE Transactions on Electron Devices* 63, no. 3 (2016): 1232–38.

[37] Bhushan, Boddepalli Santhi, Anurag Srivastava, Jyoti Bhadouria, Rinkoo Bhatia, and Pankaj Mishra. "Aromaticity Influence on Electron Transport of Molecular Single Electron Transistor: DFT Investigation." In *Nanoelectronic and Information Systems (iNIS), 2016 IEEE International Symposium on*, pp. 113–17. IEEE, 2016.

[38] Srivastava, Anurag, Kamalpreet Kaur, Ritu Sharma, Priyanka Chauhan, U. S. Sharma, and Chetan Pathak. "Orientation-dependent performance analysis of benzene/graphene-based single-electron transistors." *Journal of Electronic Materials* 43, no. 9 (2014): 3449–57.

[39] Nasri, A., A. Boubaker, B. Hafsi, W. Khaldi, and A. Kalboussi. "A comparison study of electrode material effects on the molecular single electron transistor." *Organic Electronics* 48 (2017): 7–11.

[40] Anu, Archana Sharma, Md Shahzad Khan, Anurag Srivastava, Mushahid Husain, and Mohd Shahid Khan. "High-Performance Single-Electron Transistor Based on Metal–Organic Complex of Thiophene: First Principle Study." *IEEE Transactions on Electron Devices* 64, no. 11(2017): 4628–35.

[41] Anu, Anurag Srivastava, and Mohd Shahid Khan. "First principle study of single electron transistor based on metal-organic complex of dibenzothiophene." *Organic Electronics* 53 (2018): 227–34.

[42] Anu, Anurag Srivastava, and Mohd Shahid Khan. "Charge stability diagram and addition energy spectrum for single-electron transistor based on Ni-dithiolene derivatives." *Organic Electronics* 59 (2018): 125–30.

[43] Guo, Yan-Dong, Xiao-Hong Yan, and Yang Xiao. "Computational investigation of DNA detection using single-electron transistor-based nanopore." *The Journal of Physical Chemistry C* 116, no. 40 (2012): 21609–14.

[44] Ray, S. J. "Single molecule transistor based nanopore for the detection of nicotine." *Journal of Applied Physics* 116, no. 24 (2014): 244307.

[45] Ray, S. J. "Single atom impurity in a single molecular transistor." *Journal of Applied Physics* 116, no. 15 (2014): 154302.

[46] Ray, S. J. "Humidity sensor using a single molecular transistor." *Journal of Applied Physics* 118, no. 4 (2015): 044307.

[47] Ray, S. J. "Single molecular transistor as a superior gas sensor." *Journal of Applied Physics* 118, no. 3 (2015): 034303.

[48] Jain, Barsha, K. Vinod Kumar, B. SanthiBhushan, Kumar Gaurav, Manisha Pattanaik, and Anurag Srivastava. "A tetracene-based single-electron transistor as a chlorine sensor." *Journal of Computational Electronics* 17 (2018): 1515–20.

[49] Ray, S. J. "First-principles study of MoS_2, phosphorene and graphene based single electron transistor for gas sensing applications." *Sensors and Actuators B: Chemical* 222 (2016): 492–98.

[50] Sharma, Archana, Mohd Shahid Khan, Mushahid Husain, Md Shahzad Khan, and Anurag Srivastava. "Sensing of CO and NO on Cu-Doped MoS_2 Monolayer-Based Single Electron Transistor: A First Principles Study." *IEEE Sensors Journal* 18, no. 7 (2018): 2853–60.

[51] Wang, Lin-Jun, Gang Cao, Tao Tu *et al.* “A graphene quantum dot with a single electron transistor as an integrated charge sensor.” *Applied Physics Letters* 97, no. 26 (2010): 262113.
[52] Schoelkopf, R. J., P. Wahlgren, A. A. Kozhevnikov, P. Delsing, and D. E. Prober. “The radio-frequency single-electron transistor (RF-SET): A fast and ultrasensitive electrometer.” *Science* 280, no. 5367 (1998): 1238–42.
[53] Ray, S. J., and R. Chowdhury. “Double gated single molecular transistor for charge detection.” *Journal of Applied Physics* 116, no. 3 (2014): 034307.

Chapter 4

Application of density functional theory (DFT) for emerging materials and interconnects

Kazi Muhammad Mohsin[1] *and Ashok Srivastava*[1]

4.1 Introduction

Scaling down of the critical dimensions of MOS transistors has been enabling semiconductor industry to improve the performance of electronic devices. The idea of scaling down goes back to 1965 when the founder of Intel's Gordon Moore forecasted the increase in functionality of integrated circuits (ICs), commonly known as 'Moore's Law'. Moore's law states that the number of transistors in an IC would double in every 18 months. For nearly five decades, semiconductor industry has fulfilled the prediction of this rule by constantly pushing the VLSI chip technology and maintained a tremendous effort spanning from material selection, fabrication process and novel architectures to keep the progress uncompromised. However, Moore's law may be reaching its end and a new paradigm shift with a lot more interesting things are on the way [1]. One of the most interesting trends is the exploration of novel wonder materials among the researchers. Not only experimentalists are participating in this search but also theoreticians actively participating in predicting new exotic properties of newly discovered materials. Just to name a few of these materials, carbon nanotube (CNT) [2], graphene, phosphorene, etc. For VLSI applications CNT and graphene have been studied exhaustively [3–13]. Among the various theoretical approaches, density functional theory (DFT) [14,15] is one of the widely accepted approaches in studying new materials properties for electronic applications. In this chapter, DFT will be introduced briefly and will be applied to simulate the electronic properties of a material.

4.2 Density functional theory

DFT has been widely used by physicists, material scientists and chemists as a method to understand new materials' properties utilizing first principle approach. By principle, DFT is an exact method. However, to speeding up the calculations,

[1]Division of Electrical and Computer Engineering, Louisiana State University, Baton Rouge, LA, USA

approximation for electron's exchange function is commonly used within DFT. More accurate 'wave function'-based method which solves Schrödinger equation directly without any approximation is limited only to a molecule which consists of few tens of atoms. Even using modern supercomputing technology and massive parallel resources, it is quite impossible to simulate a large atomic cluster at the scale of a present transistor technology (~10 nm). For this reason, in most of the practical cases, DFT can be utilized instead of more accurate 'wave function' approach.

To introduce DFT, a little bit of historical sketch is required. In 1928, Hartree [16] introduced a procedure to calculate approximate wave functions and energies for atoms and ions, and this is called the Hartree function. Later Fock and Slater [17], individually, proposed a self-consistent function (SCF) taking into account the Pauli's principle [18], and the multi-electron wave function (Slater-determinant) which is popular as Hartree–Fock (HF) method [14]. However, foundation of DFT was established in 1964 by Hohenberg and Kohn [15], which is known as HK theory. In a material system with n electrons, there are $3n$ (x, y, z components) variables in electrons wave functions that need to be solved which is very complicated by HF method. However, using HK theory only three variables are required as it uses functional method. In 1965, Kohn and Sham (KS) [19] simplified HK theory and made it applicable to multi-electron system. Since there were no rigorous ways to solve KS-DFT, approximations were required. Series of developments went into finding different approximation techniques. The first and the simplest approximation is the local-density approximation (LDA) [14] which by 1970 became popular and received popularity among solid-state physics community. It is now routinely used for investigating new materials and for validating experiments or to complement experimental results as a state-of-the-art theory. In 1998, Walter Kohn [19] was awarded Nobel Prize in Chemistry for his contribution in discovering DFT.

DFT started as a computational quantum mechanical modeling technique applied in physics, chemistry and material science to study electronic structure of materials in particular many body systems, atoms, molecules and condensed phases, implementation of DFT is now being extended to soft materials such as biological systems, liquids and amorphous materials. Recently, engineering disciplines are using DFT increasingly for nanoscience and nano-technological purposes. Industries are implementing this technique as a part of their next generation material search.

4.3 Theory behind DFT

Since DFT is a first principle approach it starts with the Schrödinger equation. In theory of solids, the first goal of most approaches is to find a solution of the time-independent, non-relativistic Schrödinger equation. For simplicity, all equations in this chapter are normalized with the electron mass and charge.

$$\widehat{H}\psi_i = E_i\psi_i \tag{4.1}$$

$\widehat{H}$ is the Hamiltonian for a system consisting of M nuclei and N electrons. To avoid the complexity of the development of DFT theory, we will not cover the description of Hamiltonian of the system in this chapter. Rather we will present the computationally demanding equation which will make more sense for explaining the DFT theory. By Born–Oppenheimer approximation (BOA) [18], (4.1) reduces to the following form,

$$\widehat{H_{elec}}\psi_{elec} = E_{elec}\psi_{elec} \tag{4.2}$$

Solution to this Schrödinger equation is ψ_{elec} and electronic energy is E_{elec}. The total system energy will be the sum of electronic energy and the constant nuclear repulsive energy.

$$E_{tot} = E_{elec} + E_{nu} \tag{4.3}$$

Here E_{elec} is the Eigenvalue of (4.3) and E_{nu} is the nuclear repulsive energy.

$$E_{nu} = \sum_{A=1}^{M} \sum_{B>A}^{M} \frac{Z_A Z_B}{R_{\mathrm{AB}}} \tag{4.4}$$

In (4.4), Z stands for the atomic number and R counts the inter-distance of each pair. For example, R_{AB} is the distance between the atomic centres of atom A and B. If an electronic system is in the state ψ, its expectation value of the energy is,

$$E(\psi) = \frac{\langle \psi \widehat{H} \psi \rangle}{\langle \psi \psi \rangle}. \tag{4.5}$$

Here,

$$\langle \psi \widehat{H} \psi \rangle = \int \psi^{*}\left(\vec{r}\right) \widehat{H} \psi\left(\left(\vec{r}\right)\right) d\vec{r} \tag{4.6}$$

According to the variational principle, the energy computed from a guessed ψ is an upper bound to the true ground-state energy (E_0). To obtain full minimization of the functional $E(\psi)$ with respect to all allowed N-electrons, wave functions are required. This full minimization will then give the true ground state, ψ_0 and energy (E_0). Here, expected energy (as $E(\psi_0) = E_0$) of the ground state is the minimum energy of the system.

$$E_0 = \min_{\psi \to N} E(\psi) = \min_{\psi \to N} \langle \psi \widehat{H_{elec}} \psi \rangle \tag{4.7}$$

If a material system has N electrons and given nuclear potential is V_{ext}, the variational principle formulates a procedure to obtain the ground-state wave function. In other words, the ground state energy is a function of the number of electrons N and the nuclear potential V_{ext} as described by the following (4.8).

$$E_0 = E[N,\ \mathrm{V}_{ext}] \tag{4.8}$$

Once ground state wave functions are obtained, ground state energy can be calculated from the expected value of energy. From ground state wave functions, electron density of ground state and other related properties of the material can be calculated.

So far, we have seen how to calculate ground state of a material system using (4.7). We re-write (4.7) for the minimum energy in terms of electron density as follows,

$$E_0 = \min_{\rho \to \mathrm{N}} \left(F[\rho] + \int \rho\left(\vec{r}\right) V_{nu-e}\left(\vec{r}\right) d\vec{r} \right). \tag{4.9}$$

In (4.9), the first term accounts for various electronic energy and the second one electronic interaction energy with nucleus. Here, ρ is electrons density as a function of space. Using KS theory and all necessary energy terms, the total energy can be explicitly expressed in the following equation:

$$E[\rho] = -\frac{1}{2}\sum_{i=1}^{N} \langle \psi_i \nabla^2 \psi_i \rangle + \frac{1}{2}\int\int \frac{\rho(\vec{r}_1)\rho(\vec{r}_2)}{r_{12}} d\vec{r}_1 d\vec{r}_2 + \int \rho\left(\vec{r}\right) V_{nu-e}\left(\vec{r}\right) d\vec{r} + E_{XC}[\rho] \tag{4.10}$$

The first term in (4.10) is electronic energy, the second term is the classical Columbic interaction between electrons. The third term is the energy interaction between electron and nucleus. The very last term in (4.10) corresponds to the electronic exchange correlation. All terms in (4.10) are explicit except the last term which accounts for exchange correlation. To obtain wave function dependency, all density terms can be replaced by wave functions as follows,

$$\begin{aligned} E[\rho] = &-\frac{1}{2}\sum_{i=1}^{N} \langle \psi_i \nabla^2 \psi_i \rangle + \frac{1}{2}\sum_{i}^{N}\sum_{j}^{N} \int\int \left|\psi\left(\vec{r}_1\right)\right| \frac{1}{\left|r_{12}\right|} \left|\psi\left(\vec{r}_2\right)\right|^2 d\vec{r}_1 d\vec{r}_2 \\ &- \sum_{i}^{N} \int \sum_{A}^{M} \frac{Z_A}{r_{1A}} \left|\psi\left(\vec{r}\right)\right|^2 d\vec{r} + E_{XC}[\rho] \end{aligned} \tag{4.11}$$

Using the variational principle, now the problem is to minimize (4.11) under the constraint of wave function property $\langle \psi_i \psi_j \rangle = \delta_{ij}$. From this minimization, resulting equation is the KS equation. KS equation looks like Schrödinger equation but is an approximate to the true Schrödinger equation. KS equation is described by (4.12) as follows,

$$\left(-\frac{1}{2}\nabla^2+\left[\int\frac{\rho(\vec{r}_2)}{r_{12}}d\vec{r}+V_{XC}(\vec{r}_1)-\sum_{A}^{M}\frac{Z_A}{r_{1A}}\right]\right)\psi_i=\left(-\frac{1}{2}\nabla^2+V_S(\vec{r}_1)\right)\psi_i=\epsilon_i\psi_i. \tag{4.12}$$

Once one knows the various contributions in (4.12), potential V_S can be obtained which one needs to insert into the one-particle equation, which in turn determine the wave functions and hence the ground state density and the ground state energy employing (4.11). Here, it is to be noted that V_S depends on the density, and therefore the KS equations have to be solved iteratively which is frequently referred to as self-consistent field (SCF) calculation.

4.4 Implementation of DFT

In this section, we will discuss how to implement the DFT theory using various atomic modelling open-source tools. The complete simulation flow is shown in Figure 4.1.

To solve KS equation for a given material system, QUANTUM ESPRESSO [20] has been used which is a FORTRAN code based on standard input and output options. Through text-based input and output materials, system must be defined and passed through standard input for the DFT engine to produce KS wave functions and densities. Later on, standard programming language has been used for the purpose of post processing other properties of the materials. A typical workflow

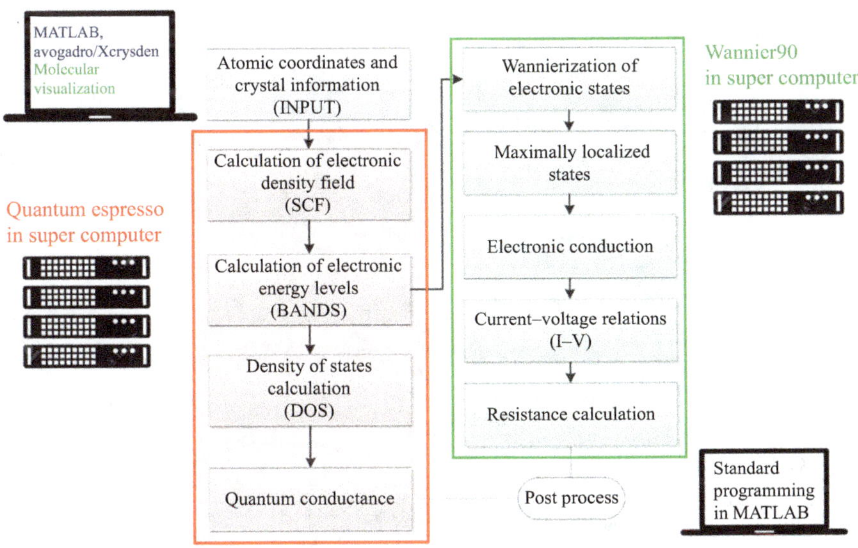

Figure 4.1 Implementation of DFT using various software and programming tools across various computational platforms

and selection of tools are shown in Figure 4.1. QUANTUAM ESPRESSO and Wannier90 have been implemented using supercomputers. Code development, testing and input/output generation were done in desktop computer.

The implementation of DFT starts with defining the crystal structure and atomic coordinates. Any programming language can be used to obtain atomic coordinates of a certain crystal structure. For example, we can take graphene's crystal structure. The first step is to define the unit cell of the graphene crystal. In the following code, Quantum Espresso input code is presented for explaining graphene crystal structure.

In Figure 4.2, the first section is to define what kind of calculation we are doing using the DFT engine which is Quantum Espresso in this example. Calculation type is used as 'SCF' which stands for 'self-consistent field' calculation. Beside calculation type, the first section also includes the computer work directory to save necessary files. The second section is to define the crystal system. The first parameter 'ibrav' is to define what kind of Bravais lattice we are working with. In this case, we are defining crystal structure of our own so it is not a predefined one. Hence the value for 'ibrav' is set to '0'. We are using only carbon atoms in this structure. So type of atom, 'ntyp' is set to '1'. Again, in a graphene crystal structure there is to lattice points, populated by two carbon atoms. So, the number of atoms 'nat' is set to '2'. Lattice constant, 'celldim (1)' of graphene structure is 2.45 Å, which is in atomic unit 4.830. After defining the crystal structure, we describe the

```
&CONTROL
      calculation = 'scf'
      restart_mode='from_scratch',
      prefix='bulk',
      pseudo_dir = '/work/',
      outdir= '/work/output/',
/
&SYSTEM
      ibrav= 0,
      celldm(1) =4.830366967101510,
      nat= 2,     ntyp= 1,
/
&ELECTRONS
      diagonalization='david',
      electron_maxstep = 100,
      mixing_beta = 0.2,
      conv_thr = 1.0d-3,
/
ATOMIC_SPECIES
    C   12.0107       C.pz-n-kjpaw_psl.0.1.UPF
CELL_PARAMETERS alat
   1.000000000000000                    0                  0
   0.500000000000000   0.866025403784439                  0
                   0                    0  10.732721359260136
ATOMIC_POSITIONS alat
C                    0                    0    0
C                    0   0.577350269189626    0
K_POINTS automatic
4 4 1 0 0 0
```

Figure 4.2 Quantum Espresso input code for 2D graphene unit cell

electronic step calculation. For linear algebra package, we used the 'Davidson' diagonalization algorithm. In SCF calculation, maximum electronic calculation iteration is set to '100' as in 'electron_maxstep'. Mixing beta is another optimization parameter. Convergence threshold is the term to define the tolerance of the iterative SCF calculation. Atomic species is to set the atom with its atomic mass in atomic unit along with the pseudopotential file where the electronic energies of a single isolated atom are pre-calculated. 'Cell parameters' are the crystal vectors of graphene expressed in the unit of lattice constant which we previously defined in 'System' section. After defining all these sections, we need to define the atomic positions of the two carbon atoms sitting at two lattice points of graphene crystal. At the end the Brillouin Zone (BZ) or 'K' points sampling scheme is to be defined. In this example, we are sampling with 4 × 4 × 1 points. The increase of 'K' points will increase the accuracy of the calculation at the price of computational speed. The output, electronic energy at different 'K' points will be generated in predefined output directory. Next step will be post-processing these energy levels to generate energy band diagram of the considered crystal system. The input code for Quantum Espresso for band diagram calculation is as shown in Figure. 4.3.

In Figure 4.4, calculated energy band diagram of 2D graphene crystal is shown along with the calculation from tight-binding method. Here, DFT accuracy has been increased by considering the van der Waals correction (vdw) along with using the plane-augmented wave (PAW) implementation [21].

```
&dos
prefix='bulk',
outdir = '/output/',
Emin = –5    , Emax = 5
DeltaE = 0.01,
fildos = '/work/kmohsi1/QE_g/dos.dat'
/
```

Figure 4.3 Quantum Espresso input code for band diagram calculation

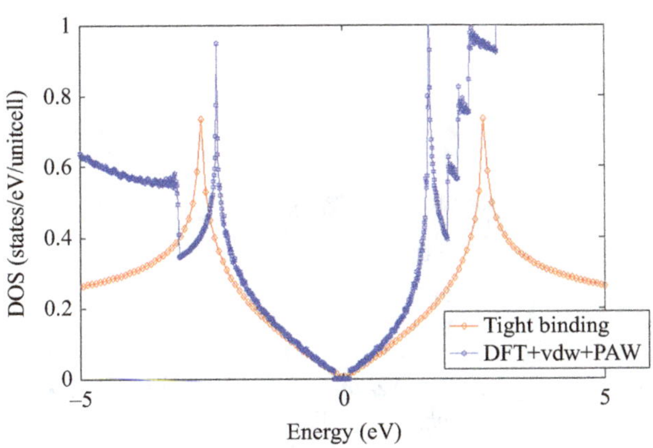

Figure 4.4 Energy band diagram for 2D graphene crystal

4.5 Hybrid material modelling with DFT

DFT can be used for modelling any material system including a hypothetical material to be synthesized in the lab. In this section, we will discuss a hybrid material of graphene and copper (G/Cu) proposed for an interconnect solution [22]. First step is to understand the crystal structure. Here, three atomic layers of copper are considered and arranged in such a way that <111> crystal plane is aligned along the *z*-axis. On top of this flat surface of copper, graphene monolayer is placed. Graphene monolayer is placed in such a way that graphene edge is zigzag towards the transport direction, which is *x*-axis in this case. Since Cu <111> plane is a hexagonal lattice with almost similar lattice constant like graphene (2% mismatch), graphene atoms are placed on <111> planes of Cu. Unit cell of bulk two-dimensional G/Cu system consists of three copper atoms in three different atomic layers along with two carbon atoms sitting on the topmost copper layer. Lattice constant and lattice vectors are shown in Figure 4.5.

For electrical transport studies, one-dimensional hybrid G/Cu nanoribbon of width 0.6 nm and height 0.8 nm is considered. For a finite length, atomistic simulation up to 10 nm of interconnect length is considered here. Since optimized structure of this hybrid material system is not known, we need to calculate the 'relaxation' step before the 'SCF' calculation. In relaxation calculation, atoms in the top two Cu layers and graphene layer were allowed to move in finding

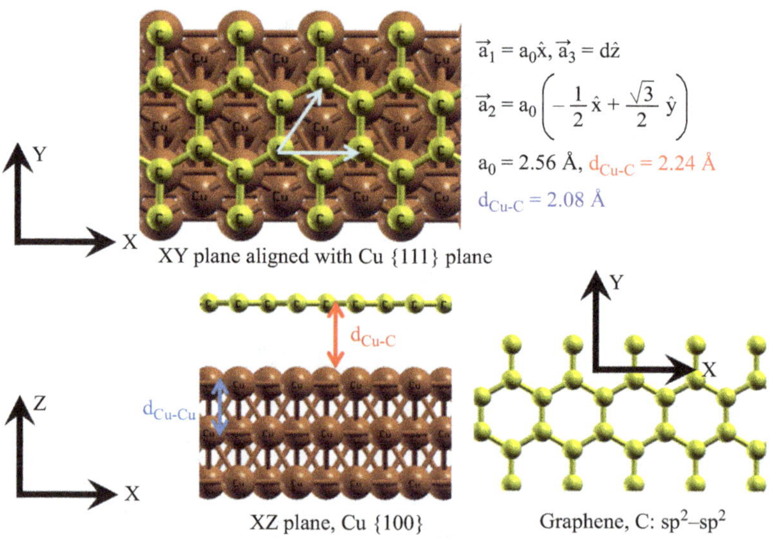

Figure 4.5 Atomic structure of graphene on copper hybrid nano-interconnect in XY plane (top) and in XZ-plane (bottom). Cu <111> plane is towards Cartesian z-axis (which is XY plane). Lattice vectors for this system are shown on the top right. Scale for the top, bottom and bottom-right are not the same

minimum energy position for the whole system in equilibrium. Bottom two Cu layers were fixed in their Cu crystal bulk position. Due to relaxation, one will find that C-Cu interlayer distance increases slightly. Relaxation calculation optimizes coordinates for all atoms in the system for finding the minimum energy state. Using these optimized atomic positions, we then performed SCF calculation.

Each SCF cycle calculates electron density field which is a minimization technique for electron density function. It tries to minimize the overall system energy for an electron density distribution. When the difference of total energies of two consecutive SCF cycles is reached as small as 10^{-9} eV, we stop the SCF calculation. In SCF calculation, we sampled BZ uniformly with $32 \times 32 \times 1$ K-grids using Monkhorst–Pack (MP) method [23] for 2D bulk system. For one-dimensional nanoribbon, we used $128 \times 1 \times 1$ K-grids. MP method ensures generation of special points in the BZ for facilitating efficient integration of periodic functions of the wave vector over entire BZ.

Electron density obtained from SCF calculation was used for another round of calculations for finding energy levels for each point of a densely sampled BZ. We used $64 \times 64 \times 1$ k-grids for BZ sampling using the MP method for 2D bulk and $256 \times 1 \times 1$ for nanoribbon. From this calculation, we obtained the electronic band structure and electrons occupations in those energy states. We performed the band structure calculation for 80 energy levels and obtained 0.7179 eV Fermi energy for the bulk case and −3.9858 eV for the hybrid nanoribbon. Later on, for all other calculations, we adjusted these Fermi energies to 0 eV when necessary for comparison or for transport calculations. From SCF calculation, we constructed band structures and calculated density of states (DOS) of this hybrid system within QE code. For DOS calculation, energy levels are adjusted in such a way that the Fermi energy becomes 0 eV. Energy spectrum is sampled with a resolution of 10 meV.

Electronic band structure of G/Cu hybrid system for bulk (2D) and nanoribbon (1D) are shown in Figures 4.6 and 4.7, respectively. From the band structure of

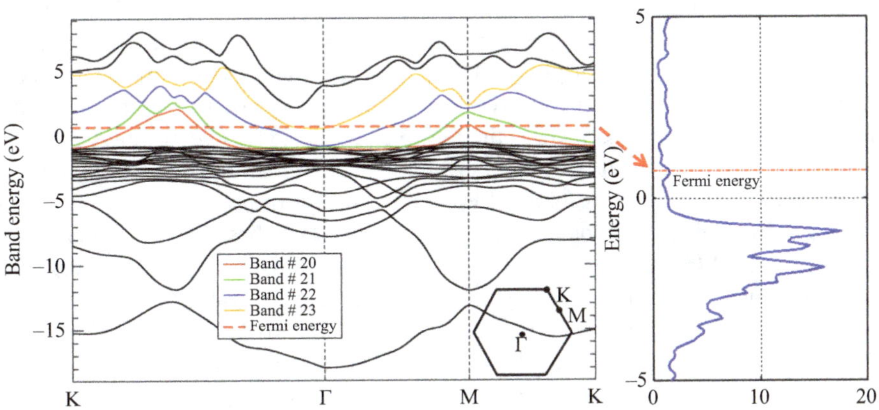

Figure 4.6 (a) Band structure of G/Cu bulk system and (b) DOS from −5 eV to 5 eV are shown

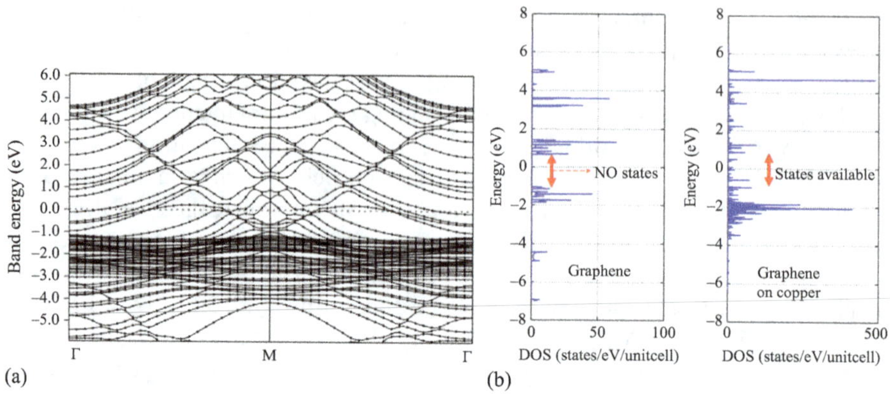

Figure 4.7 (a) Electronic band structure of G/Cu nanoribbon. Fermi level at −4.08 eV and (b) DOS of graphene interconnect are compared with graphene/copper interconnect

bulk 2D system, it is apparent that four bands are crossing Fermi level (0.7179 eV). Those Fermi level crossing bands are highlighted with red, green, blue and orange colour in Figure 4.6(a) in the order of their energy from low to high energy. We counted band index from the lowest energy one as first (near 15 eV). With this counting, bands with indices 20 to 23 are contributing in constructing the Fermi surface. For a metallic system, this multiple band crossing is expected. DOS of G/Cu bulk system is shown in Figure 4.6(b). Being an infinite two-dimensional system, DOS is continuous. Most importantly DOS is continuous and non-zero near the Fermi energy. Just below 0 eV, there is a dense crowd of bands that is consistent in DOS also.

For G/Cu nanoribbon, band structure and DOS are shown in Figure 4.7(a) and (b), respectively. Unlike GNR, there is a non-zero DOS at Fermi level for G/Cu nanoribbon. In case of nanoribbon, DOS is discrete due to one-dimensional confinement of the electron. In this hybrid system because of Cu, more states are available in an energy window near the Fermi energy. Figure 4.7(b) shows the DOS comparison of graphene and G/Cu material system. This difference of DOS in these two material systems causes their difference in current transport.

We used Wannier90 code [24] for the transport study based on the Bloch states obtained from SCF calculations. First step is to transform Bloch waves into Wannier Function and then finding maximally localized Wannier wave function (MLWF). Rest of the transport properties depend on MLWF. From MLWF, we have computed current–voltage relation (*I–V*). Typical electronic MFP of copper is 40 nm and few microns for graphene. We assumed for this hybrid system electrons MFP to be greater than 40 nm and smaller than 1000 nm. If this is the case for the MFP, then this hybrid interconnect transport should be ballistic in nature for any given interconnect length less than 40 nm. Hence, to compute current–voltage relation at different temperatures, we adopted Landauer–Buttiker (LB) formalism implemented in Wannier90 code [24],

$$I = \frac{2e^2}{h} M \frac{\mu_1 - \mu_2}{e} \tag{4.13}$$

Here, *he* is the magnitude of electronic charge and *h* is Planck constant. *M* counts the number of transport modes for a conductor, μ_1 and μ_2 are the electrochemical potentials of left and right contacts, respectively. Wannier90 code uses Bloch States obtained from QE code to obtain MLWF and construct system Hamiltonian. After obtaining Hamiltonian, Wannier90 uses non-equilibrium Green function (NEGF) for the transport calculation and transmission coefficient. In Landauer–Buttiker (LB) formalism, by definition, the transmission coefficient is quantum conductance. Due to high computational cost for first principle study, we have limited our study to a 10 nm long wire, which represents a short local interconnect and is a good example of ballistic transport. For ballistic transport, one should not use Fuchs–Sondheimer (FS) and Mayadas–Shatzkes (MS) models [25] for resistivity estimation. Therefore, in ballistic transport regime, instead of FS and MS theories we have used LB formalism.

In Figure 4.8, for graphene, no current is observed between −1.34 V and 1.34 V because of not having available states in that energy window. However, for

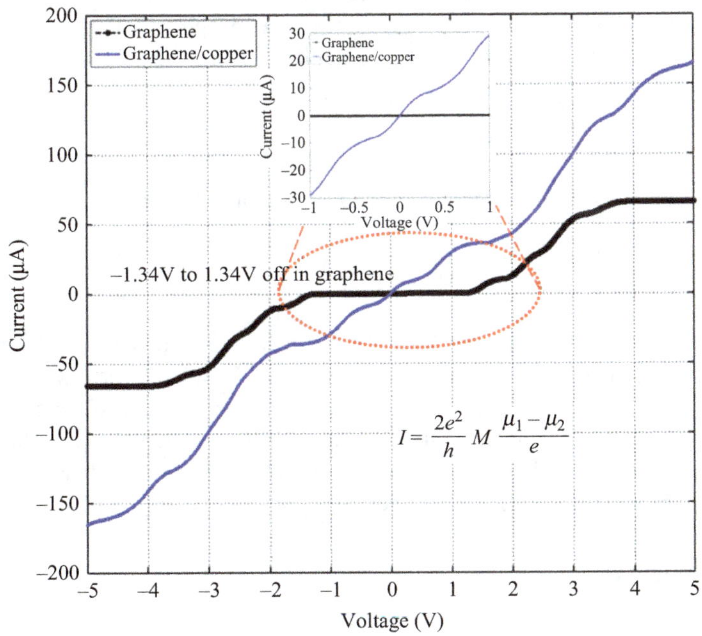

Figure 4.8 (a) Current–voltage (I–V) characteristics of G/Cu nanoribbon interconnect compared with graphene only interconnect

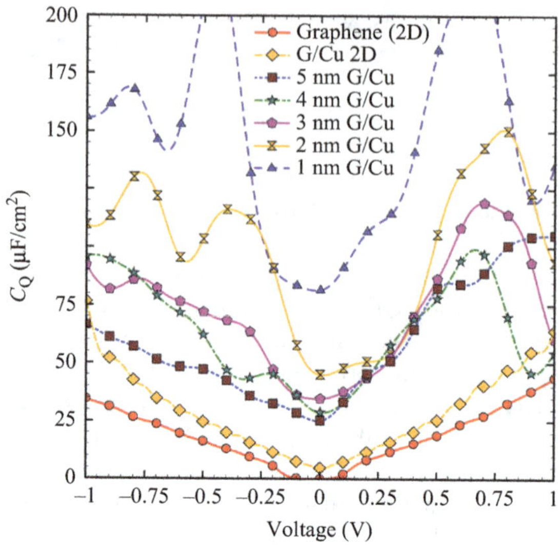

Figure 4.9 Calculated quantum capacitances at different potentials of graphene [26]

graphene on copper is still conductive in this window. Once, the *I–V* characteristics of a material are known, resistance and resistivity can be calculated from this result.

DFT can be further used for the capacitive property calculation of hybrid materials. Quantum capacitance can be calculated from the following equation,

$$C_Q = e^2 \int_{-\infty}^{+\infty} D(E)F_T(E - e\phi_G)dE \tag{4.14}$$

Here, DOS is $D(E)$ and $F_T(E)$ is thermal broadening function defined as in (4.15).

$$F_T(E) = -\frac{df}{dE} = \frac{1}{4K_BT}\operatorname{sech}^2\left(\frac{E}{2K_BT}\right) \tag{4.15}$$

Equation (4.15) will be used for estimating thermal broadening at a finite temperature for the calculation of quantum capacitance. However, for low temperatures, the calculation becomes far easier. At absolute zero temperature, F_T can be assumed as a delta function and then C_Q will be simply $e^2D(E)$. Calculated quantum capacitance is a function of applied voltage. In Figure 4.9, voltage-dependent capacitance is shown for various width of graphene–copper hybrid nanowires.

4.6 Conclusion

In this chapter, we have shown how to use DFT for modelling material and finding their electronic structure. We have shown how to calculate the band diagram and expanded it to the calculation of transport. Hybrid material modelling has been shown in the context of VLSI interconnect. We also described the calculation methods for quantum capacitance of a hybrid G/Cu nanoribbon using DFT. Although in this chapter we have shown up to transport calculation and capacitance properties, DFT can be implemented to calculate other relevant properties of the material. This calculation methodology will help finding transport properties and quantum capacitance of emerging materials. The chapter can be a good starting point in advancing an understanding of electrical performances of other nanostructures for possible interconnect and device materials.

References

[1] M. M. Waldrop, "The chips are down for Moore's law," *Nature*, vol. 530, no. 7589, pp. 144–47, 2016.

[2] S. Iijima, " Helical microtubules of graphitic carbon," *Nature*, vol. 354, no. 6348, pp. 56–58, 1991.

[3] N. Srivastava, and K. Banerjee, "Performance analysis of carbon nanotube interconnects for VLSI applications," *Proceedings of the IEEE/ACM International Conference on Computer-Aided Design (ICCAD)*, pp. 383–90, 2005.

[4] L. Hong, X. Chuan, and K. Banerjee, "Carbon nanomaterials: the ideal interconnect technology for next-generation ICs," *IEEE Design & Test of Computers*, vol. 27, no. 4, pp. 20–31, 2010.

[5] A. Srivastava, Y. Xu, and A. K. Sharma, "Carbon nanotubes for next generation very large scale integration interconnects," *Journal of Nanophotonics*, vol. 4, no. 1, pp. 041690, 1–26, 2010.

[6] S. Berber, Y.-K. Kwon, and D. Tománek, "Unusually high thermal conductivity of carbon nanotubes," *Physical Review Letters*, vol. 84, no. 20, pp. 4613–16, 2000.

[7] A. Naeemi, and J. D. Meindl, "Carbon nanotube interconnects," *Annual Review of Materials Research*, vol. 39, no. 1, pp. 255–75, 2009.

[8] M. Nihei, A. Kawabata, D. Kondo, M. Horibe, S. Sato, and Y. Awano, "Electrical properties of carbon nanotube bundles for future via interconnects," *Japanese Journal of Applied Physics*, vol. 44, no. 4A, pp. 1626, 2005.

[9] K. M. Mohsin, A. Srivastava, A. K. Sharma, and C. Mayberry, "A thermal model for carbon nanotube interconnects," *Nanomaterials*, vol. 3, no. 2, pp. 229–41, April 26, 2013.

[10] K. M. Mohsin, A. Srivastava, A. K. Sharma, and C. Mayberry, "Characterization of MWCNT VLSI interconnect with self-heating induced scatterings," *Proceedings of the IEEE Computer Society Annual Symposium on VLSI*, pp. 368–73, 2014.

[11] K. M. Mohsin, and A. Srivastava, "Characterization of SWCNT bundle based VLSI interconnect with self-heating induced scatterings," *Proceedings of the GLSVLSI'15*, pp. 265–70, 2015.

[12] Y. M. Banadaki, K. M. Mohsin, and A. Srivastava, "A graphene field effect transistor for high temperature sensing applications," *Proceedings of the SPIE, vol. 9060, Nano-, Bio-, Info-Tech Sensors and Systems*, 2014.

[13] K. M. Mohsin, Y. M. Banadaki, and A. Srivastava, "Metallic single-walled, carbon nanotube temperature sensor with self heating," *Proceedings of the SPIE, vol. 9060, Nano-, Bio-, Info-Tech Sensors and Systems*, 2014.

[14] E. K. U. Gross, and R. M. Dreizler, *Density Functional Theory*, Berlin: Springer, 1990.

[15] P. Hohenberg, and W. Kohn, "Inhomogeneous electron gas," *Physical Review*, vol. 136, no. 3B, pp. B864–B871, 1964.

[16] D. R. Hartree, "The wave mechanics of an atom with a non-Coulomb central field. part I. theory and methods," *Mathematical Proceedings of the Cambridge Philosophical Society*, vol. 24, no. 1, pp. 89–110, 1928.

[17] D. H. Kobe, "Variational principle and Slater's generalized Hartree-Fock theory for nuclei," *Physical Review*, vol. 188, no. 4, pp. 1583–89, 12/20/, 1969.

[18] D. J. Griffiths, *Introduction to Quantum Mechanics* (2nd ed.): New Jersey: Prentice Hall, 2004.

[19] W. Kohn, and L. J. Sham, "Quantum density oscillations in an inhomogeneous electron gas," *Physical Review*, vol. 137, no. 6A, pp. A1697–705, 1965.

[20] G. Paolo, B. Stefano, B. Nicola, *et al.* "QUANTUM ESPRESSO: a modular and open-source software project for quantum simulations of materials," *Journal of Physics: Condensed Matter*, vol. 21, no. 39, pp. 395502, 2009.

[21] B. Kristian, R. C. Valentino, L. Kyuho, *et al.* "van der Waals forces in density functional theory: a review of the vdW-DF method," *Reports on Progress in Physics*, vol. 78, no. 6, pp. 066501, 2015.

[22] K.M. Mohsin, A. Srivastava, A.K. Sharma, C. Mayberry and M.S. Fahad, "Temperature sensitivity of electrical resistivity of graphene/copper hybrid nano ribbon interconnect: a first principle study," *ECS J. of Solid State Science and Technology*, vol. 6, no. 4, pp. 119–24, 2017.

[23] H. J. Monkhorst, and J. D. Pack, "Special points for Brillouin-zone integrations," *Physical Review B*, vol. 13, no. 12, pp. 5188–92, 1976.

[24] A. A. Mostofi, J. R. Yates, Y.-S. Lee, I. Souza, D. Vanderbilt, and N. Marzari, "wannier90: A tool for obtaining maximally-localised Wannier functions," *Computer Physics Communications*, vol. 178, no. 9, pp. 685–99, 2008.

[25] E. H. Sondheimer, "The mean free path of electrons in metals," *Advances in Physics*, vol. 1, no. 1, pp. 1–42, 1952.

[26] K. M. Mohsin, A. Srivastava, A. Sharma, and C. Mayberry, "Capacitance of graphene/copper hybrid nano ribbon interconnect – a first principle study," paper #B06-0828 presented in session B06 on graphene and beyond 2D materials of *231st ECS Meeting*, (28 May–02 June 2017), New Orleans. Appeared in *ECS Transactions*, vol. 77, no. 11, pp. 645–50.

Chapter 5

Memristor devices and memristor-based circuits

Venkata P. Yanambaka[1], Saraju P. Mohanty[2], Elias Kougianos[3] and Dhruva Ghai[4]

There are four fundamental circuit variables: voltage, current, charge, and magnetic flux. Until the year 1971, there were only three fundamental components: resistor, capacitor, and inductor. In that year 1971, Leon O. Chua proposed a new device named "memristor" which relates charge and flux. At the time, due to lack of sophisticated fabrication facilities, the new device did not receive much attention until HP Labs successfully fabricated one in 2007. This fabrication of the new device has provided device research with a new perspective as the memristor exhibits a new hysteresis phenomenon named as "Pinched Hysteresis". The memristor can remember the voltage that passed through it even when the supply is turned off. Hence, the name memory + resistor, memristor. After the device was fabricated successfully, research has been done extensively implementing the memristor in various applications which require reconfigurability. The memristor has been used from oscillators to neural networks and logic gates to security applications giving it a wide range of applications. This chapter presents the device description, characteristics, and various applications of the memristor in analog and digital applications.

This chapter is organized as follows: different types of memristors are presented in Section 5.2. Fabrication principles of the memristor and how it works are presented in Section 5.3. For simulation purposes, various models for memristors have been proposed. Such models are presented in Section 5.4. The electrical characteristics of the memristor are presented in Section 5.5. Applications of memristors in analog and digital nanoelectronics are presented in Sections 5.6 and 5.7, respectively. Summary and future directions are presented in Section 5.8. Table 5.1 summarizes the notations and symbols used in the current chapter.

[1]School of Engineering & Technology, Central Michigan University, MI, USA
[2]Computer Science and Engineering, University of North Texas, Denton, TX, USA
[3]Electrical Engineering, University of North Texas, Denton, TX, USA
[4]Electronics and Communication Engineering, Oriental University Indore, INDIA

Table 5.1 Notations and parameters used

Parameters/ Notation	Description with units
ω	Length of the doped region
R	Generalized resistance
q	Charge
ϕ	Magnetic flux
W	Width of the doped region
D	Length of the undoped region
R_{on}	Low resistance state of memristor
R_{doped}	Resistance of the doped region
$R_{undoped}$	Resistance of the undoped region
v	Drift Velocity
l_{doped}	Length of the doped region
L_{active}	Length of the active region
R_{doped}	Resistance offered by the doped region
$M[q]$	Memristance of the device as a function of charge
μ_v	Mobility of excess ions
$X(t)$	State variable
E	Electric field across the device
$V(t)$	DC voltage at the source
$i_{Gm}(t)$	Current at the top terminal of the device in spice equivalent circuit for $I-V$ characteristics
$i_{Gx}(t)$	Current at the top terminal of the device for determining the state variable
Simscape^{TM} model	
mu	Mobility of the memristor
x	Initial W/D value
D	Width of the memristor-doped region
Ron	Minimum resistance offered by the device
$Roff$	Maximum resistance offered by the device
SPICE model	
p	Top (positive) terminal of memristor
n	Bottom (negative) terminal of memristor
XSV	External connection for plotting state variable
$x1$, $x2$, y	Parameters used for current–voltage characteristics
pV	Positive threshold voltage
nV	Negative threshold voltage
mp, mn	SV motion intensity multipliers
xp, xn	Points where SV motion is reduced
Verilog-A model	
$x0$	Initial state
uv	Dopant Mobility
d	Length of the doped region
$roff$	Minimum resistance offered by the device
ron	Maximum resistance offered by the device
vp	Positive threshold voltage
vn	Negative threshold voltage

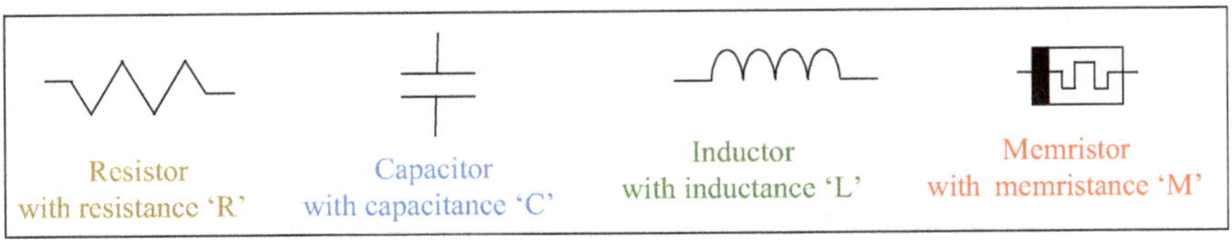

Figure 5.1 Memristor: the fourth fundamental element

5.1 Introduction

5.1.1 Brief history of memristor

There were only three fundamental circuit elements known in 1971: resistor, capacitor, and inductor. In that year, Leon O. Chua presented in his article titled "Memristor – The Missing Circuit Element", a device named memristor [1]. There are four fundamental circuit variables: voltage (v), current (i), charge (q), and magnetic flux (ϕ). Because there are four variables and three fundamental devices, Chua wanted to attain symmetry and theoretically presented the memristor. The relation between voltage and current is used by the resistor, voltage, and charge by the capacitor and current and magnetic flux by the inductor. The memristor uses the relation between charge and magnetic flux [2], as shown in Figure 5.1. It was demonstrated by Chua mathematically that the device he proposed would be able to provide a nonlinear relationship between the flux and the charge. But even before Chua published his work, there had been some current–voltage behaviors observed that could not be explained. In 2015, a new research was published by Leon O. Chua *et al.* at Hong Kong University which revealed that the first man-made memristor was actually developed in 1801 [3]. Humphry Davy conducted a carbon arc discharge experiment, which can generate light without the use of fire. The same experiment was repeated with a modern power supply and observed which revealed the fingerprint of the memristor.

In 2008, a group of scientists from HP labs successfully fabricated the memristor for the first time [2]. Since then research has been going on in various fields [4,5] to develop models and analyze the characteristics of memristors.

5.1.2 What is a memristor?

A memristor is a two-terminal device which has a variable resistance. This "memristance" depends nonlinearly on the direction and amount of the current passing through it. When the flow of current through the memristor stops, the resistance value will not change. In simple terms, it will memorize the amount of charge that passed through it before it was stopped. The *current–voltage* (*I–V*) characteristics of a memristor are similar to that of a variable resistor [6,7]. The phenomenon known as *pinched hysteresis* acts as the fingerprint of a memristor [4].

Significant research has been, and is being, conducted by the research community to study the characteristics of the device. SPICE [8–14], Verilog-A,

[11,15,16] and MATLAB/Simulink™ [11,16–18] models have been developed by the research community and designs have been proposed using various materials.

5.1.3 Applications of memristors

Figure 5.2 shows the application domains of memristors. The unique characteristics of the memristor allow it to be used for various applications. The name memory + resistor itself suggests the memory property [2,19]. The memristor has the ability to change the resistance offered to the voltage flowing through it based on the current applied to it and the direction of flow. The low power consumption of memristor allows researchers to focus on low-power application development. Currently, analog/mixed signal (AMS) design and system-on-chip (SoC) designs are more popular with the advent of new technologies and the possibility of scaling [20]. Research is being conducted to increase the scaling capability of the devices and decrease the power consumption. The memristance offered by the memristor can be changed on the fly with the current supply to the device in the right direction. With a memristive device at the core of various designs, most of the applications can be made reconfigurable. One of the main applications that can be made reconfigurable, which helps in saving chip area and power consumption, is the on-chip oscillator. A memristor-based oscillator can not only save chip area but the reconfigurability of the memristor will also allow the frequency to be changed when necessary [21,22]. Field-programmable gate array (FPGA) architectures are being proposed by various researchers to make the FPGA fabric itself reconfigurable and memristors will reduce the chip area necessary and decrease the power consumption of the FPGA [23,24]. Another application of memristors is the

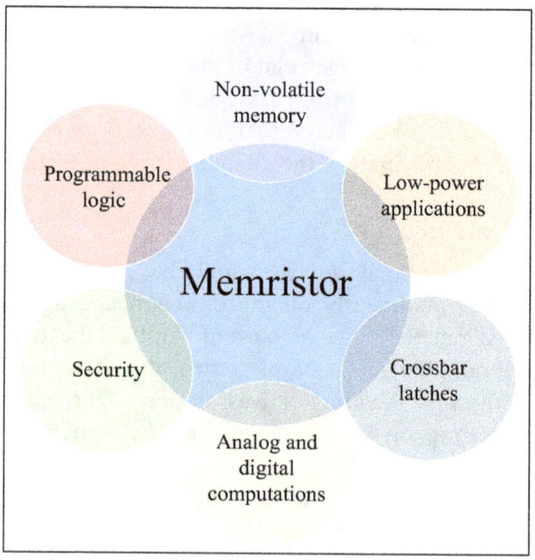

Figure 5.2 Applications of memristors

crossbar latch which can be used for various purposes including memory [25], neural networks [26], and security [27].

5.2 Types of memristors

The fabrication and structure of the memristor classifies it into various types. Such types of memristors are shown in Figure 5.3. After the memristor was rediscovered, the first memristor to be successfully fabricated was the titanium dioxide (TiO_2) type [2]. These were thin-film type memristors that were explored and various devices were developed [28–35]. Polymeric memristors make use of the dynamic doping of polymer and inorganic dielectric-type materials. Resonant memristors use specially doped quantum well diodes and manganite memristors use a substrate of bilayer oxide films based on manganite, as opposed to titanium dioxide.

Memristors are also developed using the spin of the electrons in the material, the spin-based magnetic memristors [36–40]. Based on the direction of the spin of the electron, the memristance changes. In the spin-transfer torque memristors, the relative magnetization alignment of the two electrodes affects the magnetic state of a magnetic tunnel junction changing its resistance.

5.2.1 Thin-film memristors

Thin-film memristors were the first-explored designs by HP-Laboratories [2]. In the thin-film memristor, the memristance offered by the device itself depends on the structure of the device. Usually there are two different layers of material, the active layer and the excess atoms layer. When the concentration of the atoms is higher in a single layer, the memristance is also higher; when the excess atoms are distributed across the device, the memristance is lower. The flow of current through

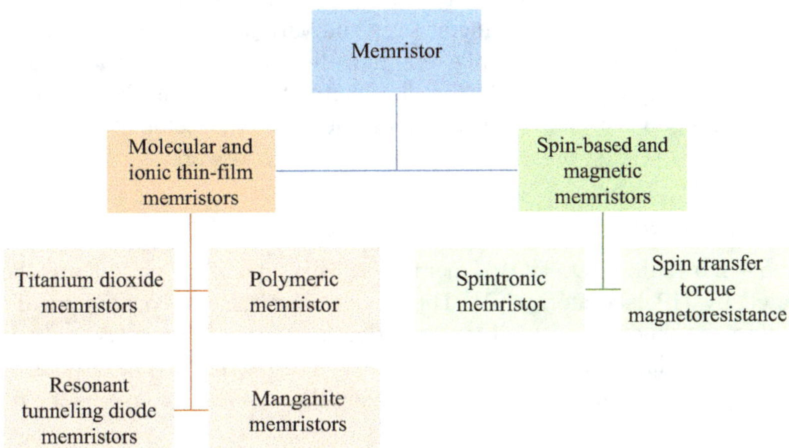

Figure 5.3 Types of memristors

the device and the direction of the voltage will determine the memristance offered by the device. A polymeric memristor is another type of thin-film memristor [41]. It also works on the principle of using a conducting layer of polymer in the device. A polyaniline (PANI) layer is responsible for the variation of memristance according to the redox state. A single passive layer between an electrode and an active thin film attempts to exaggerate the extraction of ions from the electrode.

The existing literature presents memristors with different device structure and materials used for design and fabrication [29–35,42–44].

5.2.2 Spintronic memristors

In the spintronic memristor, the spin of the electrons will decide the memristance offered by the device. They are different from thin-film memristors. A fundamental attribute of quantum electronics is the spin of the electron and other subatomic particles. Spin underlies magnetism and it is considered as the unique form of nanoworld angular momentum [45]. Spintronic memristors are being extensively explored for research and are used in various applications including neuromorphic systems [46].

There are memory designs based on the spin of the electron [47]. Various magnetic nanodevices have been engineered using the spin torque by keeping the same magnetic stack but the shape and the conditions of the devices are changed. The energy efficiency of the spintronic memristors is also comparatively high. The low-voltage fast switching property of spintronic memristors has led to extensive research and development of various applications [38,40,42].

5.3 Device structure and working of a memristor

There are various types of memristors based on the fabrication method and the structure. Thin-film and spintronic memristors are the two major types. Extensive research has been performed on memristors and various structures have been proposed using different materials like titanium dioxide, zinc oxide, and magnetic materials. In each type of memristor, the memristance offered by the device is based on the amount of current that flows through the device and the direction of the current. This section presents the different types and working of memristors.

5.3.1 Fabrication and device structure

Figure 5.4 shows the TiO_2/TiO_{2+x} memristor. This type of memristor was the first fabricated by HP laboratories [2]. The device consists of two titanium dioxide layers that are sandwiched between two electrodes. The electrodes are made of platinum or titanium. The two layers in between are active layers responsible for offering the memristance to the current flowing through it. One layer is the normal TiO_2 layer and the other is a TiO_2 layer that contains excess oxygen atoms. Based on the direction of current applied to the memristor, the excess oxygen atoms concentrate at a single layer or are distributed over the two different layers. When

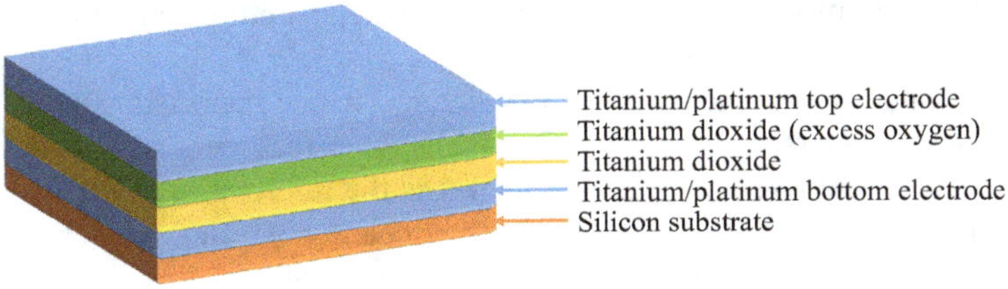

Figure 5.4 A TiO_2/TiO_{2+x}-active layer thin-film memristor

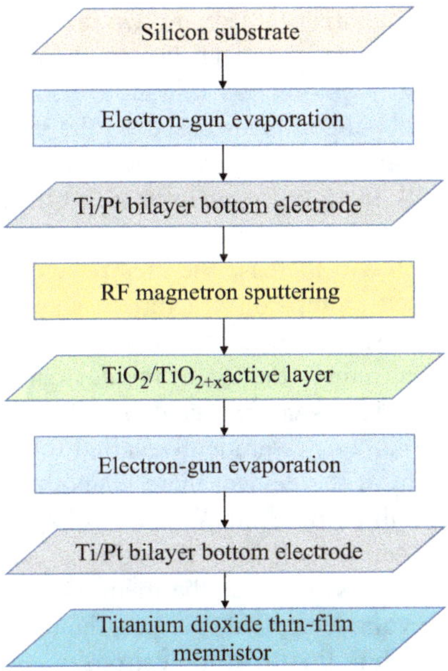

Figure 5.5 Memristor fabrication steps

the atoms are concentrated in a single layer, the memristance offered by the device will be high and when the atoms are distributed across the device, the memristance offered will be low.

The titanium dioxide thin-film memristor has the following layers: [(1)] Layer-1: the bottom titanium/platinum (Ti/Pt) bilayer electrode. Layer-2: active titanium dioxide (TiO_2) layer. Layer-3: active titanium dioxide with excess oxygen (TiO_{2+x}) layer. Layer-4: the top titanium/platinum (Ti/Pt) bilayer electrode. The process of fabricating the memristor is shown in Figure 5.5. On the silicon substrate, electron gun evaporation is performed to deposit the titanium or platinum electrodes. Then the titanium dioxide layer is deposited with radio frequency magnetron sputtering at room temperature and the titanium dioxide layer with excess atoms. The TiO_{2+x} layer is made non-stoichiometric with the addition of excess oxygen atoms by passing oxygen gas during the deposition, making it the active layer of the memristor. An additional layer of Ti/Pt bilayer is deposited for the top electrode to make it a complete memristor.

Figure 5.6 shows an example of a magnetoresistance memristor [48,49]. There are two different layers in the device. One is the fixed magnetic layer and the other is the free layer. The fixed layer is considered as the reference and the free layer is divided into two different sections each with opposite polarities. The change in the resistance occurs when the movement of the domain wall dividing the two sections is induced by the application of current.

Memristors are also manufactured based on the motion of silver dopants [29]. A cross-section of the memristor based on silver dopant motion is shown in Figure 5.6. As in the usual memristor, there are two metal electrodes on the top and bottom of the device. In between the metal layers, there is an amorphous silicon layer and amorphous silver + amorphous silicon layers. The silver ions are freely moving in this case into the silicon layer based on the voltage applied. When the top electrode is supplied with a positive voltage, the silver ions will move into the silicon layer which will decrease the memristance offered by the device. When the applied voltage direction is reversed, all the ions will move into the silver + silicon layer which will increase the memristance.

The silver dopant-based memristor structure is shown in Fig. 5.7(a) and the silicon-based memristor cross-section design is shown in Figure 5.7(b) [30]. As shown, there is a metal electrode in the top of the device which is followed by the amorphous silicon layer which is co-sputtered by silver. This layer is an active layer. The silver ratio in the layer is gradually varied toward the other end of the device itself. At the bottom of the device, there is another electrode made from a heavily doped p-type crystalline material. When a voltage is applied at the top of the device, a conduction channel will form through the active layer of amorphous silicon. This conduction layer will reduce the memristance offered by the device. When the direction of the voltage applied reverses, the memristance offered by the device will increase obstructing the flow of electrons.

The memristor structure based on silver chalcogenide is shown in Figure 5.7(c) [34,50]. In this type of memristor, tungsten electrodes are used on the top and

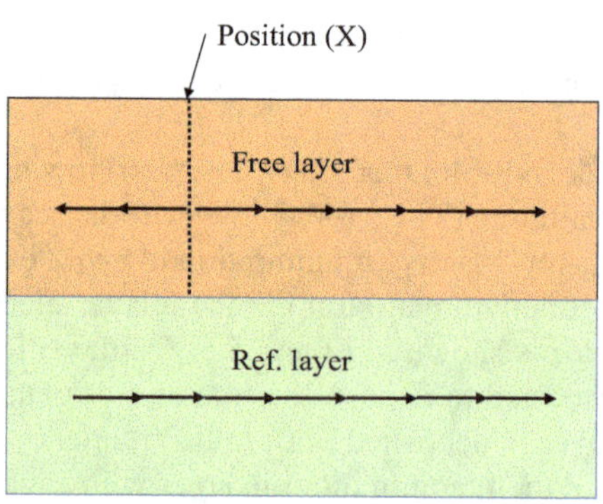

Figure 5.6 Spintronic memristor

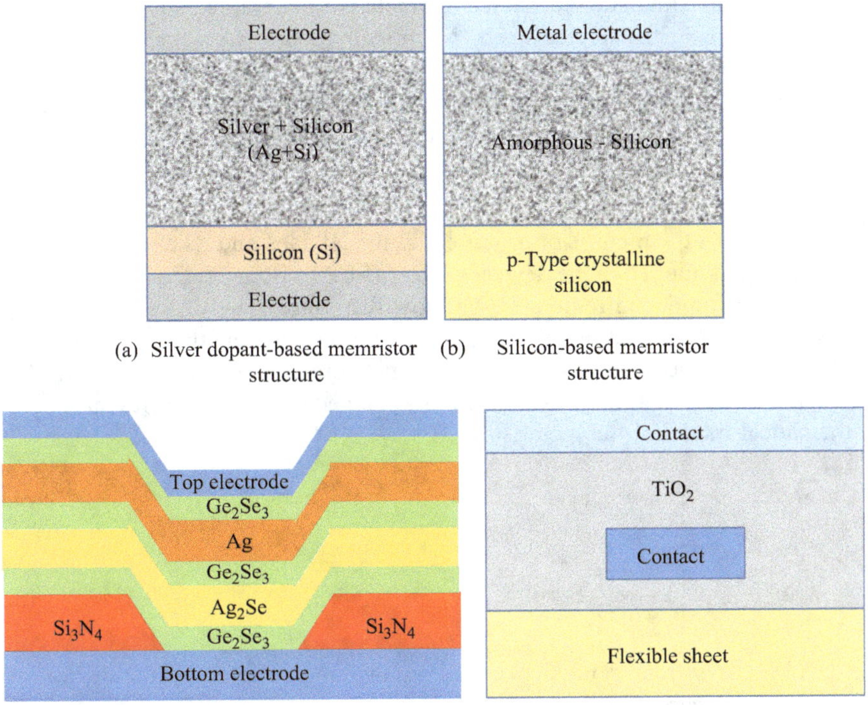

(a) Silver dopant-based memristor structure (b) Silicon-based memristor structure

(c) Silver chalcogenide-based memristor (d) Flexible solution-processed memristor

Figure 5.7 Other thin-film memristors

bottom of the device. Between the top and bottom electrodes, there are three Ge_2Se_3 layers. As shown in the figure, between the Ge_2Se_3 layers, there are Ag and Ag_2Se layers. Ag^+ ions will be able to freely move to the chalcogenide layer based on the voltage applied. If a positive voltage is applied at the top electrode, all the ions will migrate to the Ge_2Se_3 layers which will reduce the memristance offered by the device. When the polarity is reversed, the memristance of the device will increase.

Finally, the cross-section of a flexible memristor is shown in Figure 5.7(d) [42]. Laser jet transparency is used to design the flexible memristor. The electrodes used in this design are made of Al. Between the two Al electrodes, a TiO_2 layer is sandwiched. The properties exhibited by the device allow the scaling of the device in the nanometer regime.

5.4 Memristor device modeling

Before fabricating or deploying a new application design, simulations are extensively performed on that device or application to detect any failures. There are various ways to simulate the device characteristics or applications that can be implemented using the device itself. Memristor models have been proposed by various researchers in different simulating environments. This section presents various models of memristors: mathematical (analytical) model, SPICE model, Simulink model, Verilog-A(MS) model, and memristor emulators.

5.4.1 *Mathematical modeling of the memristor*

Figure 5.8 shows the memristor proposed by the HP labs and its circuit equivalent. When voltage is applied to the device, the length of the doped region (ω) is changed and so will the undoped region, D. The low resistance state of the device, R_{on} is achieved when the doped region completely occupies the length of the device, i.e., $\omega/D = 1$. When the polarity is reversed, the phenomenon will also be reversed and the high resistance of the device will be attained. HP laboratories proposed a mathematical model of the memristor [2]:

$$R(\omega) = R_{on}\frac{\omega}{D} + R_{off}\left[1 - \frac{\omega}{D}\right], \tag{5.1}$$

$$M[q] = R_{off}\left[1 - \left[\frac{1}{D^2}q(t)\mu_v R_{on}\right]\right]. \tag{5.2}$$

In (5.2), $M[q]$ represents the overall memristance offered by the device, as a function of the charge q through the device.

In a new model of memristor, the "dual-sided doped memristor", there are two layers of TiO_{2+x} between which the TiO_2 layer is sandwiched [51]. With the introduction of this new model, the noise margin and switching speed are improved

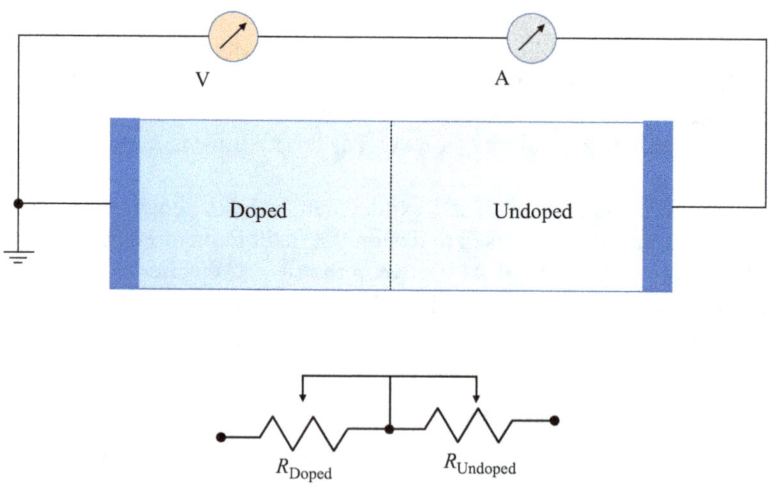

Figure 5.8 Memristor biasing

significantly. The memristance for the dual-sided doped memristor is given by:

$$M[q] = R_{off}\left[1 - \left[\frac{1}{D^2}q(t)\left[\mu_{v1}R_{on_1} + \mu_{v_2}R_{on_2}\right]R_{on}\right]\right], \tag{5.3}$$

where R_{on_1} and R_{on_2} are the on resistances of each layer.

The resistance of the memristor can be generalized as a time function [52] given by:

$$R^2 = R_o^2 - 2\kappa R_d\phi(t), \quad R \in (R_{on}, R_{off}) \tag{5.4}$$

All memristors that were initially introduced considered a periodic input for the device. A mathematical model for the memristance is provided in [53] for a DC input and symmetric periodic inputs. Based on the polarity of the input voltage, the boundary of the dopants will move in the appropriate direction, increasing or decreasing the memristance of the device. After evaluating the flux in (5.4), the resistance can be written as:

$$R^2 = R_o^2 - 2\kappa R_d V_{DC}t, \qquad R \in (R_{on}, R_{off}), \tag{5.5}$$

where V_{DC} is the DC voltage applied to the device and t is the time required to reach saturation.

5.4.2 Memristor device model using Simscape®

Memristor device characteristics can also be described using Simscape®, which is an integral part of the MATLAB® environment, capable of device- and system-level simulations [54]. Many memristor models have been developed for MATLAB® and Simulink® [18,54,55]. The availability of various libraries including system-level components allows the design and simulation of memristor circuits in Simulink®. Equations (5.14), (5.9), and (5.10) can be used to describe the working principles of the memristor in Simulink®.

Algorithm1 shows the Simulink®-based implementation of a memristor model [55]. The device parameters such as memristance and dopant mobility are described in lines 4–9 of algorithm 1. The physical boundaries of the device cannot be crossed by the doping region in the device, i.e. $0 \leq W \leq D$. dw/dt should be 0 at the boundaries if the externally applied voltage/current intends to push W beyond the limits. The memristor state variable is defined as $X = W/D \in (0, 1)$ (X_0 denotes the initial condition for X). Lines 21–29 of algorithm 1 show the memristor dynamic and implement boundary-checking to avoid boundary issues.

5.4.3 SPICE memristor device model

Various SPICE models are implemented for memristors [56–59]. Each SPICE model uses its own parameters which will represent the fingerprint of the memristor, the pinched hysteresis. Parameters are used for the depiction of $I-V$ characteristics and the state variables of the device. Figure 5.9 shows the SPICE subcircuit equivalent for the memristor. The $I-V$ characteristics of the memristor

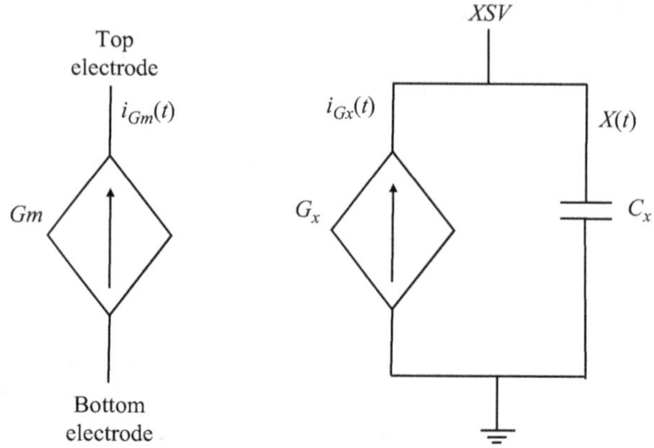

(a) Memristor SPICE equivalent for determining $I-V$ characteristics (b) Determining the state variable.

Figure 5.9 Memristor SPICE model equivalent

are modeled by a current source, G_m as shown in Figure 5.9(a). For determining the state variable, the current source is connected in parallel to a capacitor.

For determining the $I-V$ characteristics, the current is given as [59]:

Algorithm 1 Simscape® Memristor Model, Source: [55].

```
%Simscape model of a Memristor
component memristor<foundation.electrical.branch>
% mu is Mobility of Memristor
% x is the initial W/D
% D is the width of the memristor doped region
% Ron is the minimum resistance offered by the device
% Roff is the maximum resistance offered by the device.
parameters
        mu = { 1e-14,'m^2/s/V' };
        x = { .5,'1' };
        D = { 20e-9, 'm' };
        Ron = {100,'Ohm'};
        Roff = {36e3, 'Ohm'};
end
variables
        X0={.5,'1'};
        Rm ={1e3,'Ohm'};
end
function setup
        X0=x;
```

```
end
equations
        let
                az = mu * Ron / D^2;
                in
                     if(X0 <= 0 && v <= 0)||(X0 >= 1 && v >=
0)
                         X0.der == 0;
                else
                         X0.der == az & i;
                end
        end
        Rm == Ron & X0 + Roff * (1 - X0);
        v == i * Rm;
        end
    end
```

$$i_{Gm}(t) = \begin{cases} a_1 x(t)\sinh(bV(t)), & V(t) \geq 0, \\ a_2 x(t)\sinh(bV(t)), & V(t) < 0. \end{cases} \tag{5.6}$$

For determining the state variable from Figure 5.8 [59]:

$$i_{Gx}(t) = g(V(t))f(V(t), x(t)) \tag{5.7}$$

$$x(t) = \int_0^t i(t)dt \tag{5.8}$$

To plot the state variable during simulations, the port *XSV* was created in the circuit. Algorithm 2 presents a SPICE subcircuit model of the memristor [59].

Algorithm 2 SPICE memristor model, Source: [59].

```
* SPICE Model for a Memristor
*Connections:
*p - top terminal of the device.
*n - bottom terminal of the device.
*XSV - External connection for plotting state variable.
 .subckt mem_res p n XSV
*x1, x2 and y are IV characteristics parameters.
*pV, nV are Positive and Negative threshold voltages.
 .params x1=0.17 x2=0.17 y=0.05 pV=0.16 nV=0.15
*mp and mn are the SV motion intensity multipliers.
*xp and xn are the Points where the SV motion is reduced.
*alp and aln are the SV motion decay rate and x0 is the initial
value of SV.
```

```
+mp=4000 mn=4000 xp=0.3 xn=0.5 alp=1 aln=5
+x0=0.11 eta=1
*Zero State Variable Motion Functions at the boundaries of
Memristor
 .func wp(V) = \frac{xp-V}{1-xp}+1
 .func wn(V) = \frac{V}{1-xn}
 *G(V(t)) for the Threshold Voltage of Memristor
 .func G(V) = IF(V <= pV, IF(V >= -nV, 0, - mn * (e^{-+V}-e^
{nV})), mp + (e^{V}-e^{pV}))
*F(V(t),x(t)) for the SV Motion
 .func F(V1,V2) = IF(eta * V1 >= 0, IF(V2 >= xp, e^{-+alp
\times (V2-xp)} * wp(V2) ,1),
IF(V2 <= (1-xn), +e^{aln \times (V2+xn-1)} * wn(V2) ,1))
*Current - Voltage Response
 .func IVRel(V1,V2) = IF(V1 >= 0, x1 * V2 * sinh(y * V1), +x2 *
V2 * sinh(y * V1) )
*For determining the state variable
dx/dt = F(V(t),x(t)) * G(V(t))
Cx XSV 0 {1}
.ic V(XSV) = x0
Gx 0 XSV
+value={eta * F(V(p,n),V(XSV,0)) * G(V(p,n))}
*Current source for memristor Current - Voltage response.
Gm p n value = {IVRel(V(p,n),V(XSV,0))}
 .ends mem_res
```

5.4.4 Memristor device model using Verilog-A(MS)

Various models for the memristor have been proposed in Verilog-A and Verilog-AMS. SPICE uses subcircuit models whereas Verilog-A(MS) models describe the current–voltage characteristics in various fashions. Some propose a compact model [60], some use window functions [53], and others describe the characteristic equations using the language. Figure 5.10 shows a Verilog-A-based compact model for memristors [60].

A_d is the crosspoint area of the device and the length of the device is L. The structure of the device modeled is canonical where A_f is the filament area. The resistance of the device is determined by L, h, and A_f.

A Verilog-AMS memristor model is presented in [15]. The memristor described in Figure 5.8 is the device structure initially proposed by HP Labs [2]. A similar device structure was considered and the memristance equation was derived which was used for the development of the Verilog-AMS model in [15]. When the device length is D, mobility is μ_D and the on and off resistances are R_{on} and R_{off}, the memristance of the device can be given as [15]:

$$M(q) = R_o\left(\frac{\mu_D R_{on}(R_{off} - R_{on})}{D^2}\right)q(t) \tag{5.9}$$

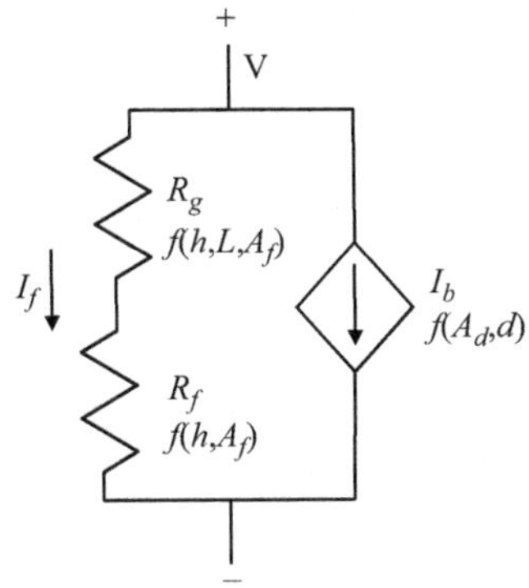

Figure 5.10 Memristor compact model using Verilog-A

Algorithm 3 presents a Verilog-AMS model [61] based on (5.9).

A polynomial metamodel-integrated Verilog-AMS design is presented in [61]. This metamodel uses a window function in the implementation of the device characteristics. When the applied source polarity changes, the window function helps the state variable in maintaining at the boundary instead of returning from the boundary. The drift velocity is given as [61]:

$$v = \begin{cases} \mu_v E & \text{if } E << E_o, \\ \mu_v E_o \exp(E/E_o) & \text{if } E \approx E_o, \end{cases} \tag{5.10}$$

where E is the electric field across the device and E_o is the characteristic high field introducing the nonlinear effect.

Algorithm 3 Verilog-AMS model of the memristor, Source: [61,62].

```
//Verilog-A Model of Memristor
'include "constants.vams"
'include "disciplines.vams"
module memristor (p1, p2);
parameter real x0 = 0.5; // Initial state
parameter real uv = 3e-18; // Dopant mobility
parameter real d = 40e-9; // Length of doped region
parameter real roff = 10e3;// Minimum Resistance
parameter real ron = 1e3; // Maximum Resistance
parameter real vp = 1.7; // Positive threshold voltage
parameter real vn = -1.7; // Negative threshold voltage
real x, xa, rm, dxdt, fz, fp, fn, integ;
inout p1, p2;
electrical p1, p2;
branch (p1, p2) mem;
analog begin
@(initial_step) begin
```

```
 fz = uv * ron / (d**2);
 fp = uv * vp / (d**2);
 fn = uv * vn / (d**2);
 xa = x0;
end
@(cross(V(mem), 0) or cross(V(mem)-vp, 0) or cross(V(mem)-
vn, 0));
         if (V(mem) >= vp)
                  dxdt = fp * limexp(I(mem) * ron / vp);
         else if (V(mem) <= vn)
         dxdt = fn * limexp(I(mem) * ron / vn);
         else dxdt = fz * I(mem);
                  if ((integ == -1 && V(mem) > 0) || (integ ==
1 && V(mem) < 0)) integ = 0;
                           @(cross(x,-1)) begin
                           integ = -1;
                           xa = 0;
                  end
                  @(cross(x-1,1)) begin
                  integ = 1;
                  xa = 1;
         end
     x = idt(dxdt, xa, integ, 1e-12);
     rm = ron * x + roff * (1 - x);
         I(mem) <+ V(mem) / rm;
end
endmodule
```

5.4.5 Memristor emulators

Modeling a memristive device with SPICE, Verilog-A(MS) and Simulink™ was possible once the $I-V$ characteristics of the device were observed and its operating principles were determined. However, when the actual device could not be implemented, the characteristics of the device could be simulated using discrete-component emulators. There are various emulators proposed [3,63–66] for memristors. Figure 5.11 [67] shows the block diagram of a memristor emulator.

The emulator uses an analog-to-digital converter (ADC), a digital potentiometer and a microcontroller as the discrete components. The resistance of the potentiometer changes continuously under the inputs from the microcontroller. The ADC output is fed to the microcontroller. The potentiometer will be updated by the microcontroller based on the preprogrammed equations of the memristor. The original memristor idea can be generalized to any class of two-terminal devices such as the memory possessing capacitor and inductor known as memcapacitor and meminductor, respectively. The three elements, memristor, memcapacitor and meminductor, are together called "memdevices". Emulators for memcapacitors and meminductors are implemented by using a memristor emulator in conjunction with operational-amplifiers (OP-AMPs) in various feedback paths [67].

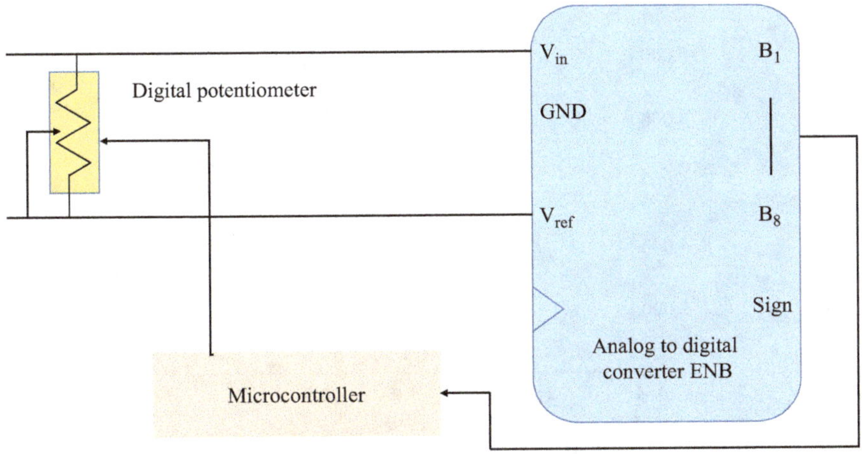

Figure 5.11 A memristor emulator

5.5 Characteristics of the memristor

Even before Chua posited the memristor, its characteristics were observed by many researchers [2–4]. Pinched hysteresis is the main fingerprint of the memristor. Pinched hysteresis was observed throughout history in various forms, such as the imperfect metal-to-metal contacts while fabricating an IC. But these effects were not significant until we reached the nanometer scale [68,69]. Resistance vs. voltage $(R-V)$ and current vs. voltage $(I-V)$ characteristics of a typical memristor are shown in Figure 5.12. As shown in Figure 5.13, the memristor provides the missing relationship between charge and flux.

Figure 5.8 shows the memristor device structure and the equivalent circuit considered to observe the current-voltage characteristics [33]. As shown in the figure, the device is divided into two regions, doped and undoped, due to the drift of dopants throughout the device based on the voltage polarity supplied. When the doped region is small, the memristance provided by the device will be high and when the doped region is large, the memristance provided will be low.

For analysis purposes, the following parameters are considered for the memristor: [(1)] L_{active}—total active length of the memristor. This is fixed once a memristor is manufactured. $l_{doped}(t)$—the doped active length of the memristor. This changes with the voltage applied across the two terminals. R_{doped}—resistance of the doped layer of length L_{active}. This is equivalent to the ON state resistance of the memristor R_{ON}. $R_{undoped}$—resistance of the undoped layer of length L_{active}. This is equivalent to the OFF state resistance of the memristor R_{OFF}. $v(t)$—the applied biasing voltage across the memristor. $q(t)$—the resulting charge in the memristor. $i(t)$—the resulting current through the memristor. μ—the average carrier mobility.

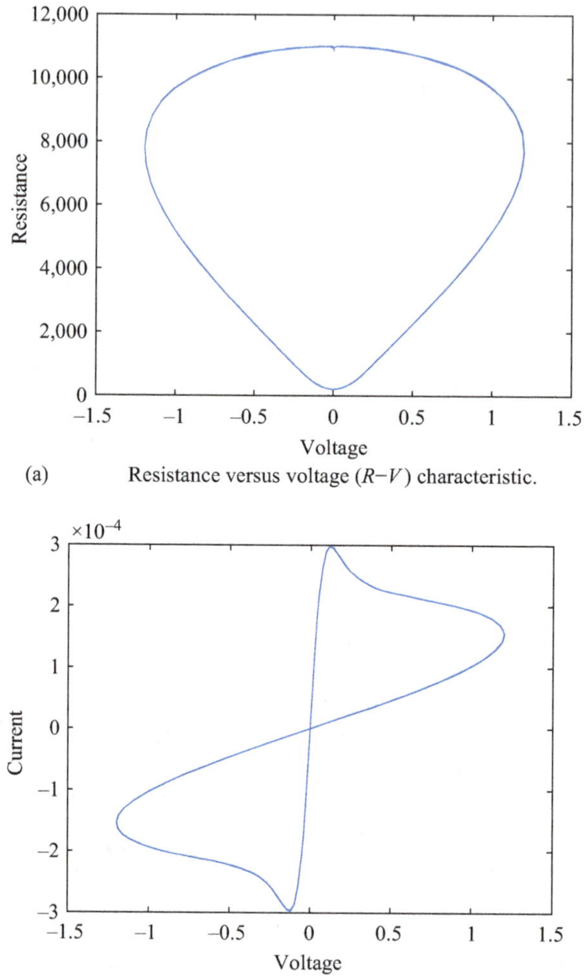

(a) Resistance versus voltage (R–V) characteristic.

(b) Current versus voltage (I–V) characteristic.

Figure 5.12 Various characteristic curves of a typical memristor

When we apply Kirchhoff's voltage law (KVL) to the circuit equivalent of the device shown in Figure 5.8, we obtain the following equation:

$$v(t) = \left(R_{doped}\left(\frac{l_{doped}(t)}{L_{active}}\right) + R_{undoped}\left(1 - \frac{l_{doped}(t)}{L_{active}}\right)\right)i(t). \tag{5.11}$$

For linear drifting with uniform field, the doped active length $l_{doped}(t)$ is obtained as the product of carrier velocity and carrier drifting time. The carrier velocity is the product of the carrier mobility (μ) and the electric field, which is given by the ratio $v(t)/L_{active}$. The drifting time is calculated as the ratio of charge

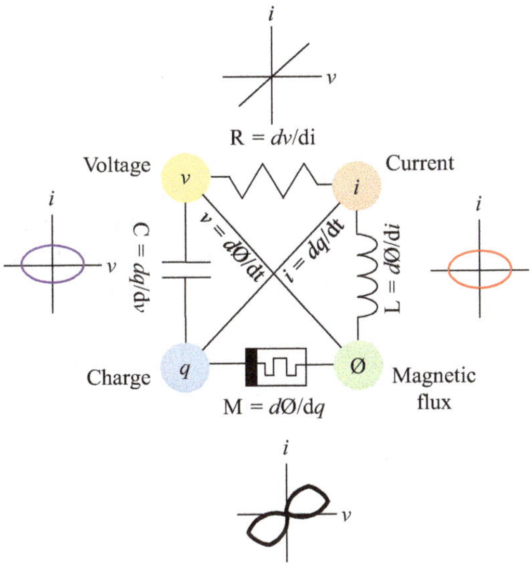

Figure 5.13 Memristor relates magnetic flux and charge; the missing connection among the four variables, v, i, q, and ϕ [33]

to current: $(q(t)/i(t))$. Thus, we obtain the following expression:

$$l_{doped}(t) = \mu\left(\frac{v(t)}{L_{active}}\right)\left(\frac{q(t)}{i(t)}\right) = \mu q(t)\left(\frac{R_{doped}}{L_{active}}\right). \tag{5.12}$$

Assuming that R_{doped} is very small compared to $R_{undoped}$, the following is derived from (5.11):

$$\frac{v(t)}{i(t)} = \left(R_{doped} - R_{undoped}\right)\left(\frac{l_{doped}(t)}{L_{active}}\right) + R_{undoped}. \tag{5.13}$$

From (5.12) and (5.13), the memristance (M) can be derived as:

$$M(q) = \frac{v(t)}{i(t)} = \left(1 - \left(\frac{\mu R_{doped}}{L_{active}^2}\right)q(t)\right)R_{undoped}. \tag{5.14}$$

5.6 Memristors in analog nanoelectronics

Analog applications are one of the main components of any electronic system. Oscillators, such as PLLs, are at the heart of the electronic devices driving many circuits. Neural network is another main focus of research that is the base of machine learning. To implement these types of designs, more analog components need to be used which tend to occupy larger chip area compared to digital electronics. This section presents some memristor-based analog implementations.

The memristance offered by the memristor can be changed on the fly by supplying current to the device in the required direction. Hence, with memristive components at the core of these designs, reconfigurability and reprogrammability can be added to the system.

5.6.1 *Memristance controlled oscillator*

One of the simplest oscillators is the Wien oscillator. Figure 5.14 shows the circuit of a Wien oscillator comprising of two capacitors and four resistors. An oscillator oscillates without any external input or oscillation source. Positive and negative feedback combined with an OP-AMP drives the Wien oscillator into an unstable state and stable oscillations are generated as a result. A frequency range of 20 Hz to 20 kHz can be generated using a Wien oscillator. The values of the resistors and capacitors are constant in a conventional oscillator. The condition for sustained oscillations can be given as:

$$\frac{C_2}{C_1} + \frac{R_1}{R_2} = \frac{R_3}{R_4}, \tag{5.15}$$

and the frequency of any Wien oscillator is calculated by the following:

$$f = \left(\frac{1}{2\pi\sqrt{R_1 R_2 C_1 C_2}}\right). \tag{5.16}$$

In the Wien oscillator, when the resistors are replaced with memristors, the memristance will drive the oscillations and based on the resistance offered by the memristance, the oscillations can be controlled [33,70] according to (5.16). It is assumed that the memristance of the memristor is not changed intentionally during the operation of the oscillator. In such case, the initial memristance of the device will determine the oscillation frequency.

Five different configurations can be attained by replacing the resistors with memristors. R_1 is replaced with the memristor M_1 in configuration 1 and the memristance is labeled $R_{1,mem}$. R_2 is replaced with memristor M_2 in configuration 2 and the memristance is labeled $R_{2,mem}$. R_3 is replaced with memristor M_3 in

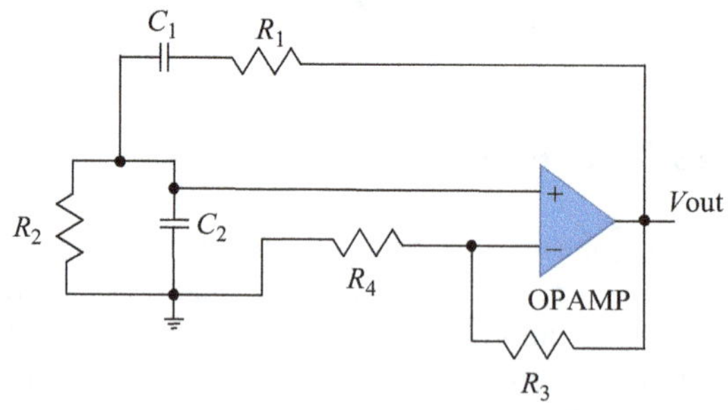

Figure 5.14 Wien oscillator schematic

configuration 3 and the memristance is labeled a $R_{3,mem}$. R_4 is replaced with memristor M_4 in configuration 4 and the memristance is labeled as $R_{4,mem}$. In the fifth configuration, all the resistors are replaced with memristors. The conditions for sustained oscillation and the oscillation frequencies for the five configurations are shown in Table 5.2. For brevity, configurations 2 and 5 are shown in Figures 5.15 and 5.16.

Table 5.2 Equations for oscillators

Configuration	Condition for oscillation	Frequency
Traditional	$\frac{C_2}{C_1} + \frac{R_1}{R_2} = \frac{R_3}{R_4}$	$\frac{1}{2\pi\sqrt{R_1 R_2 C_1 C_2}}$
Configuration 1	$1 + \frac{R_{1,mem}}{R_2} = \frac{R_3}{R_4}$	$\frac{1}{2\pi C\sqrt{R_{1,mem} R_2}}$
Configuration 2	$1 + \frac{R_1}{R_{2,mem}} = \frac{R_3}{R_4}$	$\frac{1}{2\pi C\sqrt{R_1 R_{2,mem}}}$
Configuration 3	$1 + \frac{R_1}{R_2} = \frac{R_{3,mem}}{R_{4,mem}}$	$\frac{1}{2\pi C\sqrt{R_1 R_2}}$
Configuration 4	$1 + \frac{R_1}{R_{2,mem}} = \frac{R_3}{R_{4,mem}}$	$\frac{1}{2\pi C\sqrt{R_1 R_{2,mem}}}$
Configuration 5	$1 + \frac{R_{1,mem}}{R_{2,mem}} = \frac{R_{3,mem}}{R_{4,mem}}$	$\frac{1}{2\pi C\sqrt{R_{1,mem} R_{2,mem}}}$

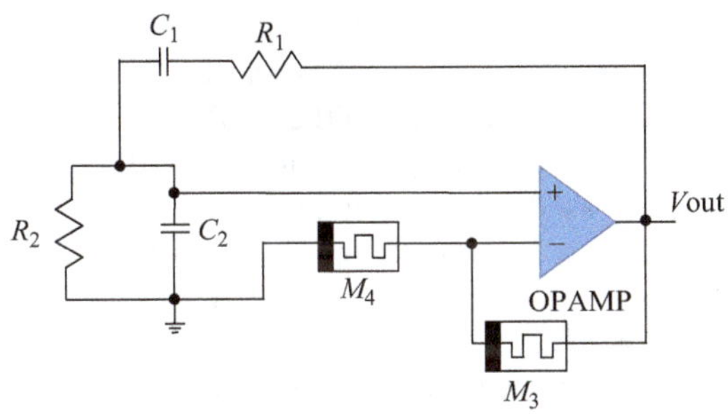

Figure 5.15 Two resistors replaced with memristors

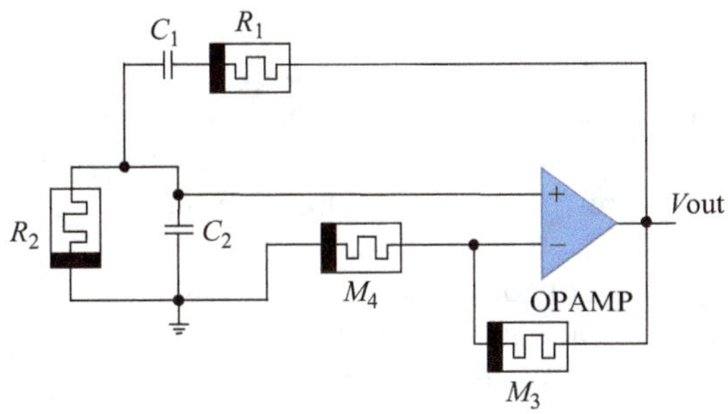

Figure 5.16 Wien oscillator schematic with five memristors

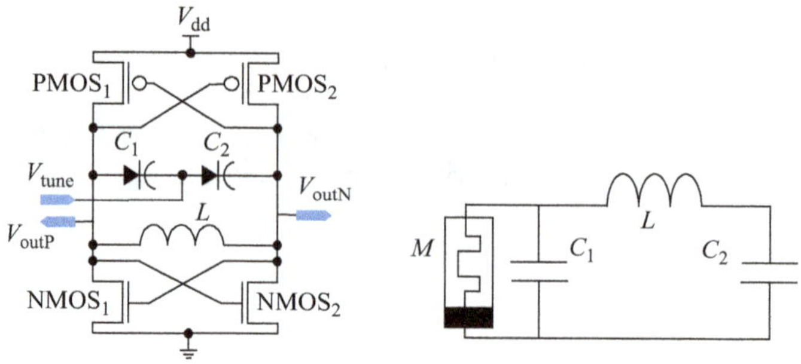

(a) Traditional LC-VCO. (b) Memristor-VCO.

Figure 5.17 Circuits of analog oscillators

5.6.2 LC-tank oscillator

Figure 5.17(a) shows an inductor–capacitor (LC)-based voltage-controlled oscillator (VCO). The LC-VCO is used as an oscillator in various electronic devices for controlling the frequency of the phase-locked loop (PLL) which is one of the main components of any synchronous circuit. A PLL also acts as a global clock in some circuits [71]. In a conventional LC-VCO, a DC voltage is applied to the input through a loop filter. Even though the VCO produces clean oscillations, the space occupied by it is very large. An alternate to the LC-VCO cannot be introduced into the circuits without compromising the oscillations or adding disturbances.

Hence, a memristor-based LC-VCO is presented in Figure 5.17(b) [21]. The memristor-based LC-VCO reduces the chip area significantly by replacing some components in the conventional design and also produces clean oscillations [21]. In the negative resistance region of the memristor, it acts as an active device, so it can produce sustained oscillations when combined with the LC tank.

5.6.3 Programmable Schmitt trigger oscillator

Memristors in the core of an electronic analog or digital system will increase its reconfigurability. Extensive research is being conducted on memristive-based devices [22,72–74] to achieve reconfigurability and reprogrammability. The introduction of the memristor into an oscillator will increase the reconfigurability of the oscillator on the fly and with minimal power consumption. Figure 5.18(b) shows the implementation of a reconfigurable Schmitt trigger oscillator with memristor.

A Schmitt trigger oscillator is one of the most commonly used oscillator designs for various applications such as inductive and capacitive sensing [75], pressure sensing [76] and frequency generators for on-chip applications [77]. A reconfigurable Schmitt trigger oscillator is very useful in such applications. Figure 5.18(a) and (b) shows conventional and memristor-based Schmitt trigger

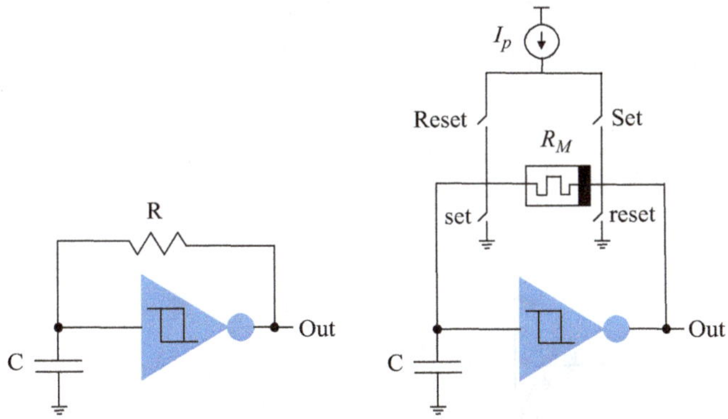

(a) Schmitt trigger oscillator (b) Memristor-based programmable oscillator.

Figure 5.18 Conventional Schmitt trigger oscillator and the memristor-based Schmitt trigger oscillator with added programmability

oscillator designs. The resistor in the conventional design is replaced by a memristor. But the memristor requires additional circuitry to be reconfigured or programmed. This consists of four switches, two "reset" and two "set" switches connected to a current source I_p.

The four switches control the flow of current to the memristor which will regulate the resistance offered by the device. Thus, the oscillation frequency can be changed and reprogrammed. There are two stages in the reprogramming: The "set" switches are opened and the "reset" switches are closed. In this state, the current from the source I_p flows to the memristor. This increases the memristance offered by the memristor which changes the oscillation frequency. When the "set" switches are closed and the "reset" switches are opened, the current flows through the memristor in the reverse direction which will decrease the memristance offered by the device. When all switches are open, the normal operation of the Schmitt trigger oscillator will continue and the memristor will act as a normal resistor. This generates sustained oscillations.

5.6.4 Neuromorphic chips

Memristors are also being used in the implementation of neuromorphic chips [5,78–82]. The friend and foe of the neuromorphic chip has always been technology. When parallel computing is necessary, neuromorphic chips are always valuable components. A very large number of processing elements are needed to perform efficient neuromorphic computing. A hardware implementation of a neural network will also help in simulating the functionality of a human brain [83]. With the advancements of technology and introduction of more low-power high-performing capabilities, neuromorphic computing has also equally advanced.

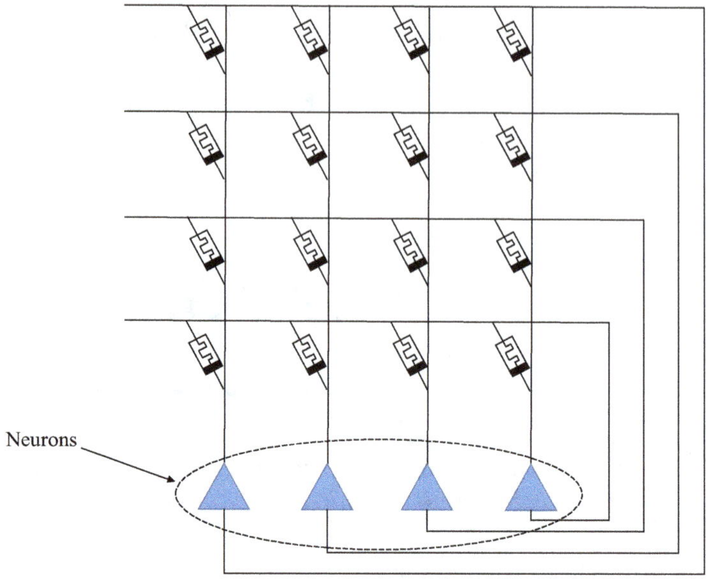

Figure 5.19 Memristor-based neural network

The winner-take-all (WTA) algorithm is one of the more efficient algorithms where one neuron will inhibit the neighbors to find the solution [78]. Figure 5.19 shows a neural network architecture based on a memristor crossbar topology implementing a WTA algorithm. A memristor-MOS technology (MMOST) array is used and the memristors and the crossbar network connectivity are used to implement the synapses [78]. As shown in the figure, a neuron can connect to every other neuron. Every neuron will fire and inhibit the neighbor neuron in a competitive learning environment. The last standing neuron will stay active.

5.7 Memristors in digital nanoelectronics

Digital nanoelectronics have come a very long way and Moore's law has been the driving force of IC technology development till now. Transistors have scaled down and reached almost their limit. Memristors are predicted to replace transistors in the near future as they are being introduced into the digital nanoelectronic domain. Digital components are a part of almost every SoC or every IC in the market. With the need for high-performance and low-power-consuming devices, research in memristors has grown exponentially. With the advantage of reconfigurability on the fly, memristors find their way into many major applications of digital electronics starting at the fundamental logic gates, through security, and FPGA implementations. This section presents memristor-based applications in digital nanoelectronics.

5.7.1 Memristor-based logic gate design

Logic gates are at the heart of many general purpose processors [20]. Memristors were initially considered for designing memory devices which can retain data even when power is shut off to the device [2]. Memristor-based logic design has brought scaling and other aspects of digital design into the forefront [84]. Logic gates using memristors have been implemented using various designs [84–88]. The NAND and NOR gates are considered to be the universal gates. The function of any other logic gate and any Boolean logic function can be expressed using only these two gates [88]. Memristors can be used to design these universal gates [86].

Figure 5.20 shows the designs of NAND and NOR gates using memristors [86]. Each of these designs contains only three memristors and with the possibilities of scaling of memristors, the universal gates can be fabricated consuming significantly less chip area and power. The NOR gate consists of two input memristors connected in parallel and the output memristor connected in series to the two input memristors. The output memristor is programmed with low resistance and the input is supplied through V_0 as shown in the figure. The NAND gate consists of three memristors: two input and an output memristor are connected in series. The output memristor in the case of a NAND gate is also programmed with a low resistance. Memristors can also be used to implement other logic operations such as implication [84].

5.7.2 Memristor-based full adder

Figure 5.21 shows an implementation of a memristor-based full adder design [89]. The full adder is one of the most used components in ALU or the processor. This design shows the implementation of the full adder with a minimal number of memristors. The full adder consists of two sets of memristors, the work memristors, which will correspond to the addition bits, a number of sum bits and a carry and the input memristors which are responsible for the inputs A and B. $2n$ memristors are needed for an n bit A and B and $n + 3$ work memristors are used. The resistance of R needs to be much higher compared to the resistance offered by the memristors and $R_{on} < R < R_{off}$.

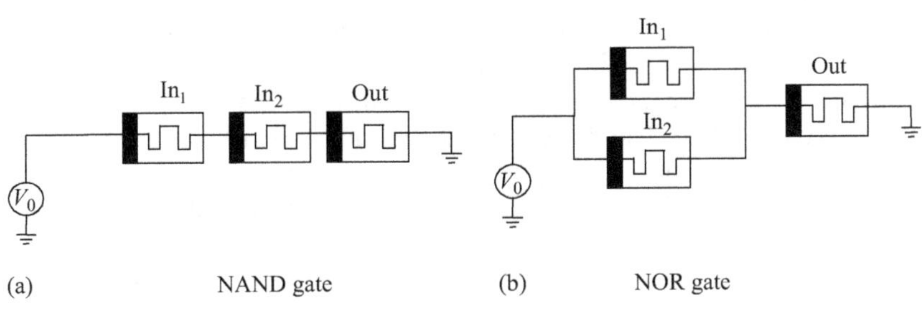

Figure 5.20 Memristor-based logic gate design [86]

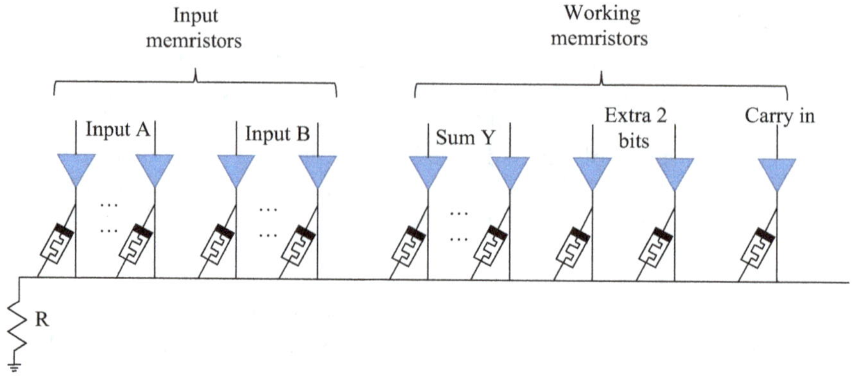

Figure 5.21 A memristor-based n-bit full adder

Besides the design shown in Figure 5.21, there are many other designs of memristor-based full adders proposed. A memristor-based XOR is used to build a full adder in [90]. In this design, four memristors will be controlling the input and four memristors will be controlling and storing the output sum and carry. An *n*-bit design can be constructed by cascading the 1-bit design [90].

5.7.3 Physical unclonable function

A physical unclonable function (PUF) takes advantage of the manufacturing variations that are introduced during the fabrication phase of the device [91]. When the device is being fabricated, due to various manufacturing steps, there will be discrepancies in the final geometry and doping profile of the device. When various instances of the same design are fabricated on the same die, the outputs will not be identical, even when driven by identical inputs. The manufacturing variations are uncontrollable, unavoidable and unpredictable. So the outputs of the devices being manufactured are also not predictable. This property is used to generate random numbers which can be used for cryptographic purposes. The input to a PUF is called a challenge and the output is called a response.

A memristor can be reprogrammed or reconfigured. When it is at the core of a PUF design, the reconfigurability of the PUF will increase giving it an exponentially high robustness [27,92,93]. Figure 5.22 shows the design of a memristor-based one-bit PUF [93]. The design is a single bit implementation of a memory design. The control signals “NEG” and “R/W” determine if the memristors are written to or read from. The write time of the memristor depends on the variations in the thickness of the device, *D*. Due to the manufacturing variations, the thickness will not be uniform even if the memristors are manufactured on the same die. When the output response is taken from the XOR, the output bit will depend on the variations in the device, which makes it random.

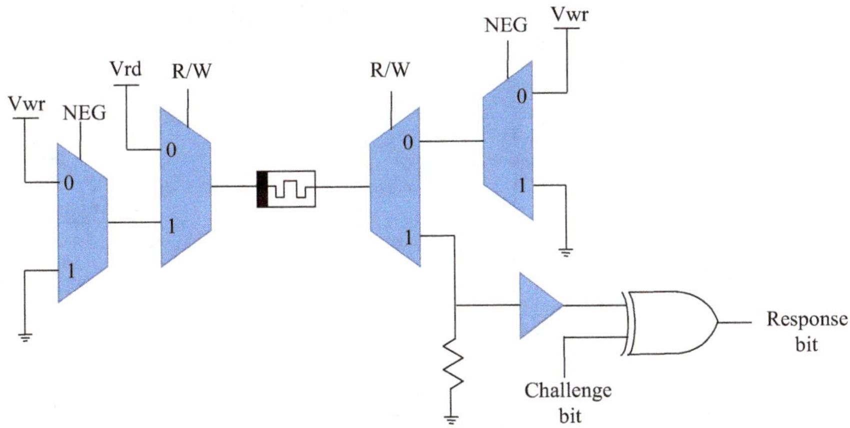

Figure 5.22 Memristor-based physical unclonable function

5.7.4 Memristor architectures for FPGAs

An FPGA offers various advantages for digital design due to its high level of the on-system configurability. Large designs can be implemented on an FPGA with the help of a hardware description language [20]. An FPGA utilizes logic elements and registers to implement the logic and the device functionality. The reconfiguration of an FPGA takes about 90% of the chip area and the power consumption [94]. Conventional FPGA components such as multiplexers, flip-flops, and SRAMs are designed using transistors. But memristors at the core of the designs provide an opportunity of reconfigurability with a very low chip area occupancy and power consumption.

Various designs of memristor-based FPGA architectures have been proposed [23,24,95]. Configurable logic blocks, switch, and connection boxes of the FPGA are combined with memristor-based basic building blocks and an FPGA architecture is presented in [95]. An emulator based on memristors is designed for the FPGA-based artificial neural network presented in [24]. Another design of FPGA based on memristors is proposed in [23] which uses memristor-based LUTs and CLBs for the design. Memristor adders are used in the look-up tables in [23].

5.7.5 Memristor crossbar

Memristors have been considered for memory design. The name "memory + resistor" was proposed due to the ability of the device to remember the voltage level even if the power to the device is cut off. Figure 5.23 shows the design of a memristor-based crossbar array memory architecture. The introduction of memristors being implemented in the design has paved the way for many new applications like reconfigurable logic [96], neural networks [26,97], and PUFs [27]. As shown in Figure 5.23, memristors are connected at the intersection of the

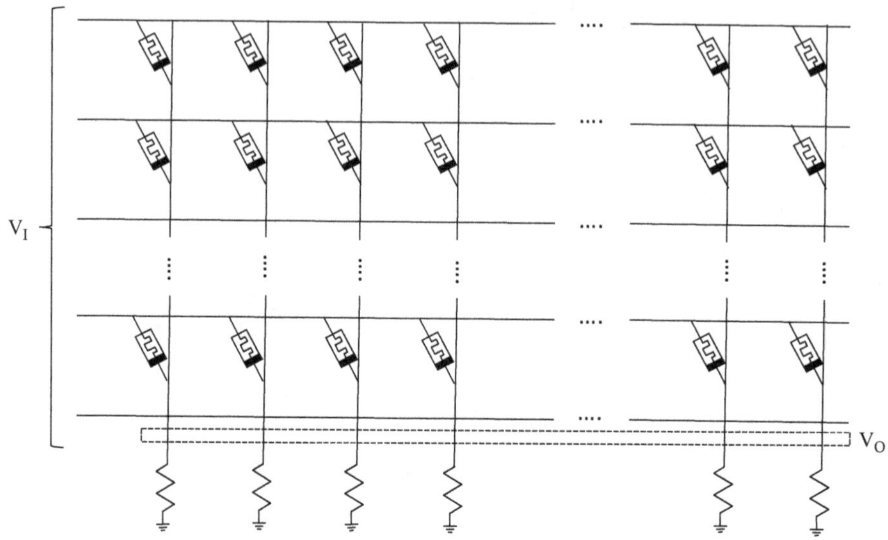

Figure 5.23 Memristor-based crossbar

horizontal and vertical wires [26]. This is much easier for the implementation of a connection matrix in neural networks.

5.8 Summary and future directions of research

The memristor provides the relation between charge and magnetic flux, two of the fundamental circuit variables. Until 1971 the device did not exist and even though the phenomenon existed and was observed, it was difficult to notice until the nano regimes were reached. When the devices got closer to the nanoscale, the memristive phenomenon and its characteristics could be observed in the imperfect metal-to-metal contact on an IC. The pinched hysteresis serves as the fingerprint of a memristor which can store the value of memristance even when the power to the device is off. With these characteristics, the application scope of memristors has become much wider. Memory design has become the main focus of research along with others such as neural networks and security.

As a future direction to the device research, the scope does not end here. Even though the device has been around for so long, a cost-effective commercialized memristor is still not available. Once a consumer level device is fabricated, implementing it in memory and other applications will be possible. There are many accurate models for memristors available and also researchers around the world have claimed to have fabricated the device, but a cost-effective device is yet to be available. Memristors are also being used in neural networks which will highly increase the effective use of artificial intelligence and machine learning.

Acknowledgments

This chapter is based on previous papers from the authors, such as the following: [33,55,61]. The authors would like to acknowledge their graduate students (Geng Zheng and Mahesh Gautam) at the University of North Texas (UNT), who helped with preliminary versions of this research.

References

[1] Chua, L.: Memristor-The Missing Circuit Element. IEEE Transactions on Circuit Theory 18, 507–19 (1971)

[2] Strukov, D.B., Snider, G.S., Stewart, D.R., and Williams, R.S.: The Missing Memristor Found. Nature 453, 80–83 (2008)

[3] Lin, D., Chua, L., and Hui, S.Y.: The First Man-Made Memristor: Circa 1801 [Scanning Our Past]. Proceedings of the IEEE 103, 131–36 (2015).

[4] Adhikari, S.P., Sah, M.P., Kim, H., and Chua, L.O.: Three Fingerprints of Memristor. IEEE Transactions on Circuits and Systems I: Regular Papers **60**, 3008–21 (2013).

[5] Sah, M.P., Kim, H., and Chua, L.O.: Brains Are Made of Memristors. IEEE Circuits and Systems Magazine 14, 12–36 (2014).

[6] Beck, A., Bednorz, J.G., Gerber, C., Rossel, C., and Widmer, D.: Reproducible Switching Effect in thin Oxide Films for Memory Applications. Applied Physics Letters 77, 139–41 (2000)

[7] Buot, F.A., and Rajagopal, A.K.: Binary Information Storage at Zero Bias in Quantum-Well Diodes. Journal of Applied Physics 76, 5552 –60 (1994)

[8] Abdalla, H., and Pickett, M.D.: SPICE Modeling of Memristors. In: Proceedings of the IEEE International Symposium on Circuits and Systems, pp. 1832–35 (2011)

[9] Biolek, D., Biolek, Z., and Biolkova, V.: SPICE Modeling of Memristive, Memcapacitative and Meminductive Systems. In: Proceedings of the European Conference on Circuit Theory and Design, pp. 249–52 (2009)

[10] Biolek, Z., Biolek, D., and Biolkova, V.: SPICE Model of Memristor With Nonlinear Dopant Drift. Radioengineering 18, 210–14 (2009)

[11] Kvatinsky, S., Talisveyberg, K., Fliter, D., Kolodny, A., Weiser, U., and Friedman, E.: Models of Memristors for SPICE Simulations. In: Proceedings of the 27th Convention of Electrical & Electronics Engineers in Israel (IEEEI), pp. 1–5 (2012). URL http://webee.technion.ac.il/people/skva/memristor.htm

[12] Mahvash, M., and Parker, A.: A Memristor SPICE Model for Designing Memristor Circuits. In: Proceedings of the 53rd IEEE International Midwest Symposium on Circuits and Systems (MWSCAS), pp. 989–92 (2010)

[13] Rak, A., and Cserey, G.: Macromodeling of the Memristor in SPICE. IEEE Transactions on Computer-Aided Design of Integrated Circuits and Systems 29, 632–36 (2010)

[14] Zhang, Y., Zhang, X., and Yu, J.: Approximated SPICE Model for Memristor. In: Proceedings of the International Conference on Communications, Circuits and Systems, pp. 928–31 (2009)

[15] da Costa, H.J.B., de Assis Brito Filho, F., and de Araujo do Nascimento, P.I.: Memristor Behavioural Modeling and Simulations Using Verilog-AMS. In: IEEE 3rd Latin American Symposium on Circuits and Systems (LASCAS), pp. 1–4 (2012).

[16] Salama, K.N.: Memristor Model. http://sensors.kaust.edu.sa/tools/memristor-model

[17] Thang, H.M.: Memristor Model-MATLAB Central. www.mathworks.com/matlabcentral/fileexchange/25082

[18] Zaplatilek, K.: Memristor modeling in MATLAB® & Simulink®. In: Proceedings of the 5th European conference on European Computing Conference, pp. 62–67 (2011)

[19] Williams, R.: How We Found The Missing Memristor. Spectrum, IEEE 45, 28–35 (2008)

[20] Mohanty, S.P.: Nanoelectronic Mixed-Signal System Design. 9780071825719. McGraw-Hill Education (2015)

[21] Fei, W., Yu, H., Zhang, W., and Yeo, K.S.: Design Exploration of Hybrid CMOS and Memristor Circuit by New Modified Nodal Analysis. IEEE Transactions on Very Large Scale Integration (VLSI) Systems PP(99) (2011).

[22] Rose, G.S., Rajendran, J., Manem, H., Karri, R., and Pino, R.E.: Leveraging Memristive Systems in the Construction of Digital Logic Circuits. Proceedings of the IEEE 100, 2033–49 (2012)

[23] Sampath, M., Mane, P.S., and Ramesha, C.K.: Hybrid CMOS-Memristor Based FPGA Architecture. In: International Conference on VLSI Systems, Architecture, Technology and Applications (VLSI-SATA), pp. 1–6 (2015).

[24] Vourkas, I., Abusleme, A., Ntinas, V., Sirakoulis, G.C., and Rubio, A.: A Digital Memristor Emulator for FPGA-Based Artificial Neural Networks. In: 1st IEEE International Verification and Security Workshop (IVSW), pp. 1–4 (2016).

[25] Kundert, K., and Zinke, O.: The Designer's Guide to Verilog-AMS, 1st edn. Kluwer Academic Publishers, Boston (2004)

[26] Hu, M., Li, H., Chen, Y., Wu, Q., Rose, G.S., and Linderman, R.W.: Memristor Crossbar-Based Neuromorphic Computing System: A Case Study. IEEE Transactions on Neural Networks and Learning Systems 25, 1864–78 (2014).

[27] Loong, J.T.H., Hashim, N.A.N., Hamid, M.S., and Hamid, F.A.: Performance Analysis of CMOS-Memristor Hybrid Ring Oscillator Physically Unclonable Function (RO-PUF). In: IEEE International Conference on Semiconductor Electronics (ICSE), pp. 304–07 (2016).

[28] Hu, M., Li, H., and Pino, R.E.: Fast Statistical Model of TiO_2 Thin-Film Memristor and Design Implication. In: IEEE/ACM International Conference on Computer-Aided Design (ICCAD), pp. 345–52 (2011)

[29] Jo, S.H., Chang, T., Ebong, I., Bhadviya, B.B., Mazumder, P., and Lu, W.: Nanoscale Memristor Device as Synapse in Neuromorphic Systems. Nano Letters 10, 1297–301 (2010)

[30] Jo, S.H., and Lu, W.: CMOS Compatible Nanoscale Nonvolatile Resistance Switching Memory. Nano Letters 8, 392–97 (2008)

[31] Khanal, G.M., Acciarito, S., Cardarilli, G.C. et al.: Synaptic Behaviour in ZnO-rGO Composites Thin Film Memristor. Electronics Letters 53(5), 296–98 (2017)

[32] Khanal, G.M., Cardarilli, G., Chakraborty, A. et al.: A ZnO-rGO Composite Thin Film Discrete Memristor. In: IEEE International Conference on Semiconductor Electronics (ICSE), pp. 129–32 (2016).

[33] Mohanty, S.P.: Memristor: From Basics to Deployment. IEEE Potentials 32, 34–39 (2013)

[34] Oblea, A., Timilsina, A., Moore, D., and Campbell, K.: Silver Chalcogenide Based Memristor Devices. In: Neural Networks (IJCNN), The 2010 International Joint Conference on, pp. 1–3 (2010)

[35] Xu, N., Fang, L., Chi, Y., Zhang, C., and Tang, Z.: Resistance Uniformity of TiO2 Memristor With Different Thin Film Thickness. In: 14th IEEE International Conference on Nanotechnology, pp. 727–31 (2014)

[36] Atasoyu, M., Altun, M., Ozoguz, S., and Roy, K.: Spin-torque Memristor Based Offset Cancellation Technique for Sense Amplifiers. In: 2017 14th International Conference on Synthesis, Modeling, Analysis and Simulation Methods and Applications to Circuit Design (SMACD), pp. 1–4 (2017)

[37] Chen, Y., Li, H., and Wang, X.: Spintronic Devices: From Memory to Memristor. In: 2010 International Conference on Communications, Circuits and Systems (ICCCAS), pp. 811–16 (2010)

[38] Fan, D., Sharad, M., and Roy, K.: Design and Synthesis of Ultralow Energy Spin-Memristor Threshold Logic. IEEE Transactions on Nanotechnology 13, 574–83 (2014)

[39] Shim, Y., Sengupta, A., and Roy, K.: Low-Power Approximate Convolution Computing Unit with Domain-Wall Motion Based – Spin-Memristor – for Image Processing Applications. In: 2016 53rd ACM/EDAC/IEEE Design Automation Conference (DAC), pp. 1–6 (2016)

[40] Wang, X., Chen, Y., Xi, H., Li, H., and Dimitrov, D.: Spintronic Memristor Through Spin-Torque-Induced Magnetization Motion. IEEE Electron Device Letters 30, 294–97 (2009)

[41] Erokhin, V.: Organic Memristors: Basic Principles. In: Proceedings of 2010 IEEE International Symposium on Circuits and Systems, pp. 5–8 (2010).

[42] Gergel-Hackett, N., Hamadani, B., Dunlap, B. et al.: A Flexible Solution-Processed Memristor. Electron Device Letters, IEEE 30, 706–08 (2009)

[43] Liu, G., Fang, L., Li, N., Sui, B.C., and Duan, Z.K.: New Behavioral Modeling Method for Crossbar-Based Memristor. In: Microelectronics and Electronics (PrimeAsia), 2010 Asia Pacific Conference on Postgraduate Research in, pp. 356–59 (2010)

[44] Prodromakis, T., Michelakis, K., and Toumazou, C.: Fabrication and Electrical Characteristics of Memristors with TiO_2/TiO_{2-x} Active Layers. In: Circuits and Systems (ISCAS), Proceedings of 2010 IEEE International Symposium on, pp. 1520–22 (2010)

[45] Bedair, S.M., Zavada, J.M., and El-Masry, N.: Spintronic Memories to Revolutionize Data Storage. IEEE Spectrum (2010)

[46] Vincent, A.F., Larroque, J., Locatelli, N. et al.: Spin-Transfer Torque Magnetic Memory as a Stochastic Memristive Synapse for Neuromorphic Systems. IEEE Transactions on Biomedical Circuits and Systems 9, 166–74 (2015).

[47] Locatelli, N., Cros, V., and Grollier, J.: Spin-Torque Building Blocks. Nature Materials 13, 11–20 (2013)

[48] Chen, Y., and Wang, X.: Compact Modeling and Corner Analysis of Spintronic Memristor. In: Nanoscale Architectures, 2009. NANOARCH '09. IEEE/ACM International Symposium on, pp. 7–12 (2009)

[49] Wang, X., and Chen, Y.: Spintronic Memristor Devices and Application. In: Design, Automation Test in Europe Conference Exhibition (DATE), 2010, pp. 667–72 (2010)

[50] Pino, R., Bohl, J., McDonald, N. et al.: Compact Method for Modeling and Simulation of Memristor Devices: Ion Conductor Chalcogenide-Based Memristor Devices. In: Nanoscale Architectures (NANOARCH), 2010 IEEE/ACM International Symposium on, pp. 1–4 (2010)

[51] Shrivastava, A., and Singh, J.: Dual Sided Doped Memristor and It's Mathematical Modelling. In: IEEE 20th International Conference on Electronics, Circuits, and Systems (ICECS), pp. 49–51 (2013).

[52] Wang, F.Y.: Memristor for Introductory. [Online] 2008. arXiv Physics, arXiv:0808.0286

[53] Radwan, A.G., Zidan, M.A., and Salama, K.N.: HP Memristor Mathematical Model for Periodic Signals and DC. In: 53rd IEEE International Midwest Symposium on Circuits and Systems, pp. 861–64 (2010).

[54] Mathworks: https://www.mathworks.com

[55] Agu, E., Mohanty, S.P., Kougianos, E., and Gautam, M.: Simscape Based Design Flow for Memristor Based Programmable Oscillators. In: Proceedings of the 23rd ACM/IEEE Great Lakes Symposium on VLSI (GLSVLSI), pp. 223–24 (2014)

[56] Biolek, D., Biolkova, V., and Kolka, Z.: Modified MIM Model of Titanium Dioxide Memristor for Reliable Simulations in SPICE. In: 14th International Conference on Synthesis, Modeling, Analysis and Simulation Methods and Applications to Circuit Design (SMACD), pp. 1–4 (2017)

[57] Ho, P.W.C., Hatem, F.O., Almurib, H.A.F., and Kumar, T.N.: Enhanced SPICE Memristor Model with Dynamic Ground. In: IEEE International Circuits and Systems Symposium (ICSyS), pp. 130–32 (2015)

[58] Mbarek, K., Rziga, F.O., Ghedira, S., and Besbes, K.: An Analysis of the Dynamics of SPICE Memristor Model. In: International Conference on Control, Automation and Diagnosis (ICCAD), pp. 54–59 (2017)

[59] Yakopcic, C., Taha, T.M., Subramanyam, G., and Pino, R.E.: Memristor SPICE Model and Crossbar Simulation Based on Devices with Nanosecond Switching Time. In: The 2013 International Joint Conference on Neural Networks (IJCNN), pp. 1–7 (2013)

[60] Nandakumar, S.R., and Rajendran, B.: Verilog-A Compact Model for a Novel Cu/SiO2/W Quantum Memristor. In: International Conference on Simulation of Semiconductor Processes and Devices (SISPAD), pp. 169–72 (2016)

[61] Zheng, G., Mohanty, S.P., Kougianos, E., and Okobiah, O.: Polynomial Metamodel Integrated Verilog-AMS for Memristor-Based Mixed-Signal System Design. In: Proceedings of the IEEE 56th International Midwest Symposium on Circuits and Systems (MWSCAS), pp. 916–19 (2013)

[62] Zheng, G., Mohanty, S.P., and Kougianos, E.: Verilog-AMS-POM: Verilog-AMS Integrated POlynomial Metamodelling of a Memristor-based Oscillator (June 2–6, 2013, Austin, Texas)

[63] Alharbi, A.G., Fouda, M.E., and Chowdhury, M.H.: Memristor Emulator Based on Practical Current Controlled Model. In: IEEE 58th International Midwest Symposium on Circuits and Systems (MWSCAS), pp. 1–4 (2015).

[64] Kim, H., Sah, M.P., Yang, C., Cho, S., and Chua, L.O.: Memristor Emulator for Memristor Circuit Applications. IEEE Transactions on Circuits and Systems: Regular Papers 59, 2422–31 (2012)

[65] Ranjan, R., Kyrmanidis, A., Hellweg, W.L. et al.: Integrated Circuit with Memristor Emulator Array and Neuron Circuits for Neuromorphic Pattern Recognition. In: 39th International Conference on Telecommunications and Signal Processing (TSP), pp. 265–68 (2016)

[66] Sah, M.P., Yang, C., Budhathoki, R.K., and Kim, H.: Features of Memristor Emulator-Based Artificial Neural Synapses. In: IEEE International Symposium on Circuits and Systems (ISCAS2013), pp. 421–24 (2013)

[67] Pershin, Y.V., and Di Ventra, M.: Memristive Circuits Simulate Memcapacitors and Meminductors. Electronics Letters 46, 517–18 (2010)

[68] Gandhi, G., Aggarwal, V., and Chua, L.O.: The First Radios Were Made Using Memristors! IEEE Circuits and Systems Magazine 13, 8–16 (2013)

[69] Johnson, D.: The Memristor's Fundamental Secrets Revealed. IEEE Spectrum (2013)

[70] Ghosal, P., Mohanty, S.P.: Power Minimization of a Memristor-Based Wien Bridge Oscillator through a Simscape Framework. In: 2015 IEEE International Symposium on Nanoelectronic and Information Systems, pp. 83–88 (2015)

[71] Garitselov, O., Mohanty, S.P., and Kougianos, E.: A Comparative Study of Metamodels for Fast and Accurate Simulation of Nano-CMOS Circuits. IEEE Transactions on Semiconductor Manufacturing 25, 26–36 (2012)

[72] Cheng, K., and Strukov, D.: 3D CMOS-Memristor Hybrid Circuits: Devices, Integration, Architecture, and Applications. In: Proceedings of the 2012 ACM international symposium on International Symposium on Physical Design, pp. 33–40. ACM (2012)

[73] Pershin, Y.V., and Di Ventra, M.: Practical Approach to Programmable Analog Circuits With Memristors. IEEE Transactions on Circuits and Systems I: Regular Papers **57**, 1857–64 (2010)

[74] Shin, S., Kim, K., and Kang, S.: Memristor Applications for Programmable Analog ICs. Nanotechnology, IEEE Transactions on 10, 266–74 (2011)

[75] Engelberg, S., Hazout, H., and Hirshowitz, J.: A Capacitive-Sensing Based Simple Serial Mouse. Instrumentation & Measurement Magazine, IEEE 13, 32–36 (2010)

[76] Moser, M., and Zangl, H.: Temperature and Pressure Monitoring of a Whipped Cream Device. In: IEEE Sensors, pp. 683–86 (2009)

[77] Arbet, D., Brenkus, J., Majer, L., and Stopjakova, V.: Oscillation-Based Built-In Self Test of Integrated Active Analog Filters. In: Applied Electronics (AE), 2011 International Conference on, pp. 1–4 (2011)

[78] Ebong, I.E., and Mazumder, P.: CMOS and Memristor-Based Neural Network Design for Position Detection. Proceedings of the IEEE 100, 2050–60 (2012)

[79] Fayyazi, A., Ansari, M., Kamal, M., Afzali-Kusha, A., and Pedram, M.: An Ultra Low-Power Memristive Neuromorphic Circuit for Internet of Things Smart Sensors. IEEE Internet of Things Journal 5, 1011–1022 (2018)

[80] Lee, W., Park, J., and Kim, J.: Electromagnetic Simulations of a Neuromorphic Hardware Using PEEC and Memristor SPICE Models. In: IEEE Electrical Design of Advanced Packaging and Systems Symposium (EDAPS), pp. 1–3 (2017)

[81] Lv, X., Cai, Y., Yang, Y. et al.: A Neural Network Circuit with Associative Learning and Forgetting Process Based on Memristor Neuromorphic Device. In: 2017 IEEE 12th International Conference on ASIC (ASICON), pp. 211–14 (2017)

[82] Wang, X., Yu, Y., Yang, N. et al.: New Synchronization Criteria for Memristor-Based Recurrent Neural Networks with Mixed Delays. In: 2017 International Workshop on Complex Systems and Networks (IWCSN), pp. 210–17 (2017)

[83] Ntinas, V., Vourkas, I., Abusleme, A., Sirakoulis, G.C., and Rubio, A.: Experimental Study of Artificial Neural Networks Using a Digital Memristor Simulator. IEEE Transactions on Neural Networks and Learning Systems 99, 1–13 (2018)

[84] Kvatinsky, S., Kolodny, A., Weiser, U.C., and Friedman, E.G.: Memristor-Based IMPLY Logic Design Procedure. In: IEEE 29th International Conference on Computer Design (ICCD), pp. 142–47 (2011)

[85] Borghetti, J., Snider, G.S., Kuekes, P.J., Yang, J.J., Stewart, D.R., and Williams, R.S.: "Memristive" Switches Enable Stateful Logic Operations via Material Implication. Nature 464, 873 (2010)

[86] Kvatinsky, S., Belousov, D., Liman, S. et al.: MAGIC - Memristor-Aided Logic. IEEE Transactions on Circuits and Systems II: Express Briefs 61, 895–99 (2014)

[87] Kvatinsky, S., Friedman, E.G., Kolodny, A., and Weiser, U.C.: TEAM: ThrEshold Adaptive Memristor Model. IEEE Transactions on Circuits and Systems I: Regular Papers **60**, 211–21 (2013)

[88] Lehtonen, E., Poikonen, J.H., and Laiho, M.: Two Memristors Suffice to Compute All Boolean Functions. Electronics Letters 46, 239–40 (2010)

[89] Lehtonen, E., and Laiho, M.: Stateful Implication Logic With Memristors. In: Proceedings of the IEEE/ACM International Symposium on Nanoscale Architectures, pp. 33–36 (2009)

[90] Wang, X., Deng, H., Feng, W., Yang, Y., and Chen, K.: Memristor-Based XOR Gate for Full Adder. In: 35th Chinese Control Conference (CCC), pp. 5847–51 (2016)

[91] Yanambaka, V.P., Mohanty, S.P., and Kougianos, E.: Novel FinFET Based Physical Unclonable Functions for Efficient Security Integration in the IoT. In: 2016 IEEE International Symposium on Nanoelectronic and Information Systems (iNIS), pp. 172–77 (2016)

[92] Loong, J.T.H., Ismail, K.A.S.C., and Hamid, F.A.: Effect of Different Memristor Window Function with Variable Random Resistance on the Performance of Memristor-Based RO-PUF. In: International Conference on Advances in Electrical, Electronic and Systems Engineering (ICAEES), pp. 445–50 (2016)

[93] Rose, G.S., McDonald, N., Yan, L.K., Wysocki, B., and Xu, K.: Foundations of Memristor Based PUF Architectures. In: 2013 IEEE/ACM International Symposium on Nanoscale Architectures (NANOARCH), pp. 52–57 (2013)

[94] Kuon, I., Tessier, R., and Rose, J.: FPGA Architecture: Survey and Challenges. Foundations and Trends in Electronic Design Automation 2, 135–253 (2008)

[95] Farooq, U., and Aslam, M.H.: Design and Implementation of Basic Building Blocks of FPGA using Memristor-Transistor Hybrid Approach. In: Fifth International Conference on the Innovative Computing Technology (INTECH 2015), pp. 142–47 (2015)

[96] Thangkhiew, P.L., Gharpinde, R., and Datta, K.: Efficient Mapping of Boolean Functions to Memristor Crossbar Using MAGIC NOR Gates. IEEE Transactions on Circuits and Systems I: Regular Papers 99, 1–11 (2018)

[97] Yakopcic, C., Rahman, N., Atahary, T., Taha, T.M., and Douglass, S.: Cognitive Domain Ontologies in a Memristor Crossbar Architecture. In: 2017 IEEE National Aerospace and Electronics Conference (NAECON), pp. 76–83 (2017)

Chapter 6

Organic–inorganic heterojunctions for optoelectronic applications

Chandra Shakher Pathak[1,2], Jitendra Pratap Singh[2] and Rajendra Singh[2]

6.1 Introduction

Conducting polymers have been attracting attention after their discovery by Shirakawa, MacDiarmid, and Heeger in 1977, who were awarded the Nobel Prize in Chemistry in 2000 for the discovery of conductive polymers [1,2]. They used organic polymer polyacetylenes, which is a conjugated polymer and insulator. They increased the conductivity of polyacetylene films by several orders of magnitude by chemical doping [1]. In recent years, there has been a lot of research activity in the field of polymer electronics. It has attracted a lot of attention because of its light-weight, high flexibility, and solution process ability [3]. Applications of conducting polymers include organic light-emitting diodes, organic thin-film transistors, solar cells, actuators and sensors, etc. [4–8]. Until the 1970s, organic polymers were considered to be nonconducting and used as insulators. After the discovery of conducting polymer, vast improvements in synthetic polymers have resulted in polymers being used for conductive applications.

Poly (3,4-ethylenedioxythiophene)-poly(styrenesulfonate) (PEDOT:PSS) is widely used conducting polymer because of its high conductivity, excellent thermal stability, transparency, structural stability, and processability [9–11]. PEDOT:PSS polymer is a promising candidate as a transparent electrode for optoelectronic devices. PEDOT is made from ethylenedioxythiophene (EDOT) monomers. PEDOT is insoluble in many common solvents, and it is unstable in its neutral state. To improve its processability, water-soluble polystyrene sulfonate (PSS), can be added and the addition of PSS causes it to become soluble. During the polymerization, PSS acts as charge-balancing dopant to yield PEDOT:PSS. PEDOT:PSS is synthesized from the EDOT monomer, and PSS acts as the counter ion. The degree of polymerization of

[1]Ben-Gurion National Solar Energy Center, Department of Solar Energy and Environmental Physics, Jacob Blaustein Institute for Desert Research, Ben-Gurion University of the Negev, Midreshet Ben-Gurion, Israel
[2]Department of Physics, Indian Institute of Technology Delhi, New Delhi, India

PEDOT is limited and the chain length of PEDOT has been estimated up to 5 ~ 10 repeating units [12]. PEDOT chains are attached to the PSS chains through the Columbic interaction and they are stabilized by the excess PSS.

The actual conductivity of PEDOT:PSS is a major limiting factor for device applications. There have been several efforts to enhance the properties of PEDOT: PSS through solvent treatments [4–6]. The solvent treatment of PEDOT:PSS films may affect the conformation of the polymer. The structure of PEDOT chain changes from benzoid to quinoid structure after solvent treatment [9]. The role of organic solvent like dimethyl sulfoxide (DMSO) is a secondary dopant, which improves the morphology of PEDOT:PSS films and improves the conductivity of the films, which depends on its concentration [9]. When DMSO is added, the interactions between the PEDOT chains and DMSO additive induce the conformational changes from coil structures to linear and expanded coil structures. The interactions are initiated by the hydrogen bond formation between the sulfonic acid groups and the polar groups of DMSO. Apart from the change in conductivity of PEDOT:PSS films, solvent treatment may also affect other properties of the films such as morphological and optical properties [13–15]. Pathak *et al.* [14] demonstrated the effect of co-solvent on the electrical, vibrational, and morphological properties of PEDOT:PSS films. They enhanced the conductivity of PEDOT:PSS films by three orders of magnitude. Graphene oxide (GO) has recently emerged as a new carbon-based material because of its low cost, large-scale production capability. The solubility of GO in water and other solvents allows it to be uniformly deposited onto wide-ranging substrates in the form of thin films.

6.2 Experimental background

6.2.1 Mechanisms of conductivity enhancement

Pristine PEDOT:PSS has lower conductivity (less than 1 S/cm) and it is a major limiting factor for device applications. There are different mechanisms reported by different groups for conductivity enhancement. Palumbiny *et al.* [5] have used glycerol and ethylene glycol and suggested that the conductivity enhancement is due to the removal of PSS and morphology change. Xia *et al.* [6] have used common organic solvents such as ethanol, isopropyl alcohol, acetonitrile, acetone, tetrahydrofuran and found that conductivity did not change remarkably. They reported that the conductivity was significantly enhanced when PEDOT:PSS film was treated with a co-solvent of water and one of these common organic solvents. They suggested that the results are different from the high boiling point organic solvents. Kim *et al.* [16] have proposed that polar solvents with higher dielectric constant induce strong screening effect between counter ions and charge carriers, which reduces the Coulombic interaction between positively charged PEDOT and negatively charged PSS dopant. Jonsson *et al.* [17] have reported that the conductivity enhancement occurs due to the reduction of the excess PSS layer that surrounds the conducting PEDOT:PSS grains. Recently, Pathak *et al.* [18,19] used graphene nano-flakes to enhance the electrical properties of PEDOT:PSS films.

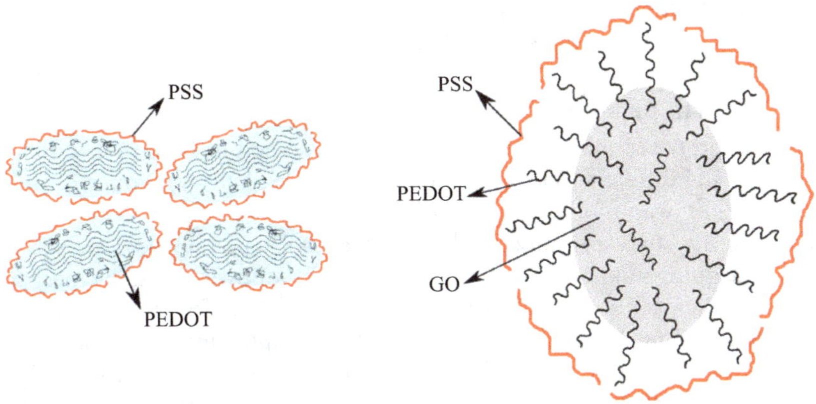

Figure 6.1 Schematic representation of conductivity enhancement mechanism of PEDOT:PSS with GO

GO has emerged as a new carbon-based material due to its low cost and large-scale production capability. Researchers have also used GO to enhance the properties of PEDOT:PSS films [20,21]. Raj *et al.* have used GO blended with PEDOT:PSS as hole-transporting layer for PTB7:PCBM bulk heterojunction solar cells [20]. GO is used as a separating agent between the conducting PEDOT and the insulating PSS to increase the conductivity of the electrode. In PEDOT:PSS, PEDOT chains are attached to insulating PSS chains through columbic attraction. The functional groups of GO separate the PEDOT and PSS chains. The columbic attraction between PEDOT and PSS chains gets weaker and improves the linear conformation of PEDOT and PSS chains. The separated PEDOT chains link up with functional groups of GO, which enhance the conductivity of PEDOT:PSS film. Figure 6.1 shows the schematic representation of conductivity enhancement mechanism of PEDOT:PSS films with GO.

6.2.2 Atomic force microscopy

Atomic force microscopy (AFM) was used to analyze the surface morphology and used to measure the force of interaction between the AFM tip and the sample. Scanning tunneling microscopy (STM) is based on the tunneling of electron between the AFM tip and sample, whereas AFM is based on the detection of attractive or repulsive forces. AFM consists of a sharp tip having a nanometer dimension which is attached to a cantilever is used to scan the sample surface. A laser beam focused on the cantilever which detects the bending of cantilever. The reflection of the laser beam is focused on photodiode detector. Deflection of the cantilever is monitored during the scanning and converted into surface image. AFM is generally operated in contact mode and tapping mode. In this chapter, topography and surface potential analyses have been done using Kelvin probe force microscopy (KPFM) in the tapping mode and current image obtained using conducting atomic force microscopy (CAFM) in the contact mode.

6.2.2.1 Kelvin probe force microscopy

Lord Kelvin proposed macroscopic Kelvin probe method in 1898 to determine the contact potential difference (V_{CPD}) between a metallic plate and sample [22]. Later in 1991, KPFM was first introduced by Nonnenmacher *et al.* [23]. The KPFM mode is basically two-scan process. In the first scan, topography of the surface is acquired in tapping mode along a single line profile. Following this, the mechanical excitation of the cantilever is turned off and in the second scan this topography is retraced at a certain lift height (LH) above the sample surface, recording local variations in contact potential difference (CPD). During the second scan, the tip-sample distance is constant and it is equal to d_{AFM} + LH, where d_{AFM} represents the tip-sample distance during the topographic scan. When the AFM tip is brought close to the sample surface, the electrostatic force is generated and it is proportional to the difference between their Fermi levels. If the work functions of the sample and the tip are different, electrons flow from the lower work function to the higher work function material to align their Fermi level and the system reach in charge equilibrium condition. The vacuum levels of the tip and sample are not the same and the surfaces of the sample and the tip are charged and an apparent V_{CPD} will form. Due to the V_{CPD}, an electrical force acts on the contact area and this force can be nullified by applying a DC voltage and it is equal to the work function difference between tip and sample. The work function of the sample can be calculated when the tip work function is known from scanning a reference sample.

6.2.2.2 Conductive atomic force microscopy

Conductive atomic force microscopy (CAFM) is usually used to analyze the local variations in the current of the sample. In this chapter work, we used Bruker's dimension ICON to record the simultaneous images of surface topography and current. The AFM controller is used for applying DC bias through the substrate during measurements. The CAFM tip is connected through a low-noise current amplifier to the AFM controllers to generate the current image.

6.2.3 Sample preparation

The silicon (Si) wafers were solvent-cleaned in an ultrasonic bath to get the surface free from contamination. Successive treatments in acetone, isopropanol and deionized (DI) water were carried out for 10 min each at room temperature. All chemicals used were of highest purity. In order to improve the wettability and to form a uniform thin film of PEDOT:PSS on a Si substrate, the addition of a suitable surfactant into the solution is required. We used zonyl fluorosurfactant as an additive. PEDOT:PSS (1.3 wt.% dispersion in H_2O) and zonyl were purchased from Sigma-Aldrich. PEDOT:PSS was filtered with polyvinylidene fluoride (PVDF) filter having pore size of 0.45 μm. PEDOT:PSS doped with 5 vol.% of organic solvents such as *n*-methyl-2-pyrrolidone (NMP), dimethyl formamide (DMF), dimethyl sulfoxide (DMSO), ethylene glycol (EG) and methanol (MeOH), GO and zonyl were spin coated on Si wafers at 2,000 rpm for 60 s. The prepared PEDOT:PSS films were then annealed at 150 °C for 10 min on a hotplate. Finally, top contacts were made with gold by thermal evaporation at a vacuum of 10^{-5} Torr. Ohmic contacts using eutectic In–Ga were made on the backside of Si.

6.3 Results and discussion

6.3.1 Thickness and morphology

Thickness of PEDOT:PSS films doped with different organic solvents was measured using stylus profilometer and found to be in the range of 70–80 nm. Figure 6.2 shows the height and line scan profile of PEDOT:PSS films doped with different organic solvents.

The average root mean square (RMS) roughness values were found to be 3.7, 2.8, 4.3, 2.8, and 2.6 nm for NMP, DMF, DMSO, EG, and MeOH doped PEDOT: PSS films, respectively. Bare PEDOT:PSS film was quite smooth with an RMS roughness of 2.5 nm. RMS roughness of PEDOT:PSS films increases after solvent doping and the variation in RMS roughness reflects the morphological change that arises from the conformation of the polymer chain. From the height line scan profile of bare and organic solvents-doped PEDOT:PSS films, we can easily see that different heights of the films which show the morphological change that arise due to the solvents. Bare PEDOT:PSS shows the uniform morphology, while organic solvents-doped PEDOT:PSS shows different heights, and this might be due to the agglomeration of PEDOT:PSS.

Figure 6.3 shows the AFM images of PEDOT:PSS films with GO concentrations. The average values of RMS roughness of the PEDOT:PSS films prepared using GO were found to be 5.5, 6.3, 6.8, 11.0, and 9.5 nm, for 0.01, 0.03, 0.05, 0.08, and 0.10 vol.% of GO, respectively.

6.3.2 Surface potential and work function

To evaluate the value of work function of bare and organic solvents-doped PEDOT: PSS films, we have used KPFM technique. Silicon tips coated with conductive Co–Cr were used to record KPFM measurements for this work. The value of ϕ_{tip} was measured and found to be 4.30 eV. The average values of work function are 4.77, 4.69, 4.78, 4.80, and 4.70 for NMP-, DMF-, DMSO-, EG-, and MeOH-doped PEDOT:PSS films, respectively. The measured work function of organic solvents-doped PEDOT:PSS thin films range from 4.69 to 4.80 eV as compared to 4.90 eV for the bare PEDOT:PSS film. The work function value of bare PEDOT:PSS is in good agreement with the reported work function 4.80 to 5.20 eV [24]. Figure 6.4 shows the potential and potential line-scan profile of PEDOT:PSS films doped with organic solvents. Potential line-scan profile shows the spatial uniformity of CPD. Bare PEDOT:PSS contains excess PSS layer that will give the higher work function of bare film around 4.90 eV. After solvents doping surface potential value

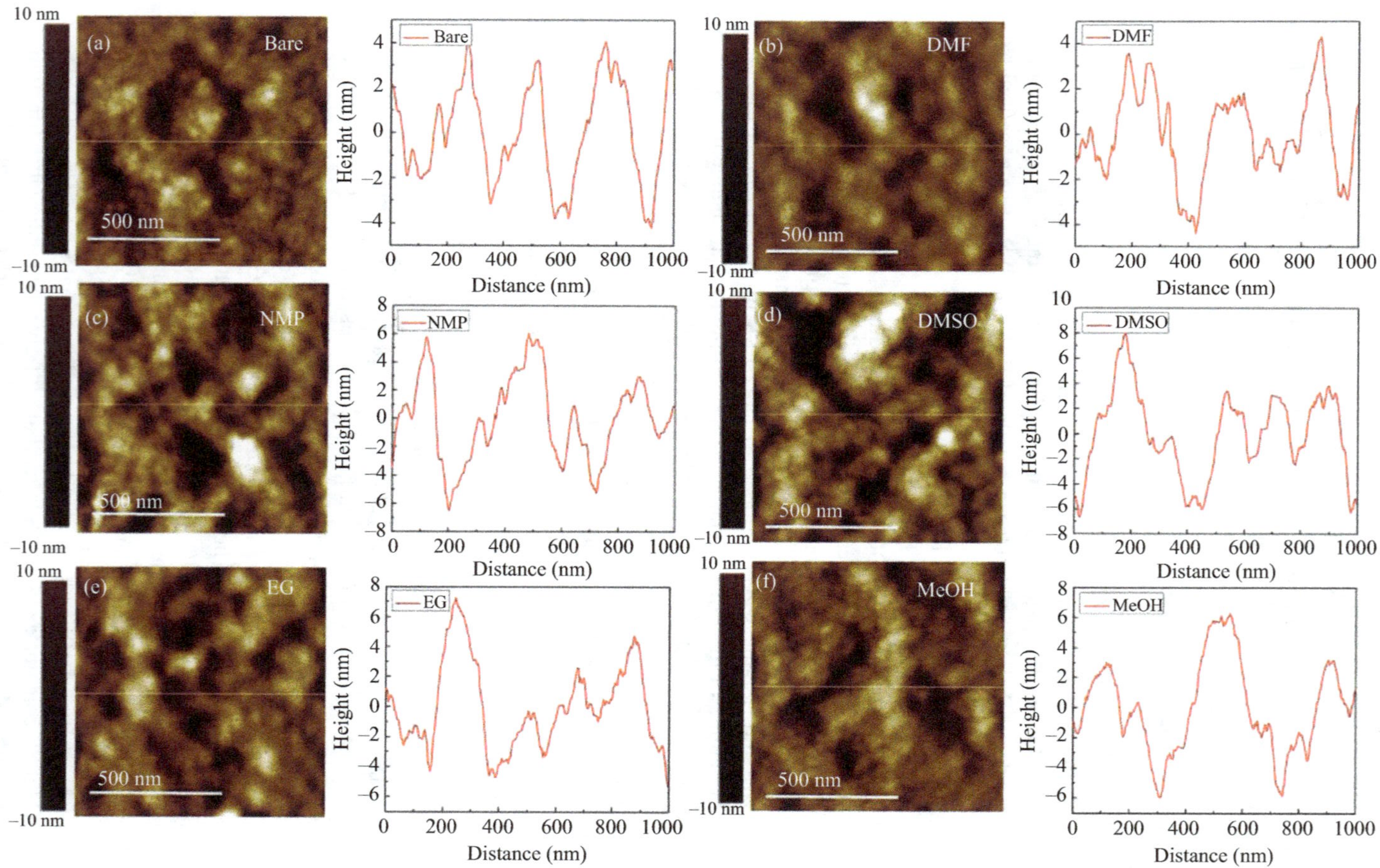

Figure 6.2 Height and respective line scan of (a) bare, (b) DMF, (c) NMP, (d) DMSO, (e) EG, and (f) MeOH organic solvents doped PEDOT:PSS films

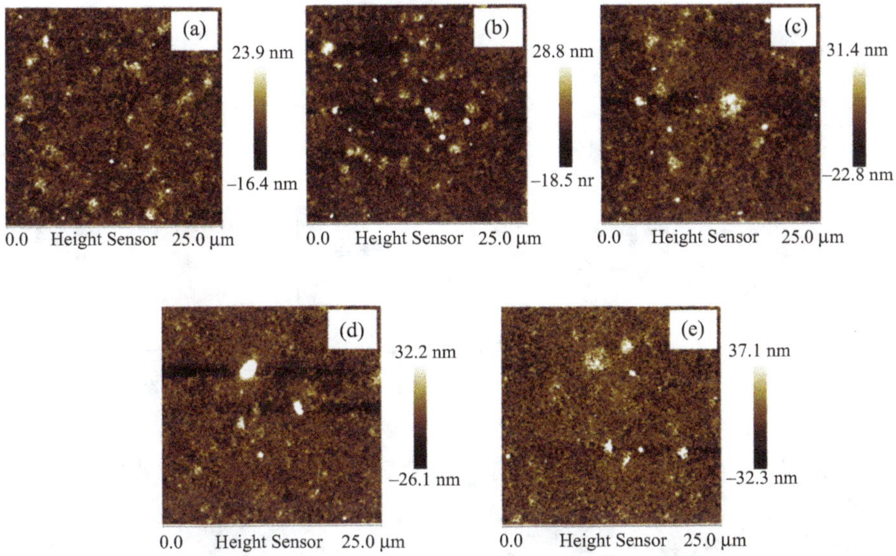

Figure 6.3 AFM images of PEDOT:PSS films with (a) 0.01, (b) 0.03, (c) 0.05, (d) 0.08, and (e) 0.10 vol.% doping concentration of GO

decreases in comparison to bare PEDOT:PSS, resulting in decrease in work function value of solvents doped PEDOT:PSS films. The work function corresponds to the energy between vacuum level and Fermi level. The variation in V_{CPD} indicates that there is a change in Fermi level of organic solvents-doped PEDOT:PSS films. The decrease in work function after solvents doping indicates that the Fermi level coming close to the conduction band.

Figure 6.5 shows the potential images of GO-PEDOT:PSS films. The average values of work function are 4.68, 4.59, 4.63, 4.57, 4.67, and 4.58 eV for 0, 0.01, 0.03, 0.05, 0.08, and 0.10 vol.% GO, respectively.

6.3.3 Conductivity

Conductivity enhancement of PEDOT:PSS films strongly depends on the chemical structure of used organic solvents for doping. The bare PEDOT:PSS film exhibited a low electrical conductivity of 0.16 S/cm, which is due to the disconnected PEDOT chain. MeOH has a low boiling point, and it has only one polar group, the electrical conductivity is not enhanced remarkably, despite having high dielectric constant. Conductivity did not change remarkably with solvents having one polar group and low boiling point [25]. However, by using polar solvents having high boiling points such as DMSO, DMF, NMP, and EG, conductivity is enhanced significantly. The high boiling point should allow more time to the PEDOT:PSS film for structural refinements during the drying process. We obtained conductivity of 40.0, 8.0, 19.0, 30.0, and 1.6 S/cm for NMP, DMF, DMSO, EG, and MeOH, respectively. MeOH has only one polar group and there is not much enhancement

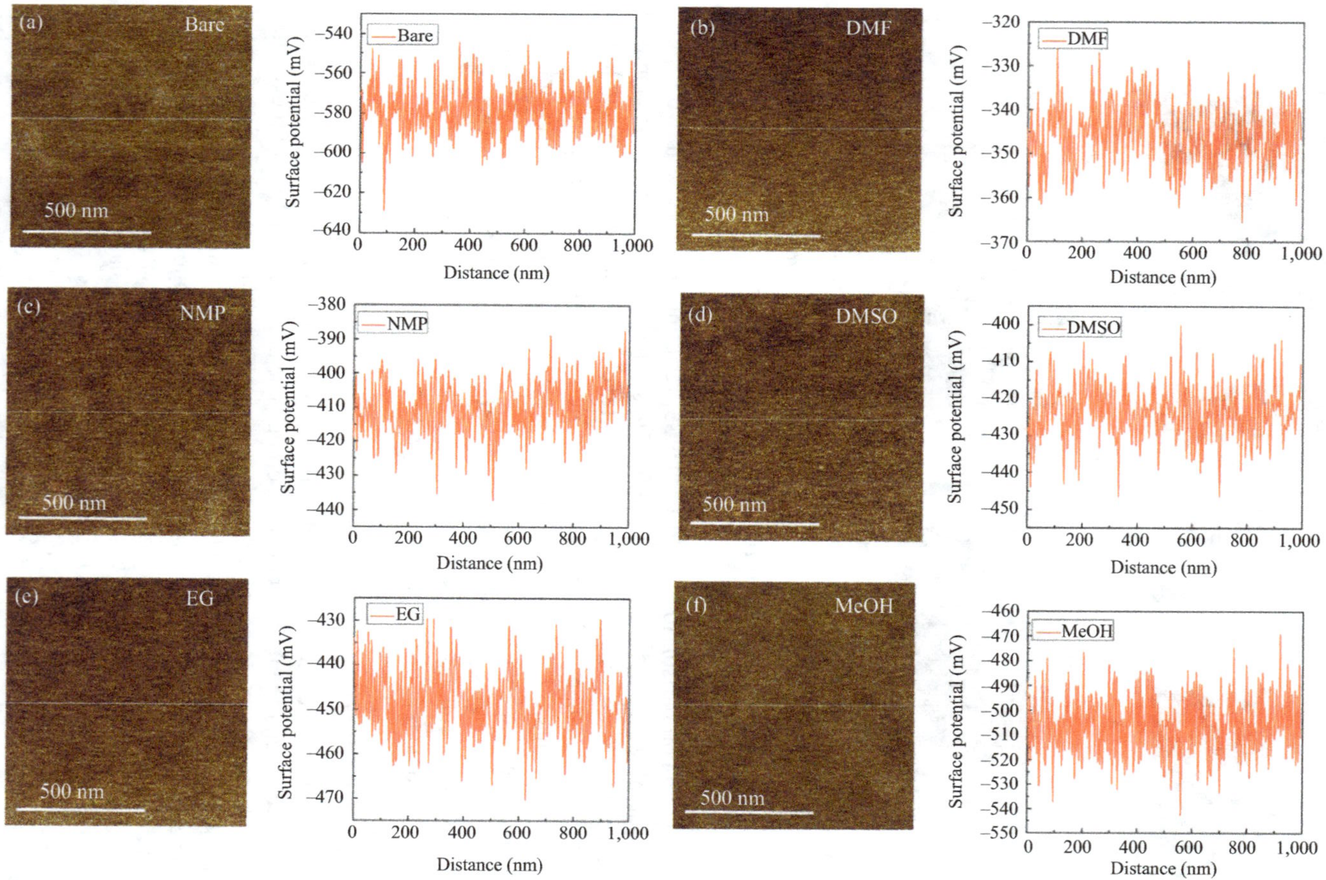

Figure 6.4 Potential and respective line scan of (a) bare, (b) DMF, (c) NMP, (d) DMSO, (e) EG, and (f) MeOH organic solvents doped PEDOT:PSS films

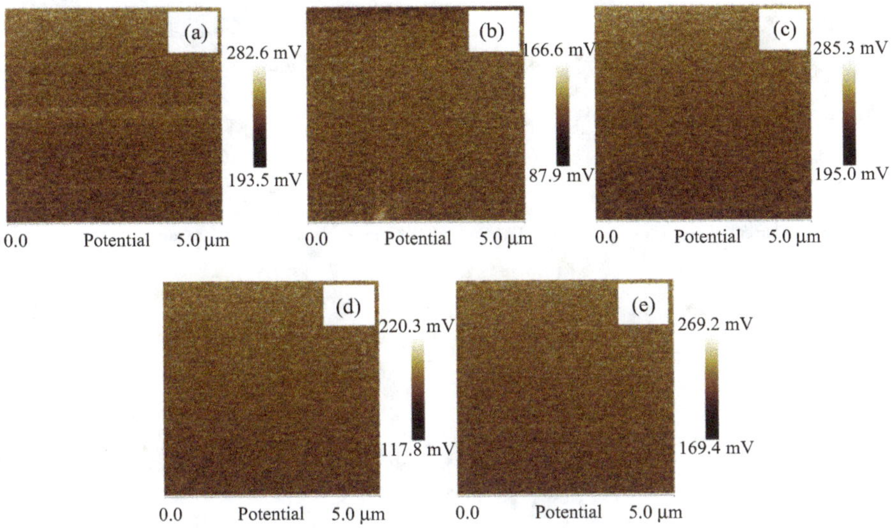

Figure 6.5 Surface potential images of PEDOT:PSS films with (a) 0.01, (b) 0.03, (c) 0.05, (d) 0.08, and (e) 0.10 vol.% doping concentration of GO

in conductivity. DMSO has only one polar group and it has strong dipole moment close to PEDOT chain so that the dipole–dipole interaction can take place after the formation of hydrogen bond with PSS [20]. From our results, we showed that the conductivity obtained with NMP doping is much higher and it has the best compromise between high boiling point and high dielectric constant [26]. Similar enhancement in the conductivity of solvent-modified PEDOT:PSS films was also observed by Kim and Jonsson [16,17]. The reduction of PSS is confirmed by the CAFM technique. CAFM current images of organic solvents-doped PEDOT:PSS films are shown in Figure 6.6. From the current image, we observed the current in nA range for bare-, DMF-, NMP-, DMSO-, EG-, and MeOH-doped PEDOT:PSS films, respectively. Bare PEDOT:PSS has the minimum current and maximum current was observed for NMP-modified PEDOT:PSS film. This shows the reduction of more PSS from the surface for NMP-doped PEDOT:PSS film. The current values obtained for all films are well correlated with the conductivity value of PEDOT:PSS films.

The conductivity of PEDOT:PSS film increases with GO addition and after the optimal value of GO conductivity starts to decrease. The average enhanced conductivity of GO-PEDOT:PSS film with 0.05 vol.% GO was found to be 118 S/cm compared to 0.16 S/cm for pristine PEDOT:PSS film. The conductivity of PEDOT:PSS films then decreases with further increase in GO vol. The decrease in conductivity after optimal doping of GO is mainly due to the reduction in structure orientation of PEDOT due to the presence of more GO on the surface. Figure 6.7 shows the variation in the conductivity of PEDOT:PSS films with GO vol.% deposited on glass substrate. In PEDOT:PSS, conductive PEDOT chains are attached to insulating PSS chains through columbic attraction.

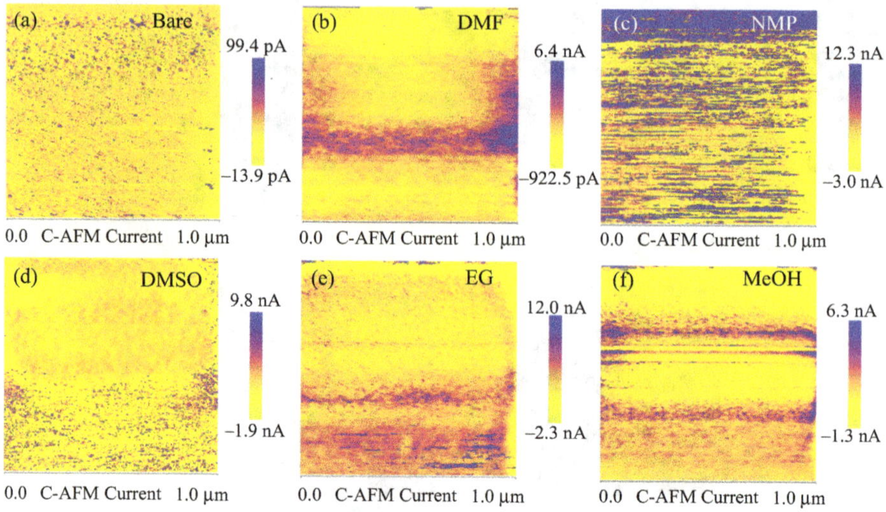

Figure 6.6 CAFM images of (a) bare, (b) DMF, (c) NMP, (d) DMSO, (e) EG, and (f) MeOH organic solvents doped PEDOT:PSS films

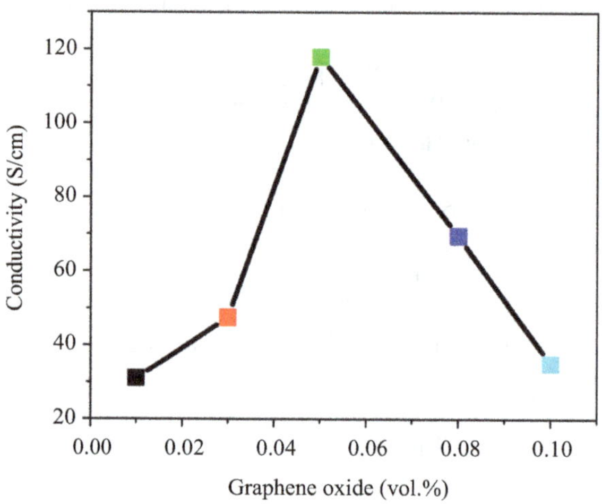

Figure 6.7 Variation in electrical conductivity of PEDOT:PSS films with doping concentration of GO varying from 0 to 0.10 vol.%

PEDOT and PSS chains are separated by the functional groups of GO such as –OH and –COOH. As a result of GO addition to PEDOT:PSS solution, the columbic attraction between PEDOT and PSS chains becomes weaker which improves the linear confirmation of PEDOT and PSS chains. The separated PEDOT chains link up with functional groups of GO that enhance the conductivity of PEDOT:PSS film as shown in Figure 6.1. We got the maximum conductivity (118 S/cm) of PEDOT:PSS film for 0.05 vol.% GO.

6.3.4 Raman spectra

Raman spectroscopy is a very sensitive technique to study the structural changes occurring in polymers films.

Figure 6.8 shows Raman spectra of organic solvents-doped PEDOT:PSS films using 514 nm Ar laser excitation. The Raman spectra show peaks at 1,367, 1,440, 1,506, and 1,568 cm^{-1}. The band at 1,440 cm^{-1} is associated with the C_α=C_β symmetric vibration [25]. The band near about 1,367 cm^{-1} is associated with the C_β–C_β stretching. Raman peaks located at 1,506 and 1,568 cm^{-1} are associated with the C_α=C_β asymmetric stretching vibrations [9]. Two kinds of resonant structures have been proposed for PEDOT, namely benzoid and quinoid structure [25]. High-dielectric constant solvents could cause phase segregation between PEDOT and PSS due to a reduction in their Coulombic interaction, which rearranges the PEDOT segments into a more linear, quinoid conformation.

Raman peaks of PEDOT:PSS film with GO contents get narrower and slightly shifted to higher wavelength (1,442 cm^{-1}) as shown in Figure 6.9. This shift is attributed to C=C stretching of the thiophene ring and transformation of the resonance structure of the PEDOT:PSS after the addition of the GO. Structural alteration of PEDOT:PSS by GO addition confirmed by shift in Raman peak. Inset shows the Raman spectra of GO film presenting a disorder-induced band at 1,350 cm^{-1} and a graphitic *G* band at 1,600 cm^{-1}. The *D* band associated with the vibration of carbon atom with dangling bond in crystal lattice plane terminations of disordered graphite and the *G* peak corresponds to sp^2 hybridized carbon in a two-dimensional hexagonal lattice [21].

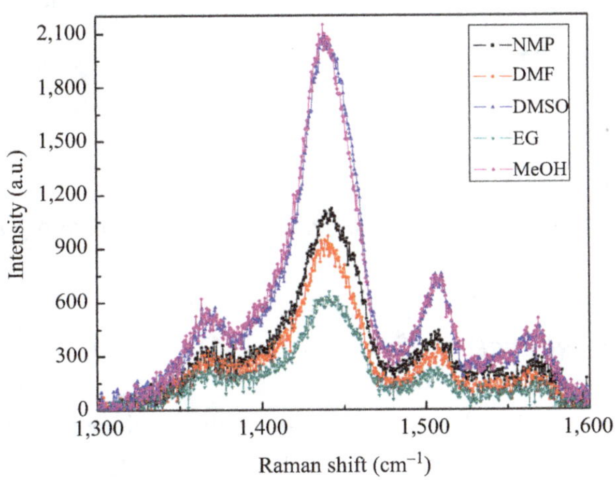

Figure 6.8 Raman spectra of organic solvents-doped PEDOT:PSS films

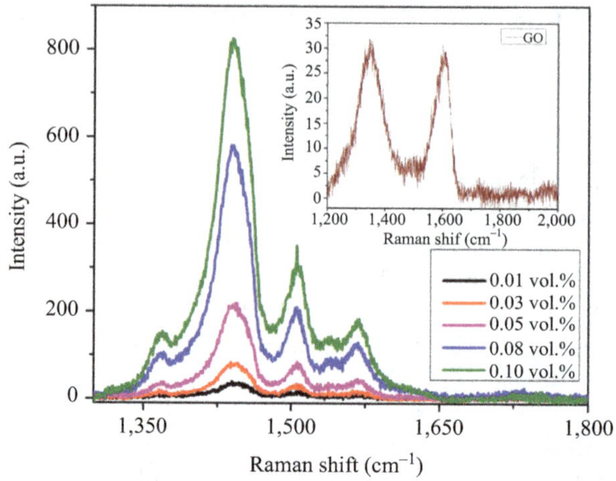

Figure 6.9 Raman spectra of PEDOT:PSS films with doping concentration of GO varying from 0 to 0.10 vol.%. Inset shows Raman spectra of GO

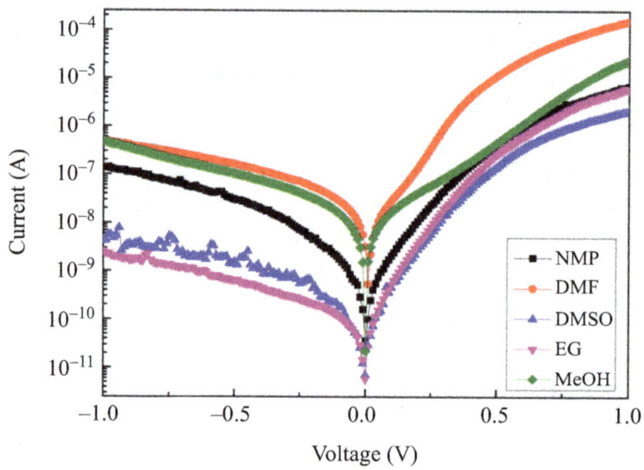

Figure 6.10 Semi-log forward and reverse I–V characteristics for Au/PEDOT:PSS/n-Si/In-Ga heterojunction diodes doped with organic solvents

6.3.5 *Electrical characteristics of PEDOT:PSS/n-Si heterojunction diodes*

Current–voltage (*I–V*) characteristics of solvents-doped PEDOT:PSS/*n*-Si heterojunction diodes are shown in Figure 6.10. The current transport mechanism in PEDOT:PSS/Si heterojunction diode is like that in a Schottky diode. The current transport mechanism in PEDOT:PSS-based diode is like that in a Schottky diode [27]. Current–voltage (*I–V*) characteristics of the Schottky junction can be expressed by the thermionic emission theory [28].

$$I = I_O \exp\left(\frac{qV}{nkT} - 1\right) \tag{6.1}$$

$$I_O = AA^{**}T^2 \exp(-q\phi_{bn}/kT) \tag{6.2}$$

where I_O is the saturation current, V is the applied voltage, A is the diode area, A^{**} is the effective Richardson constant, ϕ_{bn} is the apparent or measured barrier height, n is the ideality factor, T is the temperature, and k is the Boltzmann constant. n and ϕ_{bn} are obtained from forward characteristics as [28],

$$n = q/kT(dV/d\ln I) \tag{6.3}$$

$$\phi_{bn} = \frac{kT}{q}\ln\left(AA**T^2/I_o\right) \tag{6.4}$$

The experimental *I–V* data are plotted as log *I* vs. *V*, and *n* and barrier height (ϕ_{bn}) are calculated from the intercept and slope of the linear fit to the linear part of forward characteristics. The values of *n* and ϕ_{bn}were found as 2.6, 2.2, 2.2, 2.1, 4.5 and 0.76, 0.70, 0.82, 0.81, 0.70 eV, respectively, for NMP-, DMF-, DMSO-, EG-, and MeOH-doped PEDOT:PSS/*n*-Si heterojunction diodes.

For bare PEDOT:PSS/*n*-Si heterojunction diode, we obtained the ideality factor of 9.6 and barrier height 0.57 eV. According to the Schottky–Mott model, $\phi_{bn} = \phi_m - \chi$, where ϕ_m is the metal work function and χ is the electron affinity. Here, we have taken the work function value of organic solvents-doped PEDOT: PSS obtained with the help of KPFM as ϕ_m and χ is the electron affinity of Si. ϕ_{bn} values are close to the value predicted by Schottky–Mott model. The value of ϕ_{bn} from Schottky–Mott model were found as 0.72, 0.64, 0.73, 0.75, 0.65 eV, respectively for NMP-, DMF-, DMSO-, EG-, and MeOH-doped PEDOT:PSS/*n*-Si heterojunction diodes. High-quality junction was formed for all organic solvents-doped PEDOT:PSS/*n*-Si interfaces. The diode with EG-doped PEDOT:PSS/*n*-Si heterojunction shows the lowest value of ideality factor of 2.1. For organic solvent EG, there are two polar groups, one polar group stabilizes PSS and the other polar group interacts with dipoles or positive charge of PEDOT coil like polymer chain, rearranging them to form a linear and expanded chain network, which enhances the charge transfer across the device [14,29].

Figure 6.11 represents the *I–V* characteristics of fabricated Au/GO-PEDOT: PSS/*n*-Si/In-Ga diodes.

There is not much variation in the barrier height and the obtained ϕ_{bn} is very close to the value as predicted by Schottky–Mott model. The values of *n* and ϕ_{bn} are summarized in Table 6.1.

For PEDOT:PSS/*n*-Si heterojunction diode without GO, we got the ideality factor of 9.6 and barrier height 0.57 eV. The improvement in ideality factor indicates enhanced charge transport properties due to the GO at the interface. For 0.05 vol.% GO, we got much better ideality factor and barrier height than other GO

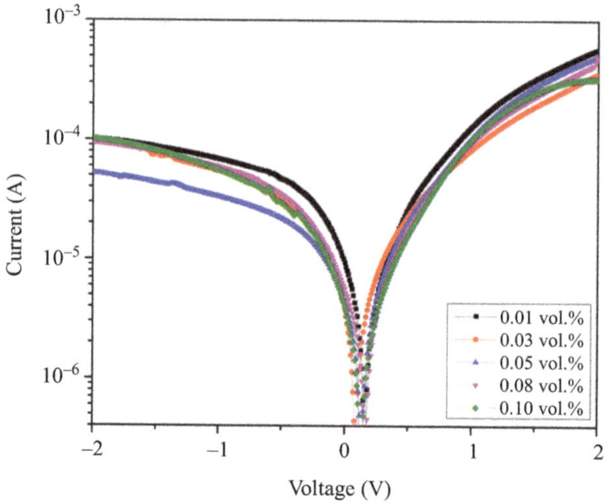

Figure 6.11 Forward and reverse current–voltage (I–V) characteristics of PEDOT:PSS/n-Si heterojunction diode with doping concentration of GO varying from 0 to 0.10 vol.%

Table 6.1 Calculated PEDOT:PSS/n-Si heterojunction diode parameters as a function of GO concentration

GO (vol.%)	Ideality factor	Barrier height (eV)
0.01	6.8	0.56
0.03	7.8	0.55
0.05	5.3	0.58
0.08	6.7	0.57
0.10	7.4	0.56

vol.%. This is due to the presence of less amount of PSS chain network on the surface and rearrangement of PEDOT, which enhances the charge transfer across the device and improves the diode parameters and the reduction of PSS is also confirmed by highest conductivity value obtained for 0.05 vol.% GO. Low ideality factor, low reverse leakage current and higher barrier height are needed for good quality diodes, and the heterojunction diodes fabricated with 0.05 vol.% GO have all these requirements, which confirms the formation of good quality junction.

After the optimal value of GO, the ideality factor starts to increase and it might be due to the decreased conductivity. The larger value of ideality factor may be attributed to secondary mechanism like generation-recombination at the interface. In this case, the charge transport across the interface is no longer due to the thermionic emission but the other mechanism like generation and recombination of charge carriers in depletion region might be the dominant current transport mechanism. The electronic conduction through the diodes affected by the charge traps in the conductive polymer and traps are locations arising from the defects, impurities, disorders, etc. The interface states between the GO-PEDOT:PSS and Si play an important role in the current transport process. Interface defects may also lead to a lateral inhomogeneous distribution of barrier heights at the interface [30–32]. The larger value of the ideality factor indicates that the forward current is governed by the recombination current.

6.3.6 Photovoltaic characteristics of PEDOT:PSS/n-Si solar cell

In order to characterize the photovoltaic behavior of the fabricated diodes, the current density–voltage (*J–V*) characteristics of the diodes were performed under solar simulator shown in Figure 6.12. Table 6.2 represents the average value of measured photovoltaic properties. We got the maximum open-circuit voltage (V_{oc}) of 510 mV for EG-doped PEDOT:PSS and minimum of 344 mV for MeOH-doped PEDOT:PSS/*n*-Si solar cell. We got the fill factor (*FF*) values between 16.67% and

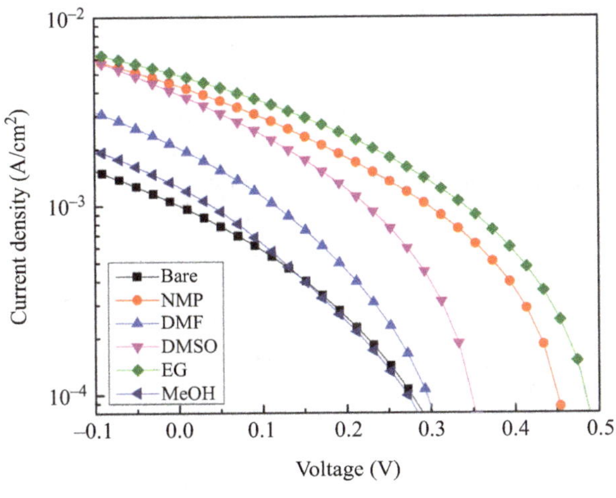

Figure 6.12 J–V characteristics of bare and organic solvents-doped PEDOT: PSS/n-Si solar cell under AM 1.5 illumination

*Table 6.2 Photovoltaic properties of organic solvents doped PEDOT:PSS/*n-Si *solar cell*

Solvent (5 vol.%)	V_{oc} (mV)	J_{sc} (mA/cm^2)	*FF* (%)	*PCE* (%)
Bare	358	1.11	17.04	0.068
NMP	472	4.80	17.73	0.403
DMF	341	2.26	16.67	0.129
DMSO	370	4.34	18.22	0.297
EG	510	5.46	18.91	0.526
MeOH	344	1.40	14.55	0.070

18.91%. The low values of *FF* indicate extra current-limiting effects which are not essentially related to PEDOT:PSS properties but arise from recombination sites present at either Si surface or Au/PEDOT:PSS interface which causes larger ideality factors and lower *FF* values [33]. The bare PEDOT:PSS sample shows the short-circuit current density (J_{sc}) of 1.11 mA/cm^2 and after doping with organic solvents J_{sc} increases up to 5.46 mA/cm^2 and well correlates with the increased conductivity of PEDOT:PSS films. The removal of PSS and enhanced conductivity has a positive effect on the value of J_{sc}. For bare PEDOT:PSS, power conversion efficiency (PCE) value is 0.068% which was found to be increased up to a maximum of 0.526% after the addition of organic solvents.

The maximum PCE is obtained for EG-doped PEDOT:PSS film. PCE has increased around 8 times by the addition of EG compared to the bare PEDOT:PSS film. Due to the low value of *FF* and J_{sc}, we obtained a low value of PCE which is attributed to the fact that the PEDOT:PSS films were deposited in air, resulting in the polymer particles being oxidized during the deposition process. The presence of oxygen has an adverse effect on the performance of polymer especially on *FF* [14]. The low value of PCE is because of higher leakage current, low value of J_{sc} and *FF* of fabricated cells. We observed maximum PCE for EG-modified PEDOT:PSS due to the low leakage current. The value of V_{OC} is in good agreement with the reported value, while the J_{sc} and *FF* values are not high. PEDOT:PSS is hygroscopic in nature and it can absorb moisture during the fabrication process. This might be one reason for low J_{sc} and *FF*. PEDOT:PSS cells without solvent have a PCE of 0.068% due to the low conductivity and organic solvents-modified PEDOT:PSS solar cells have higher PCE in comparison to bare due to better conductivity.

6.3.7 Energy band diagram

The energy band diagram of the PEDOT:PSS/*n*-Si interface is shown in Figure 6.13. The work function of pristine PEDOT:PSS film is 4.90 eV as obtained with the help of KPFM measurements.

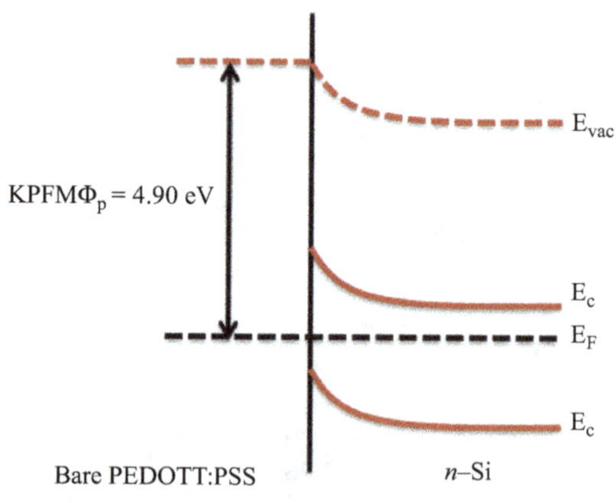

Figure 6.13 Energy band diagram of PEDOT:PSS/n-Si heterojunction diode

The band bending creates a space charge region resulting in a barrier to the charge flow at the interface of PEDOT:PSS/*n*-Si. When the two materials are brought in contact with each other, they will interact so as to achieve thermal equilibrium resulting in the constancy of the E_F.

6.4 Summary

Various properties of different optimized organic solvents and GO-doped PEDOT: PSS films, PEDOT:PSS/Si heterojunctions, and solar cells have been studied in detail. The conductivity of PEDOT:PSS films was enhanced by three orders of magnitude with GO whereas all films remain highly transparent (>85%) in the visible region. The partial removal of PSS and the formation of conducting PEDOT-connected networks contribute to the enhanced electrical conductivity of PEDOT:PSS films. The removal of PSS was also confirmed by CAFM measurements. Solar cell fabricated with EG-doped PEDOT:PSS film showed a maximum PCE as compared to other solvents-doped PEDOT:PSS film. The highly conducting and transparent material can be used in various future optoelectronic devices.

Acknowledgments

C.S. Pathak is grateful to the Department of Physics and Nanoscale research facility (NRF), Indian Institute of Technology Delhi, India, for providing characterization facilities.

References

[1] Chiang C.K., Fincher C.R., Jr., Park Y.W *et al.* 'Electrical conductivity in doped polyacetylene'. *Phys. Rev. Lett.* 1977, vol. 39 (17), pp. 1098–101

[2] Shirakawa H., Lewis E.J., McDiarmid A.G., Chiang C.K., and Heeger A.J., 'Synthesis of Electrically Conducting Organic Polymers: Halogen derivatives of polyacetylene, (CH)x'. *J. Chem. Soc. Chem. Commun.* 1977, vol. 16, pp. 578–80

[3] Groenendaal L., Jonas F., Freitag D., Pielartzik H., and Reynolds J.R., 'Poly (3,4-ethylenedioxythiophene) and its derivatives: past, present, and future'. *Adv. Mater.* 2000, vol. 12 (7), pp. 481–94

[4] Xia Y., Zhang H., and Ouyang J., 'Highly conductive PEDOT:PSS films prepared through a treatment with zwitterions and their application in polymer photovoltaic cells'. *J. Mater. Chem.* 2010, vol. 20 (43), pp. 9740–47

[5] Palumbiny C.M., Heller C., Schaffer C.J. *et al.* 'Molecular reorientation and structural changes in co solvent-treated highly conductive PEDOT:PSS electrodes for flexible indium tin oxide-free organic electronics'. *J. Phys. Chem. C.* 2014, vol. 118 (25) pp. 13598–606

[6] Xia Y., and Ouyang J., 'PEDOT:PSS films with significantly enhanced conductivities induced by preferential solvation with cosolvents and their

application in polymer photovoltaic cells'. *J. Mater. Chem.* 2011, vol. 21, pp. 4927–36

[7] Livermore P.A., Jin R., Wang X., Chen L., Bradley D. D. C., and Mello John C. de, 'High efficiency organic light-emitting diodes with PEDOT-based conducting polymer anodes'. *J. Mater. Chem.* 2008, vol. 18 pp. 4414–20

[8] Sun K., Zhang S., Li P., *et al.* 'Review on application of PEDOTs and PEDOT:PSS in energy conversion and storage devices'. *J. Mater Sci: Mater. Electron.* 2015, vol. 26 (7), pp. 4438–62

[9] Pathak C.S., Singh J.P., and Singh R., 'Effect of dimethyl sulfoxide on the electrical properties of PEDOT:PSS/*n*-Si heterojunction diodes'. *Curr. Appl. Phys.* 2015, vol. 15 (4), pp. 528–34

[10] He L., Jiang C., Wang H., and Lai D., 'High efficiency planar Si/organic heterojunction hybrid solar cells'. *Appl. Phys. Lett.* 2012, vol. 100 (7) pp. 073503

[11] Zhang Y., Zu F., Lee S.T., Liao L., Zhao N., and Sun B., 'Heterojunction with organic thin layers on silicon for record efficiency hybrid solar cells'. *Adv. Energy. Mater.* 2014, vol. 4 (2), pp. 2195–223

[12] Kok M. M. de, Buechel M., Vulto S. I. E., *et* al., 'Modification of PEDOT: PSS as hole injection layer in polymer LEDs'. *Phys. Stat. Sol. (A)* 2004, vol. 201(6), pp. 1342–59

[13] Cruz I. C., Reyes M. R., Frutis M. A. A., Rodriguez A.G., and Sandoval R.L., 'Study of the effect of DMSO concentration on the thickness of the PSS insulating barrier in PEDOT:PSS thin films'. *Synth. Met.* 2010, vol. 160 (13–14), pp. 1501–06

[14] Pathak C.S., Singh J.P., and Singh R., 'optimizing the electrical properties of PEDOT: PSS films by co-solvents and their application in polymer photovoltaic cells'. *Appl. Phys. Lett.* 2017, vol. 111 (10), pp. 102107

[15] Pathak C.S., Singh J.P., and Singh R., 'Modification of electrical properties of PEDOT:PSS/*p*-Si heterojunction diodes by doping with dimethyl sulfoxide'. *Chem. Phys. Lett.* 2016, vol. 652, pp. 162–66

[16] Kim J.Y., Jung J.H., Lee D.E., and Joo J., 'Enhancement of electrical conductivity of poly (3,4-ethylenedioxythiophene)/poly(4-styrenesulfonate) by a change of solvents'. *Synth. Met.* 2002, vol. 126(2-3), pp. 311–16

[17] Jönsson S. K. M., Birgersonb J., Crispin X., *et al.* 'The effects of solvents on the morphology and sheet resistance in poly(3,4-ethylenedioxythiophene)–polystyrenesulfonic acid (PEDOT–PSS) films'. *Synth. Met.* 2003, vol. 139 (1), pp. 1–10

[18] Pathak C.S., Singh J.P., and Singh R., 'Preparation of novel graphene-PEDOT: PSS nanocomposite films and fabrication of heterojunction diodes with n-Si'. *Chem. Phys. Lett.* 2018, vol. 694, pp. 75–81

[19] Pathak C.S., Singh J.P., and Singh R., 'A novel composite material of graphene and PEDOT:PSS' *Proceeding of DAE Solid State Physics Symposium*; Noida, India, Dec 2015. AIP; 2016, pp. 140021

[20] Raj P.G., Rani V.S., Kanwat A., and Jang J., 'Enhanced organic photovoltaic properties via structural modifications in PEDOT:PSS due to graphene oxide doping'. *Mater. Res. Bull.* 2016, vol. 74, pp. 346–52

[21] Pathak C.S., Gill S.S., Pathak P.K., and Singh R., 'Modification in electrical and morphological properties of PEDOT: PSS films with solvents and graphene oxide'. *Proceeding of National Conference on Recent Advances in Experimental and Theoretical Physics*; Jammu and Kashmir, India, April 2018. AIP; 2006 pp. 030040

[22] Kelvin L., 'Contact electricity of metals'. *Phil. Mag.* 1898, vol. 46, pp. 82–120

[23] Nonnenmacher M., Oboyle M. P., and Wickramasinghe H. K., 'Kelvin probe force microscopy'. *Appl. Phys. Lett.* 1991, vol. 58 (25), pp. 2921–23

[24] Nardes A.M., Kemerink M., Kok M.M. de, Vinken E., Maturova K., and Janssen R.A.J., 'Conductivity, work function, and environmental stability of PEDOT:PSS thin films treated with sorbitol'. *Org. Elect.* 2008, vol. 9 (5), pp. 727–34

[25] Ouyang J., Xu Q., Chu C.–W., Yang Y., Li G., and Shinar J., 'On the mechanism of conductivity enhancement in poly (3,4-ethylenedioxythiophene):poly(styrene sulfonate) film through solvent treatment'. *Polymer* 2004, vol. 45(25), pp. 8443–50

[26] Pathak C.S., Kapoor R., Singh J.P., and Singh R., 'Investigation of the effect of organic solvents on the electrical characteristics of PEDOT: PSS/p-Si heterojunction diodes'. *Thin Solid Films* 2017, vol. 622, pp. 115–21

[27] Zhang Y., Zu F., Lee S.T., Liao L., Zhao N., and Sun B., 'Heterojunction with organic thin layers on silicon for record efficiency hybrid solar cells'. *Adv. Energy Mater.* 2014, vol. 4 (2), pp. 1300923

[28] Sze, S.M., Physics of semiconductor devices. New York, Wiley; 1969. p. 255

[29] Pathak C.S, Gill S.S., Singh J.P., and Singh R., "Effect of solvents on electrical properties of PEDOT:PSS/*n*-Si heterojunction diodes", *Proceeding of* 3rd *International Conference on Emerging Electronics*, Bombay, India, Dec.2016. IEEE; 2016, pp. 1–3

[30] Werner J. H., and Guttler H. H., 'Barrier inhomogeneities at Schottky contacts' *J. Appl. Phys.* 1991, vol. 69(3), pp. 1522–33

[31] Pathak C.S., Garg M., Singh J.P., and Singh R., 'Current transport properties of monolayer graphene/ *n*-Si Schottky diodes'. *Semicond. Sci. Technol.* 2018, vol. 33 (5), pp. 055066

[32] Pathak C.S., Singh J.P., and Singh R., 'Temperature dependent electrical characteristics of PEDOT:PSS/*n*-Si heterojunction diode'. *Inv. J. Sci. Tech.*, 2016, vol. 9 (3), pp. 1–5

[33] Pietsch M., Bashouti M. Y., and Christiansen S., 'The role of hole transport in hybrid inorganic/organic silicon/poly (3,4-ethylenedioxy-thiophene): poly (styrenesulfonate) heterojunction solar cells'. *J. Phys. Chem. C* 2013, vol. 117 (18), pp. 9049–55

Chapter 7

Emerging high-κ dielectrics for nanometer CMOS technologies and memory devices

Durgamadhab (Durga) Misra[1], Md Nasir Uddin Bhuyian[2], Yi Ming Ding[3], Kolla Lakshmi Ganapathi[4], and Navakanta Bhat[5]

7.1 Introduction

The gate length of complementary metal-oxide semiconductor (CMOS) devices has been scaled to below 10 nm with a three-year delay from the predicted year from ITRS 2007 report [1]. The aggressive scaling has slowed down due to the transistor reaching its physical limits. Three approaches are being followed to continue scaling of chips [2]: (i) gate stack material, (ii) channel material, and (iii) device architecture. The dielectric layer of gate stack is one of the leading candidates for devices below 10 nm [3]. Metal gate high-κ (MGHK) has been implemented to boost the chip performance while keeping the physical thickness thick enough to prevent large direct tunneling current. Secondly, different channel materials with higher carrier mobility, other than Si, have been considered [4]. While only strained silicon has been implemented thus far materials such as Ge or III-Vs are still in their research stage. 3D structures like FinFET have been successfully implemented to improve the drain-induced barrier lowering (DIBL) related to short channel effect [5].

While CMOS technology is scaling down, the deposition process of high-κ gate dielectric and annealing has significantly improved. Atomic layer deposition (ALD) method for high-κ deposition provided several advantages over alternative deposition methods, such as chemical vapor deposition and various physical vapor deposition (PVD) techniques, due to its conformity, control over materials thickness, step coverage, and composition. These desirable characteristics originate from self-saturating nature of ALD processes [6].

[1]ECE Department, New Jersey Institute of Technology, Newark, NJ, USA
[2]Globalfoundries, Malta, NY, USA
[3]Western Digital Corporation, Shanghai, China
[4]Physics Department, Indian Institute of Technology Madras, Chennai, India
[5]Centre for Nano Science and Engineering (CeNSE), Indian Institute of Science, Bangalore, India

During or after the ALD deposition or post deposition, the quality of dielectric in terms of equivalent oxide thickness (EOT) and interface state density between the substrate and dielectric can be improved by exposing them to a slot-plane-antenna (SPA) plasma with various gases such as O_2 or inert gases [7]. The SPA plasma provides a high-density plasma at low electron temperature, where the radicals diffuse from the plasma generation region to the wafer surface. SPA plasma is also a very low-damage plasma process compared to conventional inductively coupled plasma (ICP) or electron cyclotron resonance (ECR) plasma [7]. It was observed that the SPA plasma helps better film densification as well as improved interfacial layer (IL) growth. The dielectric is prevented from crystallization at low annealing temperatures.

Several electrical characterization methods are used to evaluate the interface quality or oxide quality to investigate the defects. These methods are categorized into two major groups, i.e. evaluation of capacitance of the gate stack and leakage current through the dielectric. Because of preexisting defects, experimental results can deviate from theoretical calculations. On the other hand, these deviations are utilized to evaluate the defects in the device. Characterization methods such as conductance method [8], capacitance–voltage (CV) at various low temperatures, flicker noise, capacitance transient spectroscopy, deep-level transient spectroscopy (DLTS) [9], CV hysteresis, and time-dependent dielectric breakdown (TDDB) [10] are utilized for the dielectric quality and interface evaluations.

This chapter describes the next-generation high-κ gate dielectric for high-mobility substrates like germanium (high-κ/Ge) and for memory devices. One of the critical issues in MGHK is the high interface state density (D_{it} $\sim 10^{12}$ $cm^{-2}eV^{-1}$) [11] compared to the traditional SiO_2/Si system ($\sim 10^{10}$ $cm^{-2}eV^{-1}$) [12]. Conventional SiO_2/Si system prevailed over decades due to its perfect interface quality due to thermally grown SiO_2 on Si substrate [12]. Introducing high-κ with metal gate (HKMG) brings additional reliability issues such as threshold voltage degradation (ΔV_{th}) after the bias temperature instability (BTI) stress in both nMOS and pMOS transistors [13]. This is due to the degradation of both interface and high-κ gate dielectrics. If silicon substrate is replaced by other materials such as Ge, it is necessary to address the interface defects density before expected mobility can be achieved [14]. Therefore, it is imperative to evaluate the high-κ dielectric layer and the interface quality for next-generation devices.

The objective of this chapter is to use various electrical characterization techniques to study the interface quality and high-κ dielectrics deposited by various process conditions. This provides comprehensive information on the defects, such as density, energy level, time constant and how they interact with other parameters (like flat band voltage, V_{FB}, and dielectric lifetime). Both theoretical model and experimental work are described. Different evaluation methods can provide a good analytical approach to study the dielectrics in the gate stacks. The correlation of experimental data from different methods can enhance the understanding of the defects behavior. Since the next-generation gate dielectrics on high-mobility substrates involve nanoscale devices, it requires a detailed understanding to integrate the technology into standard CMOS technology. Furthermore, this study discusses the advantages and disadvantages of various techniques, since each method has its own limitations such as like sensitivity, range, different extracted parameters, and the difficulty of implementation.

The organization of this chapter is as follows. Section 7.2 reviews the state-of-the-art MOS-capacitor. New interface control technique is introduced. The current status of high-κ/Ge was discussed, and its subsequent interface challenge is addressed before it can be fully considered for commercial use. Section 7.3 discusses the dry and wet-processed interface layer properties for three different p type Ge/ALD 1 nm-Al_2O_3/ALD 3.5 nm-ZrO_2/ALD TiN gate. Several parameters such as EOT, flat band voltage, bulk doping, and surface potential as a function of gate voltage are reported. It is also discussed that the high-frequency capacitance of TiN/ZrO_2/Al_2O_3/p-Ge gate stacks measured in the accumulation region depends on the device area after substrate resistance correction. Section 7.4 deals with the TiN/ZrO_2/Al_2O_3/p-Ge gate stacks subjected to the different slot-plane antenna plasma oxidation (SPAO) annealing conditions, namely, (i) before high-κ ALD, (ii) between high-κ ALD, and (iii) after high-κ ALD. After XPS (X-ray photoelectron spectroscopy) and EOT (estimated by capacitance voltage) measurement, the carrier transport mechanisms on these samples were extracted at high field range to reveal how SPAO can effectively remove traps in high-κ layer. The reliability of IL (GeO_x/GeO_2) is evaluated by TDDB performance under substrate electron injection condition. Section 7.5 describes the impact of SPAO on the dielectrics when $HfZrO_2$ is used as the gate dielectric and Zr percentage is varied. Section 7.6 describes the use of the various dielectrics in next-generation resistive random-access memory (RRAM) devices. Section 7.7 summarizes the overall research of this chapter and outlines a few future challenges.

7.2 Historical perspective and current status

The first point-contact transistor was invented was in fact a germanium transistor [15]. When MOS transistors were introduced it required an excellent dielectric for field effect. SiO_2 on silicon provided that solution because Ge does not form this oxide layer on its surface so easily and GeO_2 is hydroscopic and not thermally stable [16]. In 2004, EOT (oxide thickness calculated by using dielectric constant of SiO_2) was scaled down to $\sim$1.2 nm [1]. However, ultrathin SiO_2 suffers from direct-tunneling current which increases exponentially as thickness decreases [17]. To overcome gate leakage problems, initially the addition of N into SiO_2 has been used either by post-deposition annealing in nitrogen ambient or forming a nitride/oxide stack structure. As incorporating nitrogen into SiO_2, it not only increases the dielectric constant but also acts as a better barrier preventing boron penetration from polysilicon gate. SiON served as a transition stage between high-κ and SiO_2, which has maxim dielectric constant less than 8 [18].

High-κ materials were first studied in memory devices. Before it can be implemented in CMOS Technology, the following issues has to be considered first: (a) permittivity, bandgap, and band alignment to silicon, (b) thermodynamical stability, (c) film morphology and deposition method, (d) interface quality and bulk defects, and (e) gate compatibility and process compatibility [3]. The detailed properties of various high-κ dielectrics are listed in [3]. In order to have a good insulating property, it is suggested that conduction band offset (CBO) between high-κ and substrate should be larger than 1 eV to inhibit Schottky emission, and it is same for valence band offset as well. Considering different work function between substrate and the high-κ dielectric, specifically a bandgap of 4 eV is necessary to avoid serious leakage current and breakdown. Finally, it was suggested that ZrO_2 and HfO_2 are good candidates since they have bandgap larger than 4 eV and its dielectric constant is still large enough for further EOT scaling.

In gate first CMOS processes, the gate stacks must undergo rapid thermal annealing (RTA) at temperature as high as 1,000 °C [19]. This requires that the gate oxides must be thermally and chemically stable with the contacting materials [20]. From this point of view, HfO_2 has better thermal stability than ZrO_2 [21]. Additionally, as ZrO_2 and HfO_2 thin films were grown by ALD, the structural and electrical behavior of the films were somewhat precursor-dependent, revealing better insulating properties in the films grown from oxygen-containing precursors, therefore the HfO_2 films showed lower leakage compared to ZrO_2 [22]. It is desirable to have an amorphous high-κ layer after necessary processing treatments due to several benefits of the amorphous structure. Polycrystalline gate dielectrics are not favored as gate oxide layer since grain boundaries serve as high-leakage paths. HfO_2 or ZrO_2, crystallize at much lower temperatures at ~400 °C and ~300 °C. [23]. The crystalline temperature of dielectrics can be increased by incorporating other impurities, which was first studied by van Dover [24].

Interface between high-κ and substrate must have excellent electrical property and low interface state density, D_{it}. Fixed charges present at the interface can cause flat band voltage shift. Large D_{it} degrades mobility by surface scattering mechanism. Most of the high-κ materials reported D_{it} range from 10^{11} to 10^{12} $cm^{-2}eV^{-1}$ [25,26], which is much higher than conventional thermal grown SiO_2 [27,28] on silicon. Interface treatment, therefore, is necessary before depositing high-κ layer to obtain a low D_{it} interface. Also, the overall EOT value strongly depends on the thickness of IL. Similar to interface defects, bulk defects formed in high-κ oxides during deposition also cause degraded transistor performance and it is reported that high-κ materials are intrinsically defective because of the bonding structure and cannot relax easily [29]. The bonding in high-κ oxides is ionic. The nature of intrinsic defects in ionic oxides differs from those in SiO_2. The oxygen vacancies, oxygen interstitials, or oxygen deficiency defects are due to possible multiple valences of the transition metal [29]. Moreover, high-κ oxides achieve their high-κ value because of the low-lying soft polar modes. These modes could be a limit on scattering, which does not exist in SiO_2 [30]. Conventional polysilicon as gate material is not suitable for high-κ dielectrics due to Fermi level pinning problem [31]. The solution came with the introduction of the metal gate such as TiN, a midgap metal that allows a more threshold voltage control. The state of art of EOT of HF-base dielectrics is reduced to as low as 0.42 nm [32]. IL plays a key role in EOT scaling and carrier mobility in channel [33]. For future high-speed devices high-κ

gate dielectrics can be deposited on mobility substrate like germanium. As mentioned earlier, the hydroscopic property of GeO_2 hindered the development of Ge transistor. The development of high-κ material reopened the door to Ge transistor, since it is not required to have a GeO_2 as the gate dielectrics. Research work started on Ge channel in the early 2000s again because its hole mobility is four times as high as that of silicon and its electron mobility is twice as high as that of silicon.

Table 7.1 shows material characteristics of Ge and Si [34]. Additionally, the lattice constant of Ge is close to that of GaAs is expected to facilitate the integration of III-V n-MOSFETs (GaAs) and optical devices on Ge substrates in the future [34]. However, Ge nMOSFET has not been implemented successfully as predicted electron mobility due to large density of interface states [14]. Some researchers believe this mobility degradation is due to the degraded Ge interface and is inherent to Ge [35] and Ge could only be used for pMOSFETs. Figure 7.1 shows that metal/p-Ge has Fermi level pinning problem, which is not only a problem for high-κ/p-Ge, but also it is problematic for source/drain formation as well. Although HfO_2 is widely used in Si system, it is not a good selection for Ge system. HfO_2 is unsuitable on a Ge substrate, since gate leakage current density is larger than ZrO_2 [36]. Ge can diffuse into the HfO_2 layer that results in the increase in gate leakage current if no appropriate IL presents. Al_2O_3 is typically used as an IL to prevent Ge diffusion. Nevertheless, ZrO_2, which was screened out because of low thermal stability with Si, is a good candidate for replacement metal gate integration since the thermal budget is greatly reduced compared to gate first integration. ZrO_2/Ge gate stacks can sustain and improve its electrical characteristics after annealing [36].

To replace Si by Ge in future CMOS technology, researchers must find best passivation method for Ge to reduce density of interface states. Available surface passivation methods include: epi-Si passivation, surface oxidation and/or nitridation, and S-passivation. Among these, plasma-based surface passivation followed by plasma-enhanced ALD for high-κ layer showed the highest gate stack quality [37].

Table 7.1 Material characteristics of alternative channel materials for Ge and Si [34]

Material characteristics of alternative channel materials	**Ge**	**Si**
Bandgap, E_g (eV)	0.66	1.12
Electron affinity, χ (eV)	4.05	4
Hole mobility, μ_h ($cm^2V^{-1}s^{-1}$)	1,900	450
Electron mobility, μ_e ($cm^2V^{-1}s^{-1}$)	3,900	1,500
Effective density of states in valence band, N_V (cm^{-3})	6.0×10^{18}	1.04×10^{19}
Effective density of states in conduction band, N_C (cm^{-3})	1.04×10^{19}	2.8×10^{19}
Lattice constant, a (nm)	0.565	0.543
Dielectric constant, K	16	11.9
Melting point, T_m (°C)	937	1,412

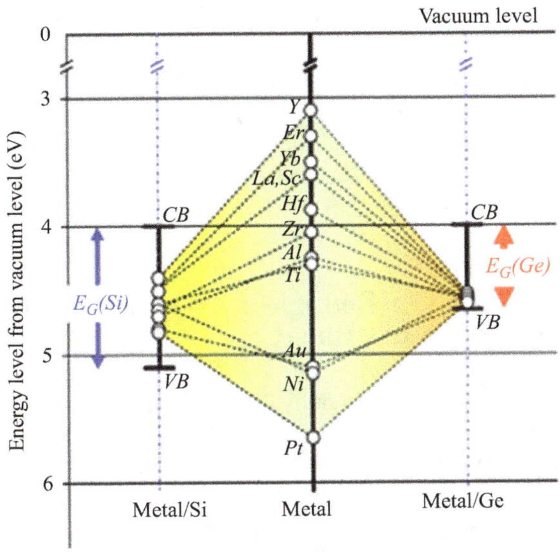

Figure 7.1 Schottky barrier heights obtained experimentally for various metals with different vacuum work-functions. In case of Ge, the Fermi level is strongly pinned near the valence band edge [38]

To achieve ultimate scaling, another solution is to deposit dielectrics directly on Ge without incorporation of an IL, which typically has a much lower capacitance value than expected. This is because, generally, an IL either intentionally or unintentionally formed during the high-κ dielectrics deposition process or during post-deposition annealing process [38]. Nevertheless, Ming Lin *et al.* reported a 0.39 nm EOT with ultrathin GeON formed by remote plasma treatment, with a D_{it} of 4×10^{12} cm^{-2}eV^{-1} [39]. Recently, GeO_2 passivation layer has been reconsidered as promising passivation layer due to its low D_{it} (6×10^{10} cm^{-2}eV^{-1}) [40]. As mentioned earlier, GeO_2 is undesirable because it is hygroscopic and water-soluble. GeO_2 is thermally unstable and converts to volatile GeO_x at approximately 430 °C. However, it is found that it is not necessary form GeO_2, instead, GeO_x can give a promising D_{it} value and it is more stable [41,42] as it can be controlled by post-deposition processes.

Since GeO is volatile [16], it is necessary to have a layer that can effectively prevent GeO volatilization and GeO growth via retarding the inter diffusion of Ge and O atoms. If GeO_x is not passivated by other elements such as nitrogen, sulfur, silicon, the interface state density increases significantly. Al_2O_3 can be considered as the first oxide layer if EOT is not aggressively scaled below 0.7 nm as it has lower intrinsic oxygen permeability [41]. Besides, Al_2O_3 has a larger bandgap and conduction band offset, which helps effectively block electron injection from Ge substrate [43]. Moreover, Houssa *et al.* used the first principle to calculate interface property and found that Al–O–Ge bond tends to give a surface states free bandgap

[44]. In this chapter, a bilayer structure 1 nm Al_2O_3/3.5 nm ZrO_2 is discussed for Ge gate stack.

In summary, although the literature reported a low D_{it} as 10^{11} cm^{-2}eV^{-1}, Ge nMOSFET still suffers low electron mobility and it can achieve approximately a maximum 1.5 times that of silicon. The underlying mechanism of low electron mobility is studied in progress and needs to be addressed before it can be deployed in future CMOS technology.

7.3 Characterization of Ge/high-κ devices with dry and wet interface treatment

It is well known that the motivation to study the Ge devices is due to their high hole and electron mobility comparable to that of silicon besides the process and integration compatibility compared to that of III-V materials. Ge devices with different high-κ gate dielectrics such as HfO_2, ZrO_2, Al_2O_3, GeON have been demonstrated. Mobility above 300 cm^{2}V^{-1}s^{-1} has been reported for Ge PMOS [45]. On the other hand, Ge NMOS has exhibited poor drive current and low mobility [46]. One possible reason is due to the quality of the gate oxide/substrate interface. Large interface defect density may pin the Fermi level, and CV and conductance–voltage (GV) data no longer behave like traditional MOS capacitor. In addition, implementing the process in 300 mm wafers for manufacturing adds to the complexity. Even though Ge/high-κ interface has been extensively studied, the high leakage current associated with these gate stacks continues to introduce frequency dispersion and hysteresis in CV and GV characteristics. These dispersions severely limit the understanding of the interface and accurate estimation of interface state density D_{it} and EOT. Low temperature measurement is, therefore, required to further enhance the measurement accuracy.

Conventionally, device information such as EOT, V_{FB}, bulk doping, and surface potential as a function of gate voltage can be obtained by CV measurement at a specific frequency, which is usually 100 kHz. However, this is carried out under a few important assumptions: (i) substrate resistance is zero; (ii) none or minimal interface defects exist; (iii) minority carrier generation rate cannot follow this specific frequency; and (iv) leakage current is small enough not to disturb the CV measurements at this specific frequency. Even Si substrate is not able to satisfy all these assumptions. Due to the scaling of technology, substrate resistance became larger, varied from a few ohms to kilo ohms, which caused frequency dispersion in the accumulation region. Interface defects cause frequency dispersion in the depletion region (in most cases). Thirdly, due to the advanced process technology (there is less bulk defects existing in Si substrate) and large bandgap (1.12 eV), minority carrier cannot follow 1 Hz frequency. Therefore, there is no frequency response in the inversion region. The DC leakage current can further disturb the CV measurement; hence, it is necessary for it to be effectively monitored before further calculation. As we had discussed, reciprocally, the dispersion of capacitance and conductance at wide range of frequency (100 Hz–1 MHz) can be utilized to calculate the R_s, the substrate resistance, and D_{it}, the interface defects density, as long as the last two assumptions are appropriate. This process is typically followed to extract the information from silicon devices.

Ge devices, however, are not yet ready for this simple evaluation process. There are two main reasons: (i) large interface density (larger than 10^{12} cm^{-2}/eV) [47] that causes Fermi level more or less pined and (ii) small bandgap (0.67 eV) that allows fast minority carrier generation and large leakage current. The first flaw of high-κ/Ge devices is a paradox when we study the interface state density of these devices. In other words, larger interface defect density changes device attributes. Both low density and high density of defects are difficult to be measured correctly. It is, therefore, important to make sure that there exist three distinctive regions (accumulation, depletion, and inversion) that can be observed in the CV measurement. The second flaw is quite significant for Ge devices, which is not as relevant in silicon. The data, as shown later, is easy to obtain in inversion region for Ge devices at moderate frequencies due to the small bandgap and low bulk defects in Ge substrate. It is, therefore, imperative to study the Ge devices at lower temperatures such that the second flaw can be more or less addressed. More importantly, a temperature scan means one additional dimension to the measurement that definitely benefits the data analysis.

In this section, we describe the measured CV and GV characteristics of Ge/ALD 1 nm-Al_2O_3/ALD 3.5 nm-ZrO_2/ALD TiN MOS capacitors on 300 mm wafers with three different interface treatments. HP4284 LCR meter was used for the measurement at ten different frequencies (1 MHz, 500 kHz, 100 kHz, 25 kHz, 10 kHz, 5 kHz, 2 kHz, 1 kHz, 500 Hz, and 100 Hz) and at five different temperatures (100 K, 150 K, 200 K, 250 K, and 300 K). The interface treatments are (i) simple chemical oxidation (Chemox); (ii) chemical oxide removal (COR) followed by 1 nm oxide by slot-plane-antenna (SPA) plasma (COR & SPAOx); and (iii) COR followed by vapor O_3 treatment (COR & O_3). The Chemox is a wet process. The other two types of samples are dry processed. The EOT, V_{FB}, bulk doping, surface potential, and interface quality are calculated, and the results were discussed with reference to the processing conditions after correcting the CV and GV data. Additionally, D_{it} was estimated by conductance method, capacitance spectroscopy, and DLTS to study the interface treatment and its impact on the defects.

In CV plots, a frequency dispersion in the negative region (accumulation region) is observed due to the substrate resistance, R_s. After a simple R_s correction the dispersion is reduced but is still present unlike silicon devices [48]. This dispersion is not acceptable for further analysis since it did not give a robust value in the accumulation region. Furthermore, the dispersion is mainly due to the interface states and large interface defects density causes Fermi level pinning before entering the accumulation region. The interface can, therefore, respond to the frequency below or equal to 10 kHz and gives a pseudo accumulation region. But can we use the value here for calculating C_{ox} or EOT? The answer is yes, since there is no major difference between interface capacitance (C_{it}) or bulk capacitance (C_B), and both of them are added to the substrate capacitance (C_s). Moreover, the measured capacitance in the accumulation region should never be above oxide capacitance (C_{ox}) unless affected by the DC leakage current. The low-frequency data, therefore, can be used to calculate EOT if we consider low-frequency capacitance in accumulation region ($C_{lf,acc}$), which is approximately equal to C_{ox}. Alternatively, Maserjian method [49] can be used to estimate an accurate C_{ox} after using (7.1). Nevertheless, the results should be comparable with $C_{lf,acc}$. EOT results of three different samples are listed in Table 7.2:

$$\frac{1}{C} = \frac{1}{C_{ox}} + \frac{2kT}{qC_{ox}} \frac{1}{V_G - V_{FB}} \tag{7.1}$$

As shown in (7.1), it is necessary to know V_{FB} before C_{ox} can be obtained. A standard way, therefore, to estimate V_{FB} is by using (7.2) below, where L_D is Debye length, ε_s is electrical permittivity of Ge substrate, V_t is the thermal energy (26 mV), N_A is the bulk doping concentration, and C_{ox} is the oxide capacitance per unit area. However, it is necessary to know the value of N_A before calculating L_D. Moreover, error in the estimation of N_A will cause uncertainty in C_{FB} result. This method is not robust to estimate V_{FB}. Therefore, we have used the method reported by Hillard *et al.* [50] to obtain $V_{FB.}$ (Table 7.2):

$$L_D = \sqrt{\frac{\varepsilon_s V_t}{qN_A}} \tag{7.2a}$$

$$C_{FB} = \frac{1}{\frac{1}{C_{ox}} + \frac{L_D}{\varepsilon_s}} \tag{7.2b}$$

Bulk concentration was calculated using (7.3). It was under the assumption that there was no interference from bulk defects and the interface was in the deep depletion region (1 MHz capacitance data at low temperature 100 K). The bulk concentration values of all three samples were around 10^{16} cm^{-3}:

$$N_A(W) = \frac{2}{q\varepsilon_s A^2 \frac{d(1/C^2)}{dV}} \tag{7.3}$$

The original purpose of measuring CV at low temperature is to remove or decrease the interference of the minority carrier generation and reduces the impact of interface defects in the inversion region. It is obvious that the capacitance significantly decreases in the inversion region when temperature is lowered [48]. However, in the range from −1.5 V to 0 V, frequency and temperature dispersion is also observed for all three samples. This is not the case in III–V GaAs devices reported by other groups [51], and Ge device reported by Kuzum *et al.* [52]. The

Table 7.2 Estimated EOT values and V_{FB} information for three samples

Sample name	**EOT (nm)**	**V_{FB} (V)**
Chemox (wet)	0.93	0.45
COR + SPAOx (dry)	1.02	0.24
COR + O_3 (dry)	0.88	0.19

measurement from their devices in the accumulation region has minimum dispersion due to R_s effects. The only possible reason for dispersion is that Fermi level can be pinned at flatband voltage. This may explain why the mobility in the p substrate of Ge is lower than expected. The calculated C_{FB} of Chemox sample using (7.2) is about 0.005 pF/μm^2. At 100 K, the capacitance measured at 1 MHz shows a constant value of around 0.0043 pF/μm^2 when the gate voltage is decreasing below around 0.5 V (V_{FB} of the Chemox sample). So, this proves that our sample is pinned at flatband due to the interface states.

Another important observation is that the capacitance value cannot reach the level 0.03 pF/μm^2 at 100 K whereas at 300 K it can. There are two possible reasons: Fermi-level statistics are changing when temperature is changing, and lower temperature causes Fermi level to be steeper instead of flat. Alternatively, as Kuzum *et al.* explained [52], this change is due to the change of emission rate of defects as stated in (7.4),

$$e_p = \frac{1}{\tau_p} = \sigma_p v_p N_v exp\left(\frac{E_V - E_T}{kT}\right) = \sigma_p v_p n_i exp\left(\frac{E_i - E_T}{kT}\right) \tag{7.4}$$

where σ_p is the defect cross section, v_p is hole thermal velocity, n_i is intrinsic carrier concentration, E_V is effective density of states of hole at valence band, E_T is valence band energy level, E_i and E_T are intrinsic energy level and defects energy level, respectively. It assumes that interface states only respond to the valence band. Also, it is further assumed that keeping all the parameters same, the time constant, τ_p will increase when temperature decreases. In other words, for a specific frequency window like 100 Hz to 1 MHz, by varying temperature, interface state information in the bandgap can be obtained if Fermi level is not pinned.

However, the second explanation is not reliable if Fermi level is not moving effectively as a function of temperature when the gate voltage is varying, especially when one considers D_{it} as a function of bandgap. One could easily observe that Fermi level in the bandgap near the flat band rarely moved under different temperatures. Because of this, no horizontal shift was observed due to capacitance interference [48]. The gate voltage was at slow rate (2 s), therefore most interface states can follow the change in DC voltage. Presence of large interface states pinned the Fermi level in the device before it entered the accumulation region. The above discussion suggests that the region of bandgap that was observed remained the same under different temperatures. However, the time constant of those measured interface states is decreased when temperature is increased, therefore they can follow the AC signal after a specific temperature. It is further observed that there is temperature dispersion at a specific frequency. It is important that the D_{it} data are plotted as function of bandgap since it is necessary to relate the gate voltage to surface potential. Software CVC.2.0 (North Carolina State University) was used to generate the surface potential [53] where C_{ox}, V_{FB}, and bulk concentration were the input. D_{it} was calculated by conductance method [8] and further verified by capacitance spectroscopy method.

As discussed earlier, we have Ge/1 nm-Al_2O_3/3.5 nm-ZrO_2/TiN gate stacks (MOSCAPs) with three different interface treatments. The COR + $SPAO_x$ and COR + O_3 are dry treatments whereas Chemox is a wet-processed interface.

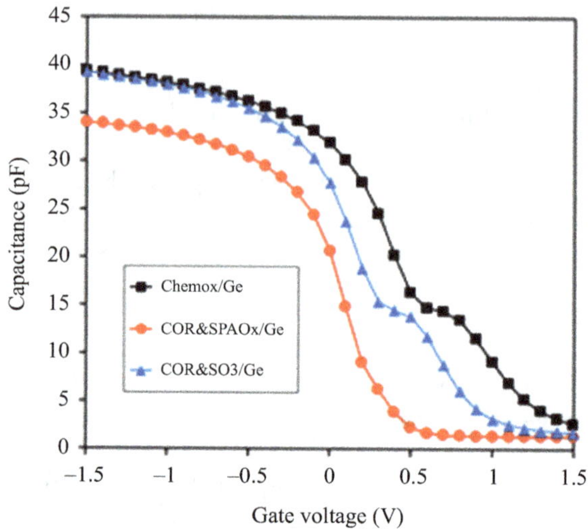

Figure 7.2 Corrected 1 MHz capacitances of three samples are plotted as a function of gate voltage at room temperature. Device area is 40 μm × 40 μm

Figure 7.2 shows corrected 1 MHz capacitances of three samples are plotted as a function of gate voltage at room temperature (device area is 40 μm × 40 μm). Considering V_{th}, the dry treated interfaces exhibit more negative V_{th} shift compared to wet treated samples since there existing positive border traps near for dry-processed samples (see DLTS section). The EOT, on the other hand, for the dry-processed interface (COR + O_3), shows a clear increase, as reported earlier (Table 7.2), COR + $SPAO_x$ shows the highest EOT because of 1 nm additional SPAO. When we plot the dry (COR + $SPAO_x$) and dry (COR + O_3) processed CV at 100 K as a function of frequency stark difference was observed between them. This indicates there exist certain bulk defects in the upper half bandgap that can follow low frequency at 100 K for dry (COR + O_3) processed sample. Even though the dry process interface exhibited excellent room temperature frequency-dependent CV, in the depletion it indicates the existence of interface states near valence band for both wet and dry-processed COR + O_3 samples. For COR + $SPAO_x$ samples, on the other hand, a reduced interface state was observed (Figure 7.2).

D_{it} was estimated by the conductance method. In Figure 7.3(a), it was clearly observed that COR & $SPAO_x$ processed interface had relatively lower interface state at room temperature than the other two different processed samples. This further suggests that SPAO samples have improved IL quality in terms of interface defects density. At low temperature (100 K), SPAO samples also show a low mid-gap D_{it}. Figure 7.3(b) compares the D_{it} values as a function of gate voltage measured by capacitance spectroscopy method and conductance method. Similar results were also observed by capacitance spectroscopy. Moreover, the difference of D_{it} estimated at 100 K and 300 K (Figure 7.3(b)) is due to their time constant variation as temperature changes. It is, therefore, imperative to understand the defect energy levels.

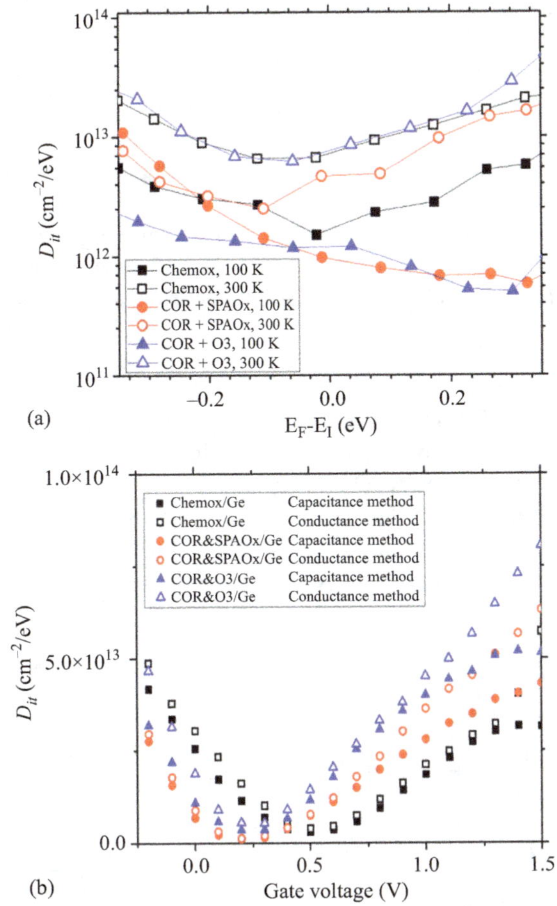

Figure 7.3 *(a) D_{it} is plotted as function of bandgap for three samples at two temperatures (100 K and 300 K), other D_{it} as a function of temperature are within this range; (b) Calculated D_{it} of three samples as a function of gate voltage by capacitance spectroscopy method and conductance method*

The low interface state density in COR + $SPAO_x$ samples indicates formation of an IL constituting GeO_x with a possible unit cell of GeO_2 layer [41]. As mentioned by Zhang *et al.*, dielectric constant decreases once IL changes from GeO_x to GeO_2 layer because of plasma oxidation. That was clearly evident in CV measurement and EOT estimation (Table 7.2) for COR + $SPAO_x$ samples. In addition, an increase in GeO_x during SPAO treatment may be possible, enhancing the EOT. In either way, interface treatment by SPAO plasma tends to passivate the interface further [41] by reducing the D_{it} (Figure 7.3) for COR + $SPAO_x$ samples at the cost of an increase in EOT. We believe the former mechanism is more responsible for D_{it} reduction. The chemical processes (both wet and dry) Chemox and COR + O_3 failed to passivate the interface with the formation of GeO_x only even though the EOT was decreased.

The interface of dry- and wet-processed Ge/High-κ MOS structures was investigated by DLTS [9] to further understand the nature of these interface defects. The MOS devices were pulsed from mid-bandgap to accumulation region, then return to mid-bandgap to obtain the majority carrier trap emission information. Figure 7.4(a)–(c) shows the DLTS data for three corresponding interface treatments. The Arrhenius plots are shown in Figure 7.4(d). Comparing the DLTS spectrum, it can be concluded that dry-processed devices have discrete border traps at the interface (H3 and H5). No such traps at the interface of wet-processed MOS devices (Chemox) were observed. For all three samples, however, interfaces such as like traps were detected (H1, H2, H4). The defects density, N_t, was estimated by (7.5),

$$N_t = \Delta V C_{ox}/q \tag{7.5}$$

where ΔV is obtained from CV plot using measured ΔC. The results suggest that it is around 10^{13} cm^{-2}/eV for all the three samples. This further confirms that the origins of interface state density, D_{it}, observed earlier for these samples (Figure 7.3) are mainly due to these traps. The time constants of these defects are in the order of

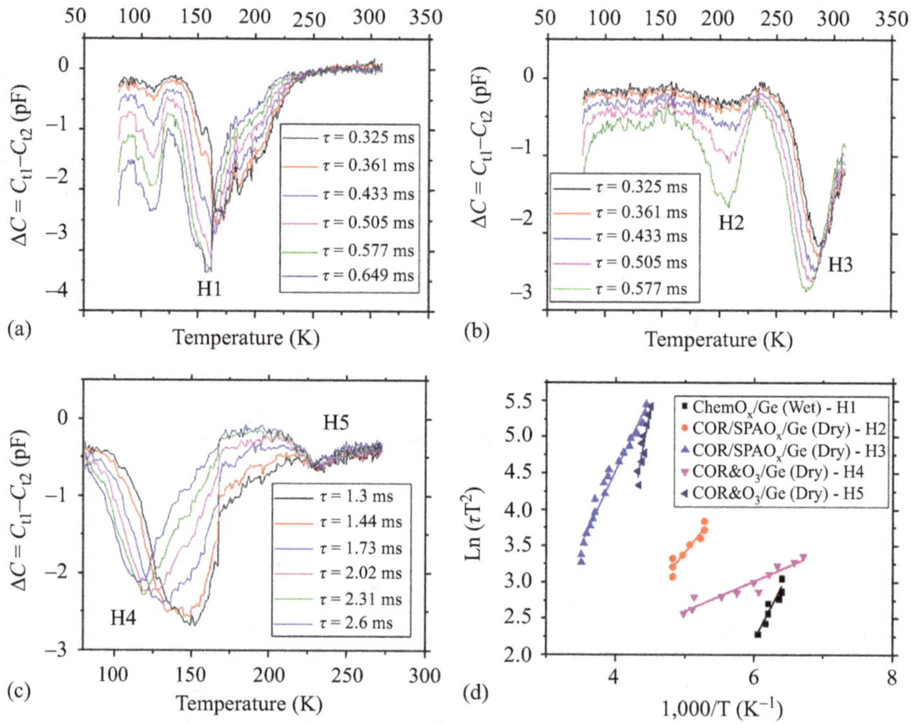

Figure 7.4 Deep-level transient spectrum (a) simple chemical oxidation (Chemox), (b) chemical oxide removal (COR) followed by 1 nm oxide by slot-plane-antenna (SPA) plasma (COR & SPAO$_x$), (c) COR followed by vapor O$_3$ treatment (COR & O$_3$), and (d) is Arrhenius plot for all three samples

Table 7.3 Arrhenius plot-fitting results

	H1	H2	H3	H4	H5
$E_T - E_V$ (eV)	0.16	0.1	0.18	0.04	0.45
σ (cm^2)	6.3×10^{-17}	1.3×10^{-19}	3.6×10^{-19}	5.8×10^{-21}	8.6×10^{-13}

milliseconds that were observed earlier. This can further explain the observed fluctuations in frequency dispersion curves at low temperatures. The relatively negative V_{th} shift in dry-processed samples compared to wet process samples can also be explained because of the presence of additional hole traps at the interface. Table 7.3 summarizes the energy levels and cross-sections [9] of these traps. The observed energy levels indicate the presence of a shallow level (H3) for COR + $SPAO_x$ sample as compared to a deep level (H5) for COR + O_3 samples. The impact of these levels on D_{it} is clearly obvious as deep levels contribute significantly to the interface state density.

When the gate leakage current densities for different types of samples were compared, the SPAO-processed interface shows lower current density [48]. The gate leakage current density increases when the EOT goes down. This is because the thinner ILs help tunneling because of the reduced tunneling barrier. The samples with SPA plasma-enhanced IL showed the lowest tunneling leakage current. It was previously reported that SPA plasma helps better oxide growth with reduced impurities [7]. In addition, the SPA plasma makes atomically flat surface and interface, which helps the reduction in leakage current density [54,55]. This further confirms that dry-processed, especially, COR + $SPAO_x$ interface is superior.

In summary, it has been demonstrated that an accurate parameter estimation method for Ge/ALD 1 nm-Al_2O_3/ALD 3.5 nm-ZrO_2/ALD TiN MOS capacitors with three different interface treatments. COR & $SPAO_x$ samples show excellent CV characteristics at room temperature. After evaluating several parameters like EOT, flatband voltage, bulk doping, and interface defects density, COR & $SPAO_x$ (dry) has better interface quality than Chemox (wet) and COR & O_3 (dry)-processed sample; however they all have larger D_{it} values in the order of 10^{13} cm^{-2}/eV that causes Fermi level more or less pinned, which is confirmed by low temperature measurements. Dry-processed sample (COR & $SPAO_x$ and COR & O_3) has more negative V_{th} shift due to the existing border trap discovered by DLTS. The levels of leakage current of three samples follow the sequence of their EOT values. Therefore, COR & $SPAO_x$ (dry) has the lowest trap-assisted tunneling effect, resulting in lowest leakage current.

7.4 Interface improvement and reliability of ZrO_2/Al_2O_3/Ge gate stack

Ge has been extensively studied to replace Si due to its higher electron and hole mobility [45,46], which can enhance metal-oxide semiconductor field effect transistor (MOSFET) speed without any physical scaling. Before the development of high-κ oxide, the hygroscopic property of GeO_2 impeded the implementation of Ge MOSFET since SiO_2 has a better stability as native oxide layer, grown on Si [37]. Nevertheless, after metal/high-κ gate stack was introduced, the formation of appropriate thickness of GeO_x or GeO_2 between high-κ and Ge substrate was reported to have low interface state density (D_{it}) [56]. Growing a few Ångström GeO_2/GeO_x intentionally is, therefore, necessary to obtain good interface quality for high-κ/Ge gate stacks [57]. The thickness of GeO_2/GeO_x layer will impact the overall EOT because of its low dielectric constant compared to high-κ layer [58]. Therefore, the tradeoff between EOT and D_{it} is an inevitable issue. Besides, interface layer treatment of Ge and controlling its thickness are also critical to further the scaling EOT below 1 nm [59]. High-κ dielectrics like HfO_2 and ZrO_2 have already been integrated into CMOS technology. It is reported that HfO_2/Ge gate stack shows a larger CV hysteresis than Al_2O_3 [8], and Al_2O_3 can block electron injection from substrate effectively by large conduction band offset related to Ge substrate [43]. Moreover, ZrO_2 showed lower leakage current than HfO_2 with similar dielectric constant [60]. Therefore, a bilayer high-κ stack (ZrO_2/Al_2O_3) was used in this study. The IL quality (interface state density, D_{it}), and EOT of these stacks were reported in a previous work [61] and it is summarized in Table 7.4. In addition, the crystalline properties of the IL need to be evaluated along with XPS analysis of IL atomic composition [62].

The reliability of high-κ layer depends on the gate leakage current density level (J_g) and charge to breakdown (Q_{BD}). The current density not only determines the performance of memory and logic circuits but also affects how fast the oxide layer degrades. After certain amount of the injection of the carriers through oxide layer, it can be irreversibly broken down. Moreover, these two parameters are somehow correlated if both of them are only dependent on the oxide layer quality. The TDDB measurement is now showing polarity dependence on thin devices since *I–V* depends on the barrier condition at the interface (gate/oxide and oxide/substrate) [63,64] and it is expected that bilayer structure will further enhance this phenomenon since different electric field across the layers and variance in dielectric quality [65]. Moreover, as EOT is scaling down to below 1 nm, IL will have more impact on the final TDDB performance [66] if a soft breakdown time is measured instead of hard breakdown. If the degradation is due to the IL, a lower current density cannot predict a longer lifetime of a dielectric layer in terms of

Table 7.4. EOT, V_{FB}, D_{it}, and IL type for three samples

Sample name	**EOT (nm)**	**V_{FB} (V)**	**D_{it} (cm^{-2}/eV) at E_i − 0.15 eV**	**IL**
Ge/SPAO/Al_2O_3/ZrO_2	0.98	0.015	2.14×10^{12}	GeO_x
Ge/Al_2O_3/SPAO/ZrO_2	1.29	−0.228	3.93×10^{11}	GeO_x
Ge/Al_2O_3/ZrO_2/SPAO	1.13	−0.362	6.87×10^{11}	GeO_2

device stability. In the first half of this section, the carrier transport mechanisms were evaluated for three different samples, namely, SPAO exposure before high-κ deposition by ALD, in between two different ALD high-κ layers, and after ALD high-κ, to understand the effect SPAO on the gate leakage current and carrier transport mechanism in dielectrics [67]. The trap distributions observed in this experiment can significantly impact the reliability of the dielectric. Therefore, samples were subjected to TDDB measurement to further understand the reliability of p-Ge/Al_2O_3/ZrO_2/TiN gate stacks under different SPAO exposure. Charge to breakdown (Q_{BD}) was estimated from the TDDB measurements which were carried out at different voltage stress conditions. Subsequently, voltage acceleration factor (AF) was extracted. TDDB and I–V characteristics of three samples were studied by using both gate electron injection (GEI) and substrate electron injection (SEI) modes. This gives an overall evaluation of the high-κ oxide and IL quality under different SPAO conditions. To evaluate the performance 1 nm Al_2O_3 and 3.5 nm ZrO_2 were deposited as gate oxide, and the samples were subjected to three SPAO annealing sequences: (i) Ge/SPAO/Al_2O_3/ZrO_2 (SPAO before high-κ), (ii) Ge/Al_2O_3/SPAO/ZrO_2 (SPAO in between two high-κ layers), and (iii) Ge/Al_2O_3/ZrO_2/SPAO (SPAO after high-κ).

Figure 7.5 shows the XRD spectra of SPAO-treated Ge MOS capacitors (TiN/ZrO_2/AlO_x/Ge) at three different stages of high-κ deposition process. XRD patterns are composed of peaks corresponds to the metal electrode (TiN) and substrate (Ge/Si) reflections [68,69]. No peaks correspond to the dielectric stack (ZrO_2/Al_2O_3) have been observed, which indicates that the dielectric stack is in amorphous phase only. No structural changes have occurred after SPAO treatment in all three cases, this reveals that the SPAO treatment is not causing any damage to the dielectric stack.

X-ray photoelectron spectroscopy was employed to examine the IL (GeO_x) formation and its oxidation states. Figure 7.6 shows the high resolution (HR) XPS

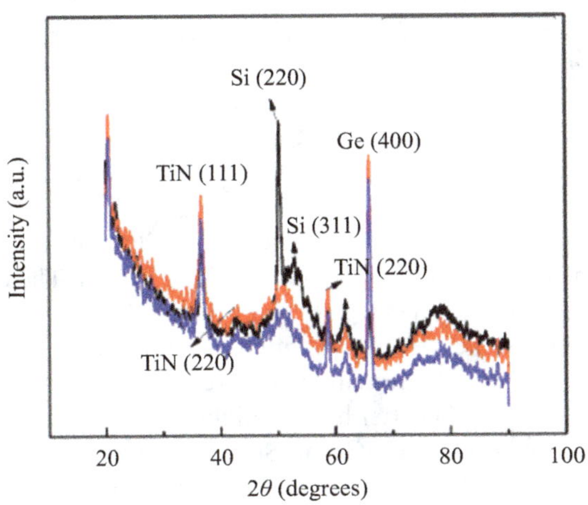

Figure 7.5 XRD spectra of all three cases

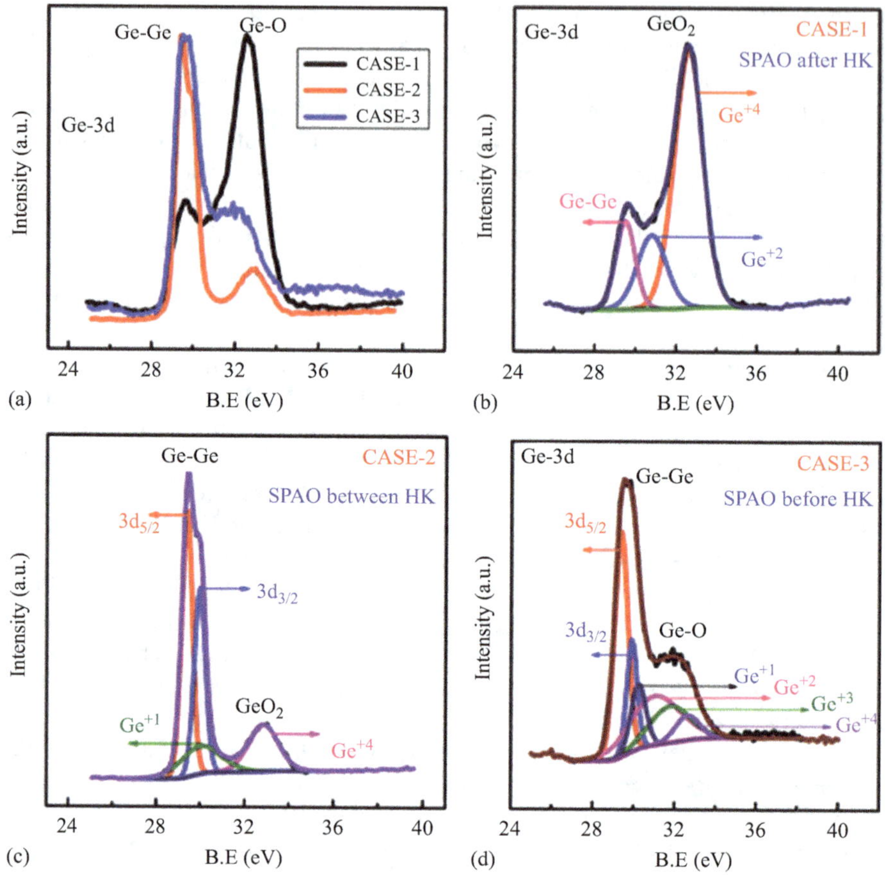

Figure 7.6 High-resolution XPS spectra of Ge-3d all three cases: (a) and deconvolution results of the Ge-3d spectra for SPAO after high-κ (b), in between high-κ layers and (c) and prior to high-κ (d)

spectra of Ge-3d in all three cases. The Ge-3d spectrum is entirely different in all three cases as shown in Figure 7.6(a). A peak is observed in Ge-3d spectra at a binding energy centered around 32.6 eV along with the Ge substrate peak (~ 29.5 eV), attributed to the formation of GeO_x in all three cases which confirms the oxidation of the high-κ stack and Ge interface [42,70,71]. The intensity of GeO_x component is more in SPAO after high-κ and is less in SPAO in between ZrO_2 and Al_2O_3 indicates the thicker IL formation in sample with SPAO after high-κ deposition and the thinner IL formation in case of SPAO in between ZrO_2 and Al_2O_3. The thicker IL in SPAO after high-κ might be due to longer duration (300 s) exposure of gate stack (ZrO_2/AlO_x/Ge) to SPAO treatment compared to other two cases, 30 s in between ZrO_2 and Al_2O_3 and 15 sec in sample with SPAO prior to high-κ deposition, which causes down diffusion of more oxygen towards the interface through the gate stack. The stability of the IL (GeO_x) depends on the oxidation state of Ge. The +4 oxidation state of Ge is more stable than the lower oxidation states, +3, +2, and +1.

The various oxidation states of GeO_x can be evaluated by the deconvolution of GeO_x peak to the corresponding to each oxidation state shown in Figure 7.6(b)–(d). The difference in binding energy of oxide (Ge–O) and substrate (Ge–Ge) peaks in SPAO after high-κ (shown in Figure 7.6(b)) is around 3.1 eV, indicates the Ge^{+4} oxidation state. The intensity of GeO_2 peak is significantly higher than the substrate peak indicates the thicker IL formation. Also, a significant component of lower oxidation state, +2 of Ge at a binding energy of ~30.8 eV is detected, indicates the suboxide formation at the interface. The difference in binding energy of oxide (Ge–O) and substrate (Ge–Ge) peaks in sample with SPAO in between ZrO_2 and Al_2O_3 (shown in Figure 7.6(c)) is around 3.2 eV, indicating a Ge^{+4} oxidation state. The intensity of GeO_2 peak is significantly lower than the substrate peak which indicates the stable thinner IL (GeO_2) formation in SPAO in between ZrO_2 and Al_2O_3 compared to samples with SPAO after high-κ. However, a very weak signal related to Ge^{+1} is detected at a binding energy of ~30 eV in SPAO in between ZrO_2 and Al_2O_3, which may influence the electrical properties of the devices. The difference in binding energy of oxide (Ge–O) and substrate (Ge–Ge) peaks in sample with SPAO prior to high-κ (shown in Figure 7.6(d)) is less than 3 eV, indicates the suboxide formation at the interface. Multi oxidation states of Ge (+1, +2, +3 and +4) are detected indicating an unstable IL formation [72,73]. This demonstrates that the direct exposure of SPAO causes damage to the Ge interface before high-κ deposition. Hence the direct exposure of SPAO to Ge is not a good choice to create superior and stable interface.

To understand the carrier transport mechanisms and/or prior to applying the physical carrier transport models, it is important to observe *I–V* characteristics at different temperatures. At negative voltage range (gate electron injection) Figure 7.7(a) and (b) shows the *I–V* dispersion at different temperatures for sample Ge/SPAO/Al_2O_3/ZrO_2 and sample Ge/Al_2O_3/SPAO/ZrO_2, respectively. On the other hand, the temperature dependence of *I–V* is greatly reduced for sample Ge/Al_2O_3/ZrO_2/SPAO (Figure 7.7 (c)). It is reported that SPAO can effectively reduce oxide traps [74]. Therefore, it is believed that reduced temperature dependence is mainly due to reduced traps density in ZrO_2 and Al_2O_3 layers. Figure 7.7(d) compares temperature dependence of gate leakage current at fix bias condition. It shows that sample Ge/Al_2O_3/ZrO_2/SPAO has the lowest leakage current, followed by sample Ge/Al_2O_3/SPAO/ZrO_2. Since *I–V* is not significantly dependent on temperature (Figure 7.7(d)) at high electric field, FN tunneling is believed to be the dominant mechanism [75] for both GEI and SEI as FN is strongly dependent on the biased voltage at high gate bias condition, which is of interest in this study. On the other hand, DT is not dependent on the applied gate bias and only dominant at low gate bias, which is beyond the scope of this study. For FN tunneling, consider the following field in the oxide layer [76] and current expressions [67]:

$$\varepsilon_{Al_2O_3} E_{Al_2O_3} = \varepsilon_{ZrO_2} E_{ZrO_2} \tag{7.6a}$$

$$E_{Al_2O_3} t_{Al_2O_3} + E_{ZrO_2} t_{ZrO_2} + V_{FB} = V_{Gate} \tag{7.6b}$$

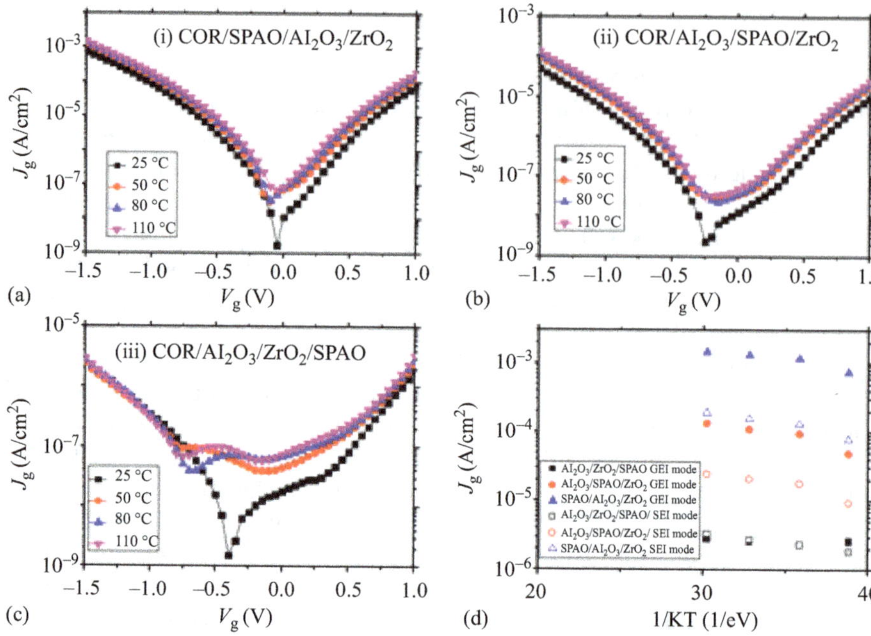

Figure 7.7 Current density (J_g) is plotted as a function of gate voltage (V_g) at four different temperatures (25 °C, 50 °C, 80 °C, 110 °C) for three different samples: (a) Ge/SPAO/Al_2O_3/ZrO_2, (b) Ge/Al_2O_3/SPAO/ZrO_2, (c) Ge/Al_2O_3/ZrO_2/SPAO, and (d) J_g as a function of V_g-V_{FB} is plotted for the above three samples at 25 °C, where V_{FB} is the flat band voltage

$$J_{FN} = \frac{q^3 m_0}{16\pi^2 \hbar m_{ox} \varnothing_b} E^2 \exp\left(- \frac{4(2m_{ox})^{1/2}}{3q\hbar} \varnothing_b^{3/2} E^{-1} \right) \tag{7.7}$$

where ε the dielectric constant, E is the electric field in oxide, and t is oxide thickness. The electric field is calculated by (7.6a) assuming the dielectric constants of ZrO_2 and Al_2O_3 as 25 and 9, respectively. The barrier heights were then extracted by fitting $\ln(J_g E^{-2})$ to $1/E$ as (2) indicates, where E is the electric field, q is the electronic charge, m_o, m_{ox}, are the electron mass in free space and in the oxide, respectively; $\hbar$ is the Planck's constant, and ϕ_b is the barrier height. The electron mass used in this calculation for ZrO_2 and Al_2O_3 are $0.5m_o$ and $0.3m_o$, respectively [77,78]. In Figure 7.8, $\ln(J_g E^{-2})$ is plotted as a function of $1/E$ and different barrier heights were obtained from the slopes in both GEI and SEI modes, respectively. Metal (TiN) to dielectric barrier height, $\phi_{b(ZrO2)}$ at high field during gate injection (barrier height of ZrO_2) is calculated as 1.3 eV. The estimated $\phi_{b(ZrO2)}$ is lower than the theoretical calculation (1.6 eV) due to the breakdown of high-κ before a sufficiently high electric field was applied. Substrate to dielectric barrier height $\phi_{b(Al2O3)}$ at high field was estimated during substrate (Ge) injection

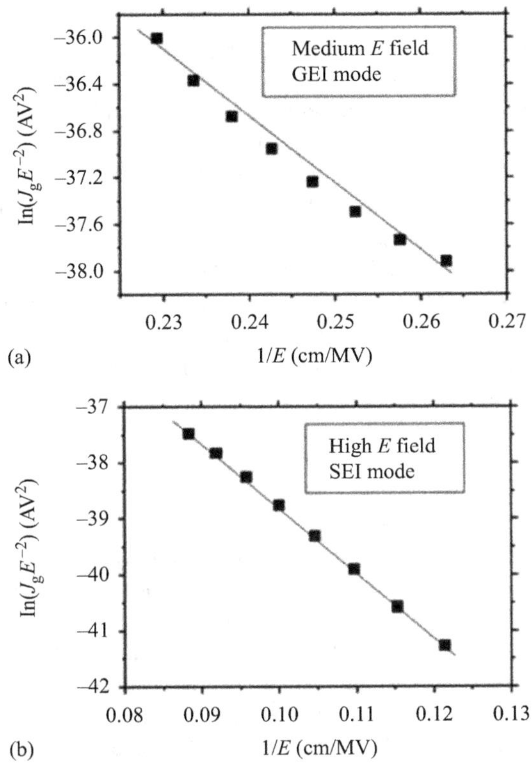

Figure 7.8 $ln(J_gE^{-2})$ is plotted as a function of 1/E as FN tunneling for sample Ge/Al_2O_3/ZrO_2/SPAO, J_g is the current density and E is the electric field: (a) gate electron injection mode (GEI) and (b) substrate electron injection mode (SEI)

(barrier height of Al_2O_3) is 2.2 eV assuming electrons tunnel through thin GeO_x. On the other hand, $\phi_{b(AlO2)}$ is very close to the expected theoretical value [43].

Note that the current–voltage characteristics in SEI mode at low field, in the range from −0.5 V to 0.5 V, low to medium *E* field range for sample Ge/Al_2O_3/ZrO_2/SPAO (Figure 7.7(c)) *I–V* is clearly a function of temperature. Therefore, Poole–Frenkel (PF) and hopping conduction (HC) mechanism is possibly the dominant mechanism in this range and fits well *I–V* characteristics. The current expression for HC is given by [67]:

$$J_{HC} = qanvexp\left(\frac{qaE}{kT} - \frac{E_a}{kT}\right) \tag{7.8}$$

In (7.8), *a* is the mean hopping distance (i.e., the mean spacing between trap sites), *n* is the electron concentration in the conduction band of the dielectric, *v* is the frequency of thermal vibration of electrons at trap sites, and E_a is the activation energy, namely, the energy level from the trap states to the bottom of conduction band.

As shown in Figure 7.9(a), HC model fits J_g-T (current temperature function) characteristic well in medium E field range. Subsequently, the slope, S, ($S = qaE - E_a$) obtained from Figure 7.9(a), can be plotted as a function of E field. In Figure 7.9(b), the fitted slope is the product of mean hopping distance (a) and electronic charge (q), and intercept of slope is the activation energy. The calculated values a and E_a are 0.3 nm and 0.16 eV, respectively, in the sample Ge/Al_2O_3/ZrO_2/SPAO.

As discussed above, sample Ge/Al_2O_3/ZrO_2/SPAO shows FN tunneling in the high E-field range since both ZrO_2 and Al_2O_3 were exposed to SPAO. On the other hand, ZrO_2 layers were not subjected to SPAO for both the samples Ge/SPAO/Al_2O_3/ZrO_2 and Ge/Al_2O_3/SPAO/ZrO_2. Therefore, both HC and PF mechanisms seem to be dominated for the entire range of measured I–V characteristics (high and low electric field range). The PF current expression is given by [67]:

$$J_{PF} = qN_C\mu E exp\left(\frac{\left(q^3/\pi\varepsilon_r\varepsilon_o E\right)^{1/2} - q\varnothing_t}{kT}\right) \tag{7.9}$$

where μ is the electronic drift mobility, N_c is the density of states in the conduction band, $q\phi_t$ is the trap energy level, and the other notations are the same as defined in (7.9). ϕ_t has the physical meaning similar to that of E_a in the HC model. The symbol difference is just to differentiate the values that are obtained using different current models. The conduction mechanism, PF is different from that of HC. In PF, electrons in trap centers are thermally excited to conductance band of oxide, and subsequently relaxed to another trap center. In case of HC, electrons transit between trap sites by trap assisted tunneling.

In Figure 7.10, both HC and PF models were applied to the sample Ge/SPAO/Al_2O_3/ZrO_2 and sample Ge/Al_2O_3/SPAO/ZrO_2. The slope value S is the fitting value from ln(J_g) versus $1/kT$ (not shown here). HC model (Figure 7.10(a)) gives same trap energy level E_{a_1} as 0.09 eV for both GEI mode (black open symbol) and

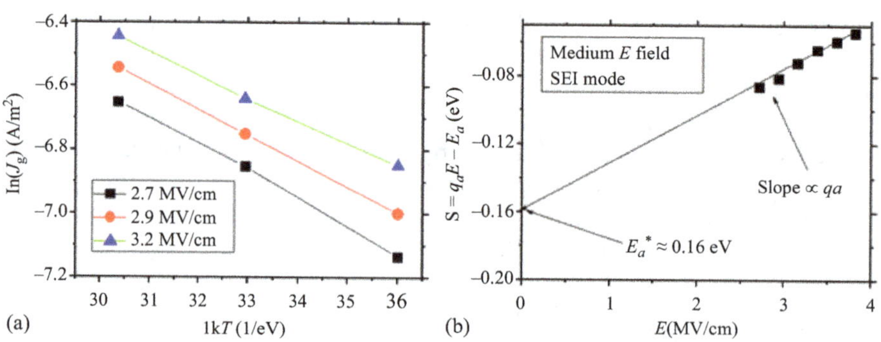

Figure 7.9 (a) Current density ln(J_g) is plotted as function of 1/kT for sample Ge/Al_2O_3/ZrO_2/SPAO, in SEI mode. (b) Slope value ($S = qaE - E_a$), which is obtained from (a) is then plotted as function electric field E (SEI mode)

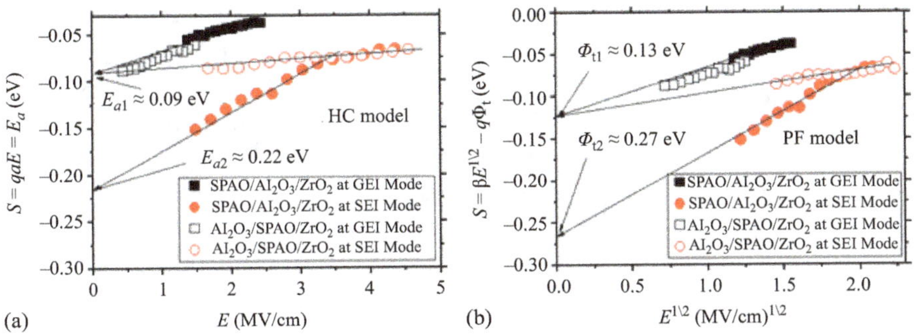

Figure 7.10 HC and Poole–Frenkel emission (PF) were used to fit the I–V characteristics for sample Ge/SPAO/Al_2O_3/ZrO_2 (solid symbol) and sample Ge/Al_2O_3/SPAO/ZrO_2 (open symbol) at both GEI mode (black symbol, squares) and SEI mode (red symbol, circles). (a) Slope value ($S = qaE - E_a$) is plotted as function electric field E. (b) Slope value ($S = \beta E^{1/2} - q\phi_t$) is plotted as function $E^{1/2}$, where $\beta = q^3/\pi\varepsilon_r\varepsilon_o$. The slope value S is the fitting value from $\ln(J_g)$ versus 1/kT

SEI mode (red open symbol) in sample Ge/Al_2O_3/SPAO/ZrO_2. On the other hand, HC model gives E_{a_1} as 0.09 eV for GEI mode (black solid symbol) and E_{a_2} as 0.22 eV for SEI mode (red solid symbol) in sample Ge/SPAO/Al_2O_3/ZrO_2. In brief, for GEI mode, electrons tunnel through ZrO_2 layer assisted by the same trap centers (E_{a_1}), since ZrO_2 layer of both samples was not subjected to SPAO. For SEI mode, electrons tunnel via two different trap centers, E_{a_1} and E_{a_2}, for sample Ge/Al_2O_3/SPAO/ZrO_2 and sample Ge/SPAO/Al_2O_3/ZrO_2, respectively. As mentioned earlier, SPAO can significantly remove the trap center E_{a_2}(0.22 eV) in GeO_x/Al_2O_3 layer for the Ge/Al_2O_3/SPAO/ZrO_2 device, when SPAO is processed after Al_2O_3 deposition. Both trap energy levels E_{a_2} and E_{a_1} were observed in Ge/SPAO/Al_2O_3/ZrO_2 sample.

When PF model (Figure 7.10(b)) was used, similar behavior was observed as that of HC model. The trap energy level, ϕ_{t1} of 0.13 eV calculated by PF model, was observed for GEI mode (black solid and open symbols) in both the samples Ge/SPAO/Al_2O_3/ZrO_2 and Ge/Al_2O_3/SPAO/ZrO_2. In SEI mode, one trap energy level ϕ_{t1} was observed with identical value of 0.13 eV. On the other hand, for SEI mode (red and black solid symbols) in Ge/SPAO/Al_2O_3/ZrO_2 a second trap energy level, ϕ_{t2} (0.27 eV), was observed in addition to ϕ_{t1} (0.13 eV). Different symbols were used as they are calculated by different models. Both HC model and PF model are similar in trap-rich oxide layers with the difference in carrier transit process between the trap sites. HC gives one more parameter, the hopping distance, *a*, of about 0.3 nm from the slope in Figure 7.10(a). Carrier transport mechanisms for GEI and SEI mode are marked in band diagram as Figure 7.11(a) and (b), respectively, and mechanisms and trap energy levels are summarized in Table 7.5. It is, therefore, clear that SPAO contributes to reduction of trap sites in

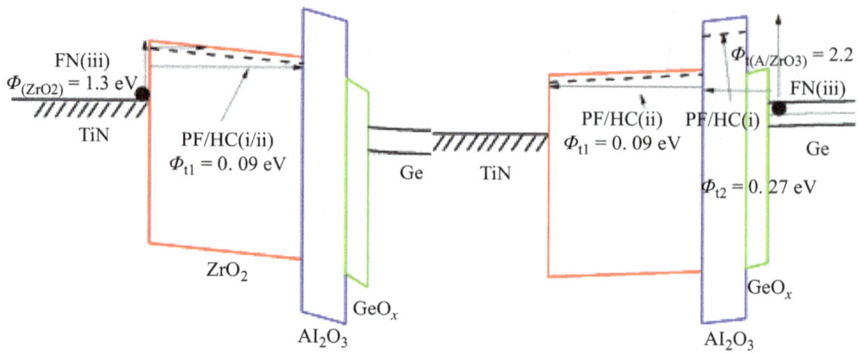

(a) Gate electron injection (GEI) (b) Substrate electron injection (SEI)

Figure 7.11 Carrier transport mechanisms are explained in band diagram for both GEI mode and SEI mode: (a) GEI mode, band diagram is simulated under $V_g = -1.5$ V and (b) SEI mode, band diagram is simulated under $V_g = 1$ V

Table 7.5 Transport mechanisms at high field region are summarized. Trap energy level ϕ_t, charge to breakdown at |1V| Q_{BD}, and AF were calculated for both GEI and SEI mode

Sample name	**Transport mechanism**‡	**Trap energy level ϕ_t (eV) GEI/SEI**	**Q_{BD} (C/μm²) at |1V| GEI/SEI**	**AF (V^{-1}) GEI/SEI**
$Ge/SPAO/Al_2O_3/ZrO_2$	PF/HC	0.13 eV/0.27 eV	$\sim 10^{-4}/10^1$	−6.1/−6.7
$Ge/Al_2O_3/SPAO/ZrO_2$	PF/HC	0.13 eV/0.13 eV	$\sim 10^2/10^1$	−10.7/−3.8
$Ge/Al_2O_3/ZrO_2/SPAO$	FN	NA	$\sim 10^2/10^{-4}$	−17.9/−1.6

‡Transport mechanisms were extracted at high electric field range.

the GeO_x/Al_2O_3 layer when SPAO is processed after Al_2O_3 deposition. This reduction process was not observed if SPAO is processed prior to Al_2O_3 deposition ($Ge/SPAO/Al_2O_3/ZrO_2$ sample). It is further observed that sample $Ge/Al_2O_3/ZrO_2/SPAO$ has the best performance for GEI mode, since most of the oxide trap centers were removed from the dielectric by SPAO [74]. The trap distribution in the dielectric and the IL will have significant impact on dielectric degradation. We have also observed difference in XPS data shown in Figure 7.6. Therefore, the performance of SEI mode is dependent on IL as will be more critical. The TDDB study was, therefore, employed to further understand the contribution of the trap distribution observed above to the degradation process, discussed in the following section.

The TDDB characteristics of all three samples were studied. The charge to breakdown, Q_{BD}, and voltage *AF* were the two main parameters used in both GEI and SEI modes to evaluate the dielectric quality and any contribution of the IL.

The oxide breakdown in this work is defined as the abrupt change of sampling gate current (I_g). Although secondary abrupt changes of I_g exist (not shown here) that were not considered. Q_{BD} was used instead of time to breakdown (T_{BD}) to further understand the trap distribution observed in the *I–V* characteristics. Weibull distribution was used to explain TDDB statistic shown in (7.10) [79], where β and η are shape parameters and they are constant:

$$F(Q_{BD}) = 1 - \exp\left(-\frac{Q_{BD}{}^{\beta}}{\eta}\right) \tag{7.10a}$$

$$W \equiv \ln(-\ln(1-F)) = \beta\ln(Q_{BD}) - \ln(\eta) \tag{7.10b}$$

$$AF = \frac{\partial \log(Q_{BD})}{\partial V} \tag{7.10c}$$

β in (7.10a) is about 0.7 for all samples and it is stress voltage-independent (not shown here). But it depends on the oxide thickness and trap sphere size (a_o) theoretically as (7.11a) indicates [80], where t_{INT} is thickness of IL [81], t_{ox} is the dielectric thickness and α is a parameter describing the correlation between t_{ox} and β:

$$\beta = \alpha\frac{t_{ox}}{a_o} \tag{7.11a}$$

Figure 7.12 shows the values of all samples for both GEI and SEI mode. The same value of 0.7 for all samples indicates that a_o is similar in all samples.

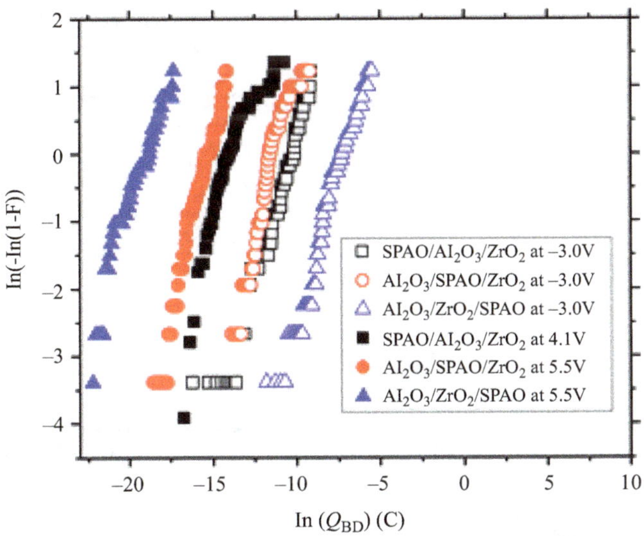

Figure 7.12 β values are around 0.7 and similar for all samples and stress conditions, which were obtained by Weibull plot

Experimentally, the observed value is linearly related to oxide thickness (7.11b), where γ is the coefficient and t_{INT} is thickness of IL [81]:

$$\beta = \gamma(t_{ox} + t_{INT}) \tag{7.11b}$$

Experimentally, it is found that the value of β tends to decrease as oxide thickness is reduced [82]. It was theoretically studied by Nigam *et al.* for thin high-κ/SiO_2 bilayer oxide and the decrease is caused by high defects generation rate in high-κ layer [83]. This low β value measured in this work also indicated that Q_{BD} has a broad statistical distribution (three orders). The final Q_{BD} is selected at 90% of its value for devices to breakdown.

Figure 7.13(a) shows that Q_{BD} values were obtained at four different gate voltages at both GEI mode (solid symbol) and SEI mode (open symbol) for all the samples. Figure 7.13(b) provides a direct comparison by plotting T_{BD} for the same samples.

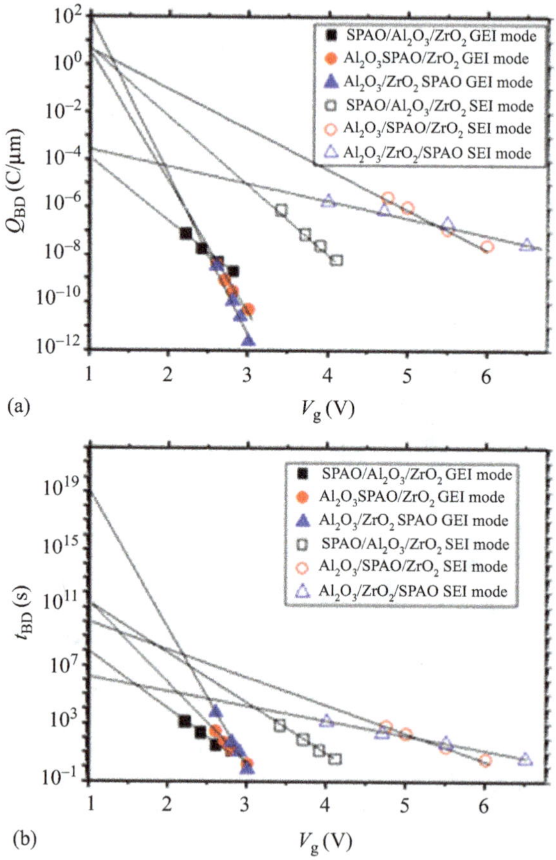

Figure 7.13 (a) Charge to breakdown value (Q_{BD}) and (b) Time to breakdown value (t_{BD}) are plotted as function of gate voltage (V_g) for both GEI mode (solid symbol) and SEI mode (open symbol)

The stress voltages were selected based on voltage ramping breakdown measurement in previous work [61]. Q_{BD} and *AF* parameters are summarized in Table 7.5, where *AF* values were obtained based on (7.10 (c)). At the GEI mode (Figure 7.11(a)), the amount of tunneling electrons is dependent on the quality of ZrO_2, which is the first layer that electron has to tunnel through. The *AF* value of −6.1 is the largest for sample Ge/SPAO/Al_2O_3/ZrO_2, followed by an *AF* value of −10.7 for sample Ge/Al_2O_3/SPAO/ZrO_2, and sample Ge/Al_2O_3/ZrO_2/SPAO has the lowest *AF* value of −17.9. Also, it was observed that Ge/Al_2O_3/ZrO_2/SPAO has the largest Q_{BD}. Therefore, it can be concluded that the SPAO can affect the TDDB characteristics GEI mode by significantly removing the traps from the dielectric layers [74] ZrO_2 and Al_2O_3. The dielectric quality was enhanced by SPAO in terms of *AF* and Q_{BD}. If traps are removed from only the Al_2O_3 layer the *AF* value increased and Q_{BD} was decreased:

$$Q_{BD} = Q_o \exp(AF \cdot V) \tag{7.12}$$

The same conclusion cannot be applied to the SEI mode, since TDDB degradation is also affected by the degradation of IL (GeO_2 or GeO_x). XPS measurement from earlier work on these devices concluded that sample Ge/Al_2O_3/ZrO_2/SPAO formed GeO_2 IL, while the other two samples formed GeO_x IL [61]. Some studies reported that ALD process will decompose GeO_2 into GeO_x [84] since GeO_2 is not thermally stable [85]. This explained why only sample Ge/Al_2O_3/ZrO_2/SPAO has GeO_2 as IL since subsequent ALD process decomposed the GeO_2 to GeO_x for other two samples. As open symbols show in Figure 7.13, sample Ge/Al_2O_3/ZrO_2/SPAO has the largest *AF* and lowest Q_{BD} among three samples due to the formation of GeO_2. On the other hand, samples Ge/SPAO/Al_2O_3/ZrO_2 and Ge/Al_2O_3/SPAO/ZrO_2 show better Q_{BD}. The formation of unstable fragmented IL causes an increase in *AF* values (Table 7.5), and only sample Ge/SPAO/Al_2O_3/ZrO_2 has similar *AF* values at both GEI and SEI mode, which indicates a similar breakdown process. Note that GeO_2 shows the worst TDDB result among three different SPAO treatments. This suggests that GeO_2 has the worst resistance to stress in terms of device stability compared to GeO_x. Electrons transit through thinner GeO_x IL, and TDDB measured the quality of Al_2O_3 rather than IL. After taking into account of D_{it} and EOT (Table 7.5), it can be concluded that formation of GeO_x or GeO_2 is helpful for improvement of D_{it} values at the cost of increased EOT values and formation of GeO_x is better for device reliability rather than GeO_2 because of rapid degradation of GeO_2 can cause gate stack instability, although sample Ge/Al_2O_3/ZrO_2/SPAO has the best performance of *I–V*.

Gate leakage current (I_g) is plotted as function sampling time for all samples.

Figure 7.14 shows that typical stress-induced leakage current (SILC) in TDDB measurement, which was not observed in GEI mode. This SILC is a process of increasing gate tunneling current via trap-assisted-tunneling. As oxide is degraded, traps are created, and enhanced trap-assisted-tunneling increases the leakage current. Another phenomenon, observed at SEI mode, was that leakage current was decreased initially before SILC process for Ge/Al_2O_3/ZrO_2/SPAO. This is possibly due to GeO_2 instability, and initially GeO_2 transformed to GeO_x, which has larger conduction band offset to block electron from substrate [86,87]. An initial decrease

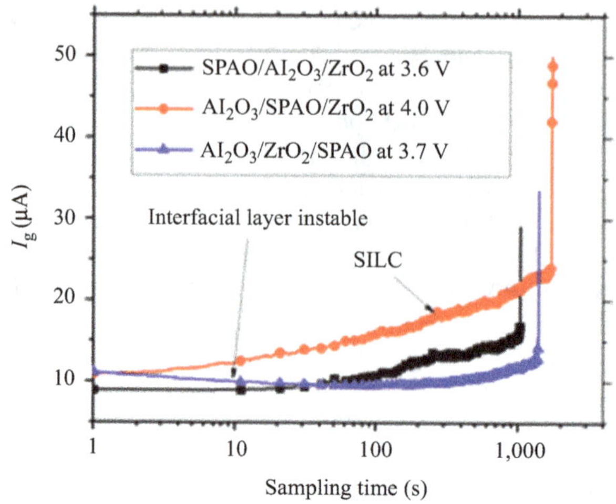

Figure 7.14 Gate leakage current (I_g) is plotted as function sampling time for all the samples

in I_G could also simply be due to electron trapping in the existing traps, followed by hole trapping and subsequently increasing SILC by trap-assisted-tunneling. It was also observed that there exists secondary breakdown in sample Ge/Al_2O_3/ZrO_2/SPAO, which is another evidence of IL degradation (not shown here). Therefore, TDDB measurements at SEI mode were representing the quality of IL rather than overall oxide breakdown (Figure 7.13).

In summary, the TiN/ZrO_2/Al_2O_3/p-Ge gate stacks were studied for their temperature-dependent carrier transport mechanisms and reliability in terms of oxide breakdown. Different SPAO annealing conditions reveal that although SPAO can effectively remove the traps in high-κ dielectrics and subsequently reduce the leakage current, the formation of GeO_2/GeO_x inevitably impacts the reliability. Trap energy levels in ZrO_2 and Al_2O_3 are found to be 0.13 eV and 0.27 eV respectively by fitting Poole-Frenkel emission model. TDDB characteristics suggest that GeO_2 can be degraded faster than GeO_x since in GeO_x electron can transit through IL even though both of the layers can give similar D_{it} values of around 5×10^{11} cm^{-2}/eV. If the ALD process decomposes GeO_2 into a stable GeO_x, the overall performance of the gate stack is more stable as shown in case of sample Ge/Al_2O_3/SPAO/ZrO_2.

7.5 Enhancement of dielectric constant with HfZrO

This section investigates the dielectric quality and interface properties of TiN/Hf1-xZrxO2/Al_2O_3/Ge gate stacks with six different Zr content (0%, 25%, 33%, 50%, 75%, and 100%). The dielectrics were subjected to SPAO after the ALD deposition process prior to metal deposition.

Figure 7.15 shows the impact of Zr incorporation in HfO_2 on EOT, flat-band voltage (Figure 7.15(a)), and gate leakage current density (Figure 7.15(b)). It is observed that with the addition of Zr in HfO_2, the EOT goes down for Zr percentage ranged from 0% to 75%, while 100% Zr (ZrO_2) showed a dramatic increase in EOT (Figure 7.15(a)). The flat-band voltage shift showed an inverse relationship with the EOT downscaling, i.e., when EOT decreases with Zr incorporation, the flat-band voltage increases (Figure 7.15(a)) with a corresponding shift of CV characteristic towards positive gate voltage [88]. The ZrO_2 showed the lowest flat-band voltage shift with the largest EOT among the samples studied in this work (Figure 7.15(a)). It was found previously for ALD $Hf_{1-x}Zr_xO_2$ on Si substrate that the addition of Zr in HfO_2 enhances EOT downscaling when the dielectrics are subjected to SPA plasma exposure [89]. We have observed enhancement in EOT for ALD $Hf_{1-x}Zr_xO_2$ Ge substrate by increasing Zr percentage up to 75% and subsequently subjecting the dielectrics to SPAO oxidation (Figure 7.15(a)). However, for ZrO_2 with no Hf present, the formation of GeO_2 IL between the Ge and Al_2O_3 due to SPAO exposure might be responsible for the increase in EOT for these dielectrics [61]. It is known that ALD Hf-based high-κ dielectrics are oxygen deficient contain many oxygen vacancies. Therefore, when ALD $Hf_{1-x}Zr_xO_2$ is subjected to SPAO oxidation, the oxygens from SPAO source mostly fill the oxygen vacancies and the contribution to GeO_2 growth at the interface is less. When the percentage of Zr is increased in $Hf_{1-x}Zr_xO_2$, the oxygen vacancy defect type modifies from doubly charged V^{2+} to single charge V^{+} type [74].

This contributes to the flat-band voltage shift towards positive direction with more Zr added to $Hf_{1-x}Zr_xO_2$ (Figure 7.15(a)). In contrast, for ZrO_2 with 100% Zr, there is a significant reduction in oxide capacitance C_{ox}, which can be translated to a negative flat-band voltage shift [88]. From Figure 7.15 (b), dielectrics with 0% to 75% Zr content showed almost identical gate leakage current density, while ZrO_2 with 100% Zr showed more than one order less gate leakage current density.

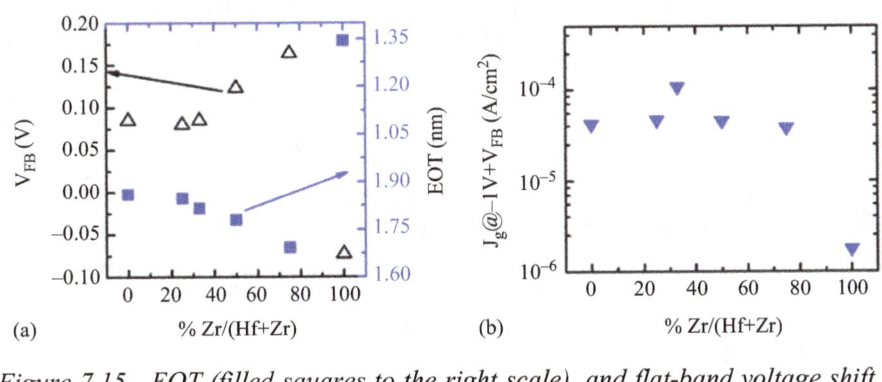

Figure 7.15 EOT (filled squares to the right scale), and flat-band voltage shift (open triangles to the left scale) (a), gate leakage current density at $-1\ V + V_{FB}$ (b) as a function of zirconium percentage for TiN/ $Hf_{1-x}Zr_xO_2/Al_2O_3$ gate stacks on germanium substrate

This can be attributed to the reduced direct tunneling because of thicker dielectric barrier. This observation further supports the observed CV characteristics [88] and corresponding EOT shown in Figure 7.15(a).

7.6 Dielectric stacks for next-generation memory devices

Non-volatile resistive random-access memory (RRAM) devices are currently being investigated as a low power, high density- and high-speed alternative [90–93]. A transition metal oxide dielectric layer is typically used as an insulating layer in a metal-insulator-metal (MIM) structure. The switching is controlled by the formation and dissolution of a conductive filament within the insulating dielectric layer that switches the device between low resistance, R_{on}, and high resistance, R_{off}, state. Initially, a forming voltage is required to create the conducting filament from the pristine insulating state. The conducting filament is being partially dissolved or repaired during post-forming switching process rather than entirely dissolving or rebuilding. In addition, during the switching process, a compliance current is set to avoid a thermodynamic (hard) breakdown of the dielectric. To reduce the large variability of switching parameters and high operation current in these RRAM devices bilayer dielectric structures are used. This second thin lower dielectric constant layer is used to vary the electric field across the switching layer [94].

In this section, several different bilayer structures using different dielectrics, electrode materials, and processing conditions are explored. While the bottom electrode (BE) and a thin lower dielectric constant of 1 nm Al_2O_3 were common in most of the devices, the second dielectric layer, the switching layer, composition was varied. By comparing the forming voltage, R_{off}/R_{on} values and both set and reset power it was observed that $HfAlO_2$ as the switching layer has the superior characteristics.

Si wafers of 12-in size were used to fabricate the metal-insulator-metal (MIM) devices. Figure 7.16(a) shows the device configurations. Two groups of devices with various switching layers were prepared for comparison. In each device, the BE has a 10 nm Ti/50 nm TiN which was followed by a low dielectric constant material of 1 nm Al_2O_3. The second dielectric layers, the switching layers in Group-I were 7-nm $HfAlO_2$ (*R1*), HfO_2 (*R3*) or $HfZrO_2$ (*R4*). This was followed by the top electrode (TE) constituting of 8 nm Ti/6 nm ALD TiN+50 nm PVD TiN. In Group-II, while the switching layer was $HfZrO_2$, the top electrode was varied to 2 nm ALD TiN/8 nm Ti/6 nm ALD TiN + 50 nm PVD TiN (*R5*) by adding a 2 nm ALD TiN layer prior to 8 nm Ti. Some devices with $HfZrO_2$ (*R2*), where the dielectric layers were subjected to a post-deposition anneal (PDA) at 700 °C for 60 s. In all cases, to enhance the switching characteristics on the top electrode 8 nm Ti was used as the cap layer material except in the case of *R5* where 2 nm of ALD TiN was deposited prior to 8 nm of Ti. The use of thin dielectric layer of 1 nm Al_2O_3 also suppresses the sneak-path problem [92] while at low resistance state (R_{on}).

When the forming voltages of the Group-I samples are compared, it was found that Al_2O_3/HfO_2 (R3) stacks have the largest forming voltage whereas much lower forming voltages were observed for both $Al_2O_3/HfAlO_2$ (*R1*) and $Al_2O_3/HfZrO_2$

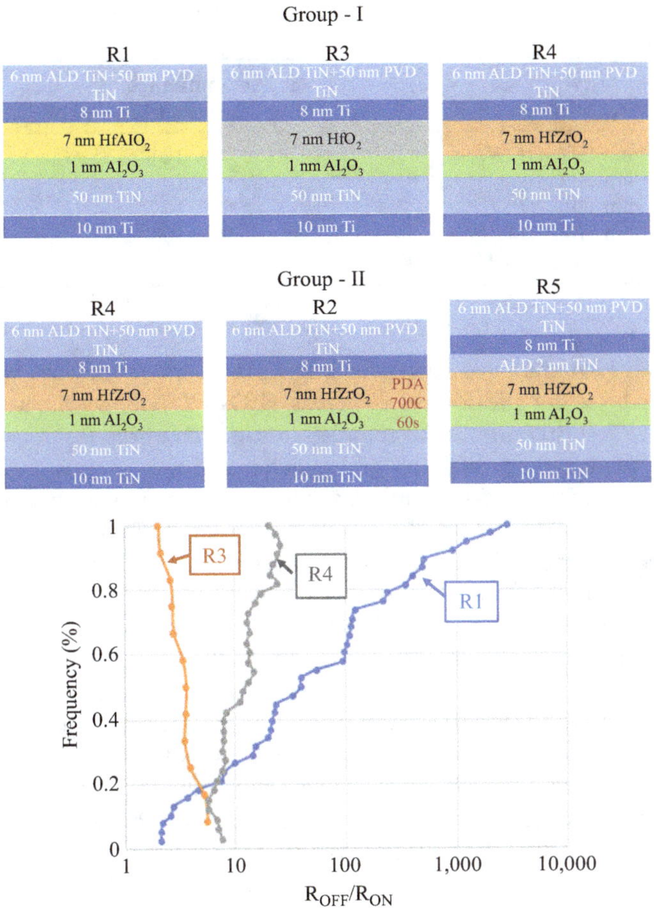

Figure 7.16 Device configurations showing the switching layer variation and process condition variations (a); Cumulative distribution of R_{off}/R_{on} ratios of all the Group-I devices (b)

(R4) for identical compliance current. It is known that in stoichiometric HfO_2 the forming voltage is usually rather high (5). The initial filament formation starts with a trigger location with either a defect site, grain boundary or oxygen vacancy location. On the other hand, $HfAlO_2$ and $HfZrO_2$ show more suitability because of the lower forming voltage. But when samples with $HfZrO_2$ were subjected to a post-deposition annealing (R2) the forming voltage increases. Whereas modifying the top electrode (TE) metal (R5) has a little effect on the forming voltage values. Table 7.6 shows the different devices, compliance currents and forming voltages.

Furthermore, by comparing the different parameters of the RRAM devices of Group-I, where dielectric variation of the switching layer is listed, it was observed

Table 7.6 Devices with compliance currents and forming voltages

Sample name	Compliance current (A)	Forming voltage (V)
1 nm Al_2O_3/7 nm $HfAlO_2$ (*R1*)	5.0×10^{-6}	2.2
1 nm Al_2O_3/7 nm $HfZrO_2$ + PDA (*R2*)	1.0×10^{-6}	4.8
1 nm Al_2O_3/7 nm HfO_2 (*R3*)	5.0×10^{-6}	12.3
1 nm Al_2O_3/7 nm $HfZrO_2$ (*R4*)	5.0×10^{-6}	1.4
1 nm Al_2O_3/7 nm $HfZrO_2$ +TE var (*R5*)	2.0×10^{-5}	1.75

that the average set and reset power requirements for the *R1* devices with $HfAlO_2$ as switching layer are the lowest. Whereas HfO_2 (*R3*) shows the worst performance. Figure 7.16(b) outlines the cumulative variation of the R_{off}/R_{on} ratios of all the Group-I devices. For reliability and endurance R_{off}/R_{on} ratios are significant. Comparing the variation of switching layer it was observed that the device 10 nm Ti/50 nm TiN/1 nm Al_2O_3/7 nm $HfAlO_2$/8 nm Ti/6 nm ALDTiN+50 nm PVD TiN (*R1*) provided the superior average R_{off}/R_{on} values and both set and reset power compared to HfO_2 (*R3*), and $HfZrO_2$ (*R4*) devices. When the $HfZrO_2$ (*R4*) devices were compared with the identical device with PDA (*R2*) R_{on} increased and R_{off} decreased for the PDA device, reducing the R_{off}/R_{on} value. While the reset power increases the set power for PDA devices moderately decreased. The impact of top electrode configuration shows that with a variation of cap layer to TiN instead of Ti decreases set power but increases reset power. When all the RRAM devices for the set power and reset power were compared it was observed that devices *R3* and *R2* behaved irregularly for set and reset powers respectively. While reset power is the lowest for the *R1* devices with $HfAlO_2$ as switching layer, the set power decreased for $HfZrO_2$ when the top electrode configuration was modified. RRAM operation and power requirement, therefore, depends on the switching layer and electrode configuration.

In this section, several dielectric stacks for enhancing RRAM operation and power requirements for a low-power operation were discussed. It was observed that that $HfAlO_2$ as the switching layer has superior characteristics while $HfZrO_2$ showed comparable characteristics. When $HfZrO_2$ as the switching layer was subjected to a PDA, the characteristics degraded because of reduced trigger points. With the top electrode metal configuration, the characteristics improved.

7.7 Summary

Since high mobility substrates such as Ge have attracted increasing attention to enhance the devices, performance for the next-generation CMOS devices, the chapter investigated the high-κ dielectrics on germanium substrates for EOT scaling and interface performance. In addition, we have looked at some of the applications of high-κ dielectrics for next-generation resistive random-access memory devices. In this study, several MOS structures, prepared by advanced ALD process and with pre and post treatment by plasma generated by slot plane antenna (SPA) were investigated. The chapter reviewed the oxide/substrate interface quality and the dielectric quality of metal-oxide semiconductor (MOS) gate stack structures as they are critical to future CMOS technology. Different electrical characterization methods such as CV, GV, DLTS, IV and reliability issues like TDDB were used throughout this chapter. Even though DLTS measurements were more difficult to implement it complements the conductance and low-frequency CV methods for device characterization. CV and GV are not able to provide true energy levels of the trap/defects and it is usually fulfilled by simulation software like CVC 7.0. The detection of trap energy level is usually limited in the depletion region and can be extended to inversion region if substrate is not doped (no minority carrier response). To extract the energy level, direct DLTS measurements are required.

The impact of different interface treatments on Ge/high-κ gate stacks was outlined. Simple interfacial treatment leads to higher D_{it} values in the order of 10^{13} cm^{-2}/eV that causes Fermi level more or less pinned. This was confirmed by low temperature measurements. More negative V_{th} shift due to existing border trap was characterized using DLTS method. The IL quality (D_{it}), and quality of high-κ layer in terms of leakage current density and TDDB under different SPAO annealing were also discussed. Controlling SPAO decomposed the unstable GeO_2 into GeO_x, which turned out to be more reliable than GeO_2 as observed in TDDB results. SPAO removes high-κ layer defects and different SPAO strategy also formed different IL thickness. Improvement and understanding of post annealing and ALD process is very critical to the future implementation of sub-nm EOT on high-κ layers regarding that they impact the IL quality.

References

[1] H. Iwai, "Roadmap for 22 nm and Beyond," *Microelectron. Eng.*, vol. 86, p. 1520, 2009.

[2] maltiel-consulting.com, "International Technology Roadmap for Semiconductors: 2012 Update," 2012, [Online]. Available: http://www.maltiel-consulting.com/ITRS-2012-Update-Overview.pdf. [Accessed: 28-Jun-2016].

[3] G. D. Wilk, R. M. Wallace, and J. M. Anthony, "High-K Gate Dielectrics: Current Status and Materials Properties Considerations," *J. Appl. Phys.*, vol. 89, p. 5243, 2001.

[4] A. Lubow, S. Ismail-Beigi, and T. P. Ma, "Comparison of Drive Currents in Metal-Oxide-Semiconductor Field-Effect Transistors Made of Si, Ge, GaAs, InGaAs, and InAs Channels," *Appl. Phys. Lett.*, vol. 96, p. 122105, 2010.

[5] D. Hisamoto, W.-C. Lee, J. Kedzierski, *et al.*, "FinFET—A Self-Aligned Double-Gate MOSFET Scalable to 20 nm," *IEEE T. Electron. Dev.*, vol. 47, 2000.

[6] R. W. Johnson, A. Hultqvist, and S. F. Bent, "A Brief Review of Atomic Layer Deposition: from Fundamentals to Applications," *Mater. Today.*, vol. 17, p. 236, 2014.

[7] T. Tanimura, Y. Watanabe, Y. Sato, Y. Kabe, and Y. Hirota, "Effect of Microwave Plasma Treatment on Silicon Dioxide Films Grown by Atomic Layer Deposition at Low Temperature," *J. Appl. Phys.*, vol. 113, p. 064102, 2013.

[8] E. H. Nicollian and A. Goetzberger, "The Si-SiO_2 Interface-Electrical Properties as Determined by the Metal-Insulator-Silicon Conductance Technique," *AT&T Tech. J.*, vol. 46, p. 1055, 1967.

[9] D. V. Lang, "Deep-Level Transient Spectroscopy: a New Method to Characterize Traps in Semiconductors," *J. Appl. Phys*, vol. 45, p. 3023, 1974.

[10] J. F. Verweij and J. H. Klootwijk, "Dielectric Breakdown I: A Review of Oxide Breakdown," *Microelectr. J*, vol. 27, p. 611, 1996.

[11] D. Misra, R. Garg, P. Srinivasan, N. Rahim, and N. A. Chowdhury, "Interface Characterization of High-K Dielectrics on Ge Substrates," *Mat. Sci. Semicon. Proc.*, vol. 9, p. 741, 2006.

[12] S. C. Vitkavage, E. A. Irene, and H. Z. Massoud, "An Investigation of Si–SiO_2 Interface Charges in Thermally Oxidized (100), (110), (111), and (511) Silicon," *J. Appl. Phys.*, vol. 68, p. 5262, 1990.

[13] A. Kerber, K. Maitra, A. Majumdar, M. Hargrove, R. J. Carter, and E. A. Cartier, "Characterization of Fast Relaxation During BTI Stress in Conventional and Advanced CMOS Devices With Gate Stacks," *IEEE Trans. Electron Devices*, vol. 55, p. 3175, 2008.

[14] D. Kuzum, T. Krishnamohan, A. Nainani, *et al.*, "High-Mobility Ge N-MOSFETs and Mobility Degradation Mechanisms," *IEEE T. Electron. Dev.*, vol. 58, p. 59, 2011.

[15] aps.org, "November 17–December 23, 1947: Invention of the First Transistor", 2000, [Online]. Available: http://www.aps.org/publications/apsnews/200011/history.cfm. [Accessed: 14-Jul 2016].

[16] K. Kit, S. Suzuki, H. Nomura, T. Takahashi, T. Nishimura, and A. Toriumi, "Direct Evidence of GeO Volatilization from GeO_2/Ge and Impact of Its Suppression on GeO_2/Ge Metal–Insulator–Semiconductor Characteristics," *Jpn. J. Appl. Phys.*, vol. 47, p. 2349, 2008.

[17] M. Fukuda, Watarumizubayashi, A. Kohno, S. Miyazaki, and M. Hirose, "Analysis of Tunnel Current through Ultrathin Gate Oxides," *Jpn. J. Appl. Phys.*, vol. 37, p. L1534, 1998.

[18] J.-L. Everaert, T. Conard, and M. Schaekers, "SiON Gate Dielectric Formation by Rapid Thermal Oxidation of Nitrided Si," in *13th IEEE International Conference on Advanced Thermal Processing of Semiconductors*, 2005, p. 135.

[19] J. Robertson, "Band Offsets of Wide-Band-Gap Oxides and Implications for Future Electronic Devices," *J. Vac. Sci. Technol. B*, vol. 18, p. 1785, 2000.

[20] A. P. Huang, Z. C. Yang, and P. K. Chu, Advances in Solid State Circuit Technologies, IntechOpen, 2010, p. 333.

[21] M. C. Zeman, C. C. Fulton, G. Lucovsky, R. J. Nemanich, and W.-C. Yang, "Thermal Stability of TiO_2, ZrO_2, or HfO_2 on Si(100) by Photoelectron Emission Microscopy," *J. Appl. Phys.*, vol. 99, p. 023519, 2006.

[22] K. Kukli, J. Niinistö, A. Tamm, *et al.*, "Atomic Layer Deposition of ZrO_2 And HfO_2 on Deep Trenched and Planar Silicon," *Microelectron. Eng.*, vol. 84, p. 2010, 2007.

[23] S. Stemmer, Y. Li, B. Foran, *et al.*, "Grazing-Incidence Small Angle X-Ray Scattering Studies of Phase Separation in Hafnium Silicate Films," *Appl. Phys. Lett.*, vol. 83, p. 3141, 2003.

[24] R. B. V. Dover, "Amorphous Lanthanide-Doped TiO_x Dielectric Films," *Appl. Phys. Lett.*, vol. 74, p. 3041, 1999.

[25] Y.-S. Lin, R. Puthenkovilakam, and J. P. Chang, "Dielectric Property and Thermal Stability of HfO_2 on Silicon," *Appl. Phy. Lett.*, vol. 81, p. 2041, 2002.

[26] P. K. Hurley, K. Cherkaoui, E. O'Connor, *et al.*, "Interface Defects in HfO_2, $LaSiO_x$, and Gd_2O_3 High-K/Metal–Gate Structures on Silicon," *J. Electrochem. Soc.*, vol. 155, p. G13, 2008.

[27] C. H. Bjorkman, J. T. Fitch, and G. Lucovsky, "Correlation between Midgap Interface State Density and Thickness-Averaged Oxide Stress and Strain at Si/SiO_2 Interfaces Formed by Thermal Oxidation of Si," *Appl. Phys. Lett.*, vol. 56, p. 1983, 1990.

[28] K. Choi, T. Ando, E. Cartier, *et al.*, "The Past, Present and Future of High-K/Metal Gates," *ECS Trans.*, vol. 53, p. 17, 2013.

[29] J. Robertson, "Interfaces and Defects of High-K Oxides on Silicon," *Solid State Electron.*, vol. 49, p. 283, 2005.

[30] M. V. Fischetti, D. A. Neumayer, and E. A. Cartier, "Effective Electron Mobility in Si Inversion Layers in Metal–Oxide–Semiconductor Systems with a High- Insulator: The Role of Remote Phonon Scattering," *J. Appl. Phys.*, vol. 90, p. 4587, 2001.

[31] C. C. Hobbs, L. R. C. Fonseca, A. Knizhnik, *et al.*, "Fermi-Level Pinning at the Polysilicon/Metal–Oxide Interface—Part II," *IEEE T. Electron. Dev.*, vol. 51, p. 978, 2004.

[32] T. Ando, M. M. Frank, K. Choi, *et al.*, "Ultimate EOT Scaling (<5Å) Using Hf-Based High-K Gate Dielectrics and Impact on Carrier Mobility," in ECS Trans., vol. 28, p. 115, 2010.

[33] K. Tatsumura, M. Goto, S. Kawanaka, and A. Kinoshita, "Correlation between Low-Field Mobility and High-Field Carrier Velocity in Quasi-Ballistic-Transport MISFETs scaled down to $L_g = 30$ nm," in 2009 IEEE International Electron Devices Meeting (IEDM), Baltimore, MD, 2009, p. 1.

[34] Y. Kamata, "High-k/Ge MOSFETs for Future Nanoelectronics," *Mater. Today*, vol. 11, p. 30, 2008.

[35] A. Dimoulas, P. Tsipas, A. Sotiropoulos, and E. K. Evangelou, "Fermi-Level Pinning and Charge Neutrality Level in Germanium," *Appl. Phys. Lett.*, vol. 89, p. 252110, 2006.

[36] Y. Kamata, Y. Kamimuta, T. Ino, R. Iijima, M. Koyama, and A. Nishiyama, "Influences of Annealing Temperature on Characteristics of Ge p-Channel Metal Oxide Semiconductor Field Effect Transistors with ZrO_2 Gate Dielectrics," *Jpn. J. Appl. Phys.*, vol. 45, p. 5651, 2006.

[37] Q. Xie, S. Deng, M. Schaekers, *et al.*, "Germanium Surface Passivation and Atomic Layer Deposition of High-K Dielectrics—a Tutorial Review on Ge-Based Mos Capacitors," *Semicond. Sci. Technol.*, vol. 27, p. 074012, 2012.

[38] A. Toriumi, T. Tabata, C. H. Lee, T. Nishimura, K. Kita, and K. Nagashio, "Opportunities and Challenges for Ge CMOS – Control of Interfacing Field on Ge is a Key," *Microelectron. Eng.*, vol. 86, p. 1571, 2009.

[39] C.-M. Lin, H.-C. Chang, I.-H. Wong, S.-J. Luo, C. W. Liu, and C. Hu, "Interfacial Layer Reduction and High Permittivity Tetragonal ZrO_2 on Germanium Reaching Ultrathin 0.39 nm Equivalent Oxide Thickness," *Appl. Phys. Lett.*, vol. 102, p. 232906, 2013.

[40] Y. Fukuda, T. Ueno, S. Hirono, and S. Hashimoto, "Electrical Characterization of Germanium Oxide/Germanium Interface Prepared by Electron-Cyclotron-Resonance Plasma Irradiation," *Jpn. J. Appl. Phys.*, vol. 44, p. 6981, 2005.

[41] R. Zhang, T. Iwasaki, N. Taoka, M. Takenaka, and S. Takagi, "Al_2O_3/GeO_x/ Ge Gate Stacks with Low Interface Trap Density Fabricated by Electron Cyclotron Resonance Plasma Postoxidation," *Appl. Phys. Lett.*, vol. 98, p. 112902, 2011.

[42] A. Delabie, F. Bellenger, M. Houssa, *et al.*, "Effective Electrical Passivation of Ge (100) for High-K Gate Dielectric Layers using Germanium Oxide," *Applied Physics Letters*, vol. 91, p. 082904, 2007.

[43] H. Seo, F. Bellenger, K. B. Chung, *et al.*, "Extrinsic Interface Formation of HfO_2 and Al_2O_3/GeO_x Gate Stacks on Ge (100) Substrates," *J. Appl. Phys.*, vol. 106, p. 044909, 2009.

[44] M. Houssa, G. Pourtois, M. Caymax, M. Meuris, and M. M. Heyns, "First-Principles Study of the Structural and Electronic Properties of (100) Ge/Ge(M) O_2 Interfaces (M = Al, La, or Hf)," *Appl. Phys. Lett.*, vol. 92, p. 242101, 2008.

[45] H. L. Shang, H. Okorn-Schmidt, K. K. Chan, *et al.*, "High Mobility p-channel Germanium MOSFETs with a Thin Ge Oxynitride Gate Dielectric," Presented at the International Electronic Devices Meet, 2002.

[46] H. Shang, K.-L. Lee, P. Kozlowski, *et al.*, "Self-Aligned n-Channel Germanium MOSFETs with a Thin Ge Oxynitride Gate Dielectric and Tungsten Gate," *IEEE Electr. Device L.*, vol. 25, p. 135, 2004.

[47] D. Misra, R. Garg, P. Srinivasan, N. Rahim, and N.A. Chowdhury, "Interface Characterization of High-k Dielectrics on Ge Substrates" *Mat. Sci. Semicon. Proc.*, vol. 9, p. 741, 2006.

[48] Y. M. Ding and D. Misra, "Oxide Structure-Dependent Interfacial Layer Defects of HfAlO/SiO_2/Si Stack Analyzed by Conductance Method," *J. Vac. Sci. Technol. B*, vol. 33, p. 021203, 2015.

[49] E. Vincent, G. Ghibaudo, G. Morin, and C. Papadas, "On the Oxide Thickness Extraction in Deep-Submicron Technologies," in Microelectron. Test Struct: *Proc. 1997 IEEE Int. Conf.*, Monterey, USA, 1997, pp.105–110.

[50] R. Hillard, J. Heddleson, D. Zier, P. Rai-Choudhury, and D. Schroder, "Diagnostic Techniques for Semiconductor Materials and Devices", Proceeding of Electrochem. Soc., Pennington, USA, 1992, pp. 261–274.

[51] R. Engel-Herbert, Y. Hwang, and S. Stemmer, "Comparison of Methods to Quantify Interface Trap Densities at Dielectric/III-V Semiconductor Interfaces," *J. Appl. Phys*, vol. 108, p. 124101, 2010.

[52] D. Kuzum, A. J. Pethe, T. Krishnamohan, and K. C. Saraswat, "Ge (100) and (111) N- and P-FETs With High Mobility and Low-T Mobility Characterization," *IEEE Trans. Electron Dev.*, vol. 56, p. 648, 2009.

[53] J. R. Hauser and K. Ahme, "Characterization of Ultra-Thin Oxides Using Electrical C-V and *I-V* Measurements,"*AIP Conf. Proc.*, vol. 449, p. 235, 1998.

[54] Y. Kabe Y, R. Hasunuma, and K. Yamabe, "Oxidation of Silicon Utilizing a Microwave Plasma System: Electric-Stress Hardening of SiO_2 Films by Controlling the Surface and Interface Roughness" *Jpn. J. Appl. Phys.*, vol. 51, p. 041104, 2012.

[55] S. Takagi and M. Takenada, "Prospective and Critical Issues of III-V/Ge CMOS on Si Platform," *ECS Trans.*, vol. 35, p. 279, 2011.

[56] X. Yang, S.-K. Wang, B. Sun, *et al.*, "A Comparative Study of Ge MOSFET using Al_2O_3GeOx/Ge Stacks-Forming High Quality GeO_x Interface Layer to Boost Device Performance," in Solid-State and Integrated Circuit Technology (ICSICT), 2014 12th IEEE International Conference, Guilin, 2014, p. 1.

[57] H. Le, W. Sheng-Kai, Z. Xiong, *et al.*, "Effect of Ultrathin GeO_x Interfacial Layer Formed by Thermal Oxidation on Al_2O_3 Capped Ge," Chin. Phys. B, vol. 23, p. 046804, 2014.

[58] S. N. A. Murad, P. T. Baine, D. W. McNeill, *et al.*, "Optimisation and Scaling of Interfacial GeO_2 Layers for High-K Gate Stacks on Germanium and Extraction of Dielectric Constant of GeO_2," *Solid State Electron.*, vol. 78, p. 136, 2012.

[59] R. Zhang, W. Chern, X. Yu, M. Takenaka, J. L. Hoyt, and S. Takagi, "High Mobility Strained-Ge pMOSFETs with 0.7-nm Ultrathin EOT using Plasma Post Oxidation $HfO_2/Al_2O_3/GeO_x$ Gate Stacks and Strain Modulation," presented at the 2013 IEEE International Electron Devices Meeting, Washington, DC, 2013.

[60] Y. Kamata, Y. Kamimuta, T. Ino, and A. Nishiyama, "Direct Comparison of ZrO_2 and HfO_2 on Ge Substrate in Terms of the Realization of Ultrathin High-K Gate Stacks," *Jpn. J. Appl. Phys.*, vol. 44, p. 2323, 2005.

[61] S. Mukhopadhyay, S. Mitra, Y. M. Ding, *et al.*, "Effect of Post Plasma Oxidation on Ge Gate Stacks Interface Formation," in ECS Trans., San Diego, US, 2016, p. 303.

[62] L.G. Kolla, Y. Ding, D. Misra and N. Bhat, "Interface states reduction in atomic layer deposited TiN/ZrO2/Al2O3/Ge Gate stacks," *J. Vac. Sci. Technol. B* vol 36, p. 021201, 2018.

[63] P. P. Apte and K. C. Saraswat, "SiO_2, Degradation with Charge Injection Polarity," *IEEE Electr. Device L.*, vol. 14, p. 512, 1993.

[64] E. Hasegawa, A. Ishitani, K. Akimoto, M. Tsukiji, and N. Ohta, "SiO_2/Si Interface Structures and Reliability Characteristics," *J. Electrochem. Soc.*, vol. 142, pp. 273–282, 1995.

[65] R. Degraeve, T. Kauerauf, A. Kerber, *et al.*, "Stress Polarity Dependence of Degradation and Breakdown of SiO_2/High-k Stacks," presented at the IEEE 41st Annual International Reliability, Dallas Texas, 2003.

[66] W. L. Chang, J. H. Stathis, and E. Cartier, "Role of Interface Layer in Stress-Induced Leakage Current in High-k/Metal-Gate Dielectric Stacks," presented at the IEEE International Reliability Physics Symposium (IRPS), Anaheim, CA, 2010.

[67] F.-C. Chiu, "A Review on Conduction Mechanisms in Dielectric Films," *Adv. Mat. Sci. Eng.*, vol. 2014, p. 1, 2014.

[68] N. K. Ponon, D. J. R. Appleby, E. Arac, *et al.* "Effect of deposition conditions and post deposition anneal on reactively sputtered titanium nitride thin films," *Thin Solid Films*, vol. 578, pp. 31–37, 2015.

[69] F. Vaz, J. Ferreira, E. Ribeiro, *et al.* "Influence of Nitrogen Content on the Structural, Mechanical and Electrical Properties of TiN Thin Films," *Surf. Coat. Technol.*, vol. 191, pp. 317–323, 2005.

[70] L. Zhang, H. Li, Y. Guo, *et al.* "Selective Passivation of GeO_2/Ge Interface Defects in Atomic Layer Deposited High-k MOS Structures," *ACS Appl. Mater. Interfaces*, vol. 7, pp. 20499–20506, 2015.

[71] R. Zhang, T. Iwasaki, N. Taoka, M. Takenaka, and S. Takagi, "High-Mobility Ge pMOSFET With 1-nm EOT Al_2O_3/GeO_x/Ge Gate Stack Fabricated by Plasma Post Oxidation," *IEEE Trans. Electron Devices*, vol. 59, pp. 335–341, 2012.

[72] R. Zhang, P.-C. Huang, J.-C. Lin, M. Takenaka, and S. Takagi, "Atomic Layer-By-Layer Oxidation of Ge (100) and (111) Surfaces by Plasma Post Oxidation of Al_2O_3/Ge Structures," *Appl. Phys. Lett.*, vol. 102, p. 81603, 2013.

[73] M. Ke, X. Yu, R. Zhang, *et al.*, "Fabrication and MOS Interface Properties of ALD $AlYO_3$/GeO_x/Ge Gate Stacks with Plasma Post Oxidation," *Microelectron Eng*, vol. 147, pp. 244–248, 2015.

[74] M. N. U. Bhuyian, D. M. S. Poddar, K. Tapily, *et al.*, "Impact of Cyclic Plasma Treatment on Oxygen Vacancy Defects in TiN/HfZrO/SiON/Si Gate Stacks," *Appl. Phys. Lett.*, vol. 106, p. 193508, 2015.

[75] N. M. Ravindra and J. Zhao, "Fowler–Nordheim Tunneling in Thin SiO_2 Films," *Smart Mater. Struct.*, vol. 1, p. 197, 1992.

[76] D. K. Cheng, Field and Wave Electromagnetics. Canada: Addison-Wesley, 1989, p. 116.

[77] S. Monaghan, P. K. Hurley, K. Cherkaoui, M. A. Negara, and A. Schenk, "Determination of Electron Effective Mass and Electron Affinity in HfO_2 using MOS and MOSFET Structures," *Solid State Electron.*, vol. 53, p. 438, 2009.

[78] M. L. Huang, Y. C. Chang, C. H. Chang, *et al.*, "Energy-Band Parameters of Atomic-Layer-Deposition Al_2O_3/InGaAs Heterostructure," *Appl. Phys. Lett.*, vol. 89, p. 012903, 2006.

[79] D. K. Schroder, Semiconductor Material and Device Characterization. Hoboken: Wiley, 2006, pp. 693, 694.

[80] J. S. Suehle, "Ultrathin Gate Oxide Reliability: Physical Models, Statistics, and Characterization," *IEEE T. Electron. Dev.*, vol. 49, p. 958, 2002.

[81] P. E. Nicollian, A. T. Krishnan, C. A. Chancellor, *et al.*, "The Current Understanding of the Trap Generation Mechanisms that Lead to the Power Law Model for Gate Dielectric Breakdown," in IEEE International Reliability Physics Symposium Proceedings. 45th Annual, Phoenix, AZ, 2007 p. 197.

[82] E. Y. Wu, J. Sune, and W. Lai, "On the Weibull shape factor of intrinsic breakdown of dielectric films and its accurate experimental determination. Part II: experimental results and the effects of stress conditions," *IEEE Trans. Elec. Dev.*, vol. 49, p. 2141, 2002.

[83] T. Nigam, A. Kerber, and P. Peumans, "Accurate Model for Time-Dependent Dielectric Breakdown of High-k Metal Gate Stacks," in 2009 IEEE International Reliability Physics Symposium, 2009, p. 523.

[84] C.-C. Cheng, C.-H. Chien, G.-L. Luo, *et al.*, "Characteristics of Atomic-Layer-Deposited Al_2O_3 High-K Dielectric Films Grown on Ge Substrates," *J. Electrochem. Soc.*, vol. 155, p. G203, 2008.

[85] P. S. Goley and M. K. Hudait, "Germanium Based Field-Effect Transistors: Challenges and Opportunities," *Mat. Sci. Semicon. Proc.*, vol. 7, p. 2301, 2014.

[86] W. F. Zhang, T.Nishimula, K. Nagashio, K. Kita, and A. Toriumi, "Conduction Band offset in GeO_2/Ge Stack Determined by Internal Photoemission Spectroscopy," *ECS Trans.*, vol. 50, p. 91, 2012.

[87] S. Rangan, E. Bersch, R. A. Bartynski, E. Garfunkel, and E. Vescovo, "GeO_x Interface Layer Reduction upon Al-Gate Deposition on a HfO_2/GeO_x/Ge (001) stack," *Appl. Phys. Lett.*, vol. 92, p. 172906, 2008.

[88] M.N. Bhuyian, P. Shao, A. Sengupta, *et al.* "Post Plasma Oxidation Processed ALD $Al_2O_3/Hf_{1-x}Zr_xO_2$ Thin Films on Ge Substrates: Reliability," *ECS J. Sol. State Sci. Technol.*, vol. 7, p. N1–N6, 2018.

[89] M. N Bhuyian, D. Misra, K. Tapily, *et al.* "Cyclic Plasma Treatment during ALD $Hf_{1-x}Zr_xO_2$ Deposition," *ECS J. Sol. State Sci. Technol.*, vol. 3, N83–N88, 2014.

[90] D. Ielmini, "Resistive Switching Memories Based on Metal Oxides: Mechanisms, Reliability and Scaling," *Semicond. Sci. Technol.* vol. 31, p. 063002, 2016.

[91] Xiaorong Chen and Jie Feng, "Improvement in Roff/Ron Ratio and Reset Current via Combining Compliance Current with Multilayer Structure in Tantalum Oxide-Based RRAM," *Appl. Phys. A*, vol. 120, p. 67. 2015.

[92] U. Chand, K-C. Huang, C-Y. Huang, and T-Y. Tseng, "Mechanism of Nonlinear Switching in HfO2-Based Crossbar RRAM With Inserting Large Bandgap Tunneling Barrier Layer," *IEEE Trans. Electron. Dev.*, vol. 62, p. 3665, 2015.

[93] D. C. Gilmer, G. Bersuker, H-Y. Park, *et al.* "Effects of RRAM Stack Configuration on Forming Voltage and Current Overshoot," IEEE International Memory Workshop (IMW), 2011.

[94] K.G. Young-Fisher, G. Bersuker, B. Butcher, A. Padovani, L. Larcher, D. Veksler, and D.C. Gilmer, "Leakage Current-Forming Voltage Relation and Oxygen Gettering in HfOx RRAM Devices," *IEEE Electron Dev. Lett.*, vol. 34, p. 750, 2013.

Chapter 8

Technology and modeling of DNTT organic thin-film transistors

Sushil Kumar Jain[1], *Amit Mahesh Joshi*[1], *and Arun Dev Dhar Dwivedi*[2]

8.1 Introduction

Dinaphtho[2,3-*b*:2′,3′-*f*]thieno[3,2-*b*]thiophene (DNTT) is a popular material for the fabrication of organic field-effect transistors (OTFTs), better stability in air, with higher mobility, and better reproducibility. DNTT-based devices exhibit very less degradation and hysteresis loss in the transfer characteristics of device over a long period of time. These DNTT OTFT features are even better than pentacene, which is also used in organic electronics fabrication field in the last decade. Due to its high performance and stability, DNTT would be the next generation of the new standard organic semiconductor (OSC). The air stability, good performance very small molecule of DNTT can be used for the implementation of a flexible amplifier circuit. Low-temperature polymers as a gate dielectric and air-stable DNTT as a semiconductor active matrix (AM) backplanes can be successfully fabricated on any flexible substrates.

From the past decade, efforts have been increased significantly to implement and develop the electronic circuit on stretchable and flexible substrates. With the advent of different soluble OSCs and conductors, printed electronic components have become feasible. Printed electronic products are usually connected to organic electronic products. They are characterized by the use of organic materials to implement devices and circuits. Devices containing organic materials are on the verge of commercialization. In recent years, the research in organic thin-film transistors (OTFTs) has increased dramatically and has been proven for various applications such as display with low cost [1], memory using OSC [2], components used in RFID tags [3], low-end electronics products, circuits, electronic components, and sensors using polymer [4]. Flexible electronics is a new era technology that establishes electronic circuits with storing electronic products on flexible, stretchable substrates such as paper, plastic and even on cloth. Compared with inorganic electronic products, organic or flexible electronic products have the

[1]Department of Electronics and Communication, Malaviya National Institute of Technology (MNIT), Jaipur, India
[2]Department of Electrical and Electronics, Poornima University, Jaipur, India

following distinct advantages: (i) it can be manufactured at low cost; (ii) it is lightweight, thin, bendable, foldable and has strong absorbance and will not be crushed. It has mechanical flexibility, low energy consumption, high emission efficiency; and (iii) it has cheaper materials and lower cost deposition processes, the cost is lower [5], therefore it is also used for large-area applications. In fact, the stacking and rapid annealing of OSC and low-temperature polymer gate dielectrics are suitable for high-throughput, low-cost printing manufacturing [6]. The researchers used (DNTT) [7], poly(3-hexylthiophene) P3HT, poly(3-alkylthiophene), poly(3-octylthiophene) P3OT and as organic materials instead of semiconductors, and used dielectric layers to create complete flexibility.

8.2 Motivation

First field-effect transistor based on organic material (OFET) was reported by Koezukaand in 1987 and it opened a new door for researchers all over the world in this field. Since then, great progress has been made in the synthesis of organic materials, manufacturing techniques, processing methods, and equipment. Organic electronics refers to new era electronics that use polymers or carbon-based small molecules as functional materials for conductors, semiconductors, and dielectrics. Main reasons for this choice are as follows [5,6]:

1. Low process temperature
2. Low-cost manufacturing
3. Roll-to-roll manufacturing
4. Large-area electronics
5. Large material choice
6. Solution processable

Despite these advantages, organic electronics have some drawbacks which do not allow them to compete with inorganic material (silicon)-based electronic devices. The disadvantages are as follows:

1. Low carrier mobility
2. Mono-type circuits
3. Large variability
4. Limited controllability

However, the mentioned distinct properties of DNTT have been made it one of the potential candidates for various applications.

8.2.1 Potential applications of flexible organic electronics

Various applications of OTFTs have been developed and cover a wide range of technologies and products, including memory, displays, sensors, lighting equipment, radio frequency identification (RFID) tags, flexible batteries, OPVs, and OLED lighting.

From Figure 8.1, it can be seen that organic electronics have great potential in the future by introducing products to the market in new forms.

The research community for integrating smart systems would explore opportunities to develop circuits and systems with OTFT. Figure 8.2 shows the comparison of organic and inorganic semiconductors because organic technology has various advantages over inorganic technology [8].

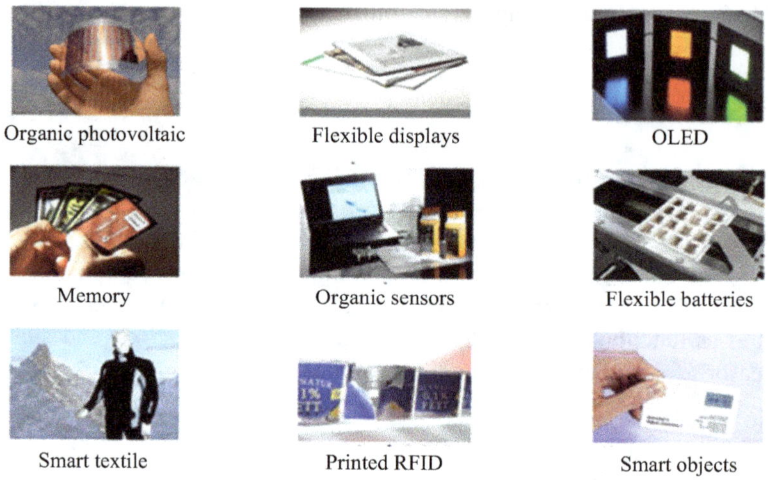

Figure 8.1 Applications using organic electronics [1–3,8]

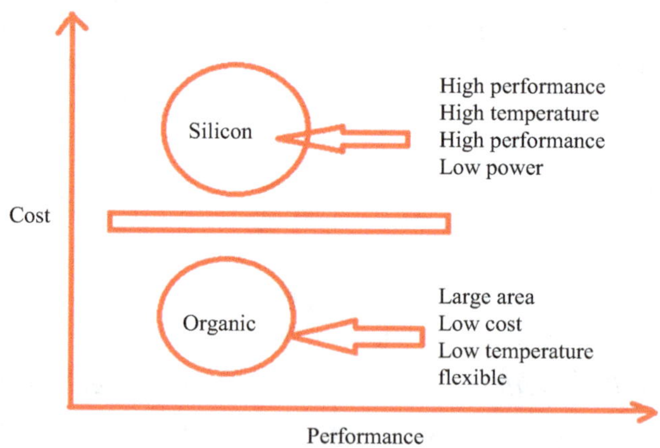

Figure 8.2 Graphical representation of cost versus performance of inorganic and OSCs [8]

8.3 Organic thin-film transistors (OTFTs)

In 1962, Weimer mainly proposed and developed the transistor using a thin film by producing a cadmium sulfide semiconductor film (inorganic semiconductor). After some years in 1979, Le Comber *et al.* [9] reported the possibility of film formation at relatively low temperatures using TFT. Subsequently, in the 1980s, several scientists and researchers including Tsumura [10], Kudo *et al.* [11] and Ebisawa *et al.* [12] suggested organic transistors on plastic foils and glass. Afterward, there were several efforts to improve the performance of thin-film transistors using organic material. Basically, four types of TFT structures are defined based on the positions of the source, gate and drain and contacts with respect to the layer of OSC. OTFTs are broadly classified into staggered and coplanar structures. Coplanar structure (bottom contact) consists of the conductive channel, source terminal and drain terminal is on the same side of OSC layer. In contrast, electrically conductive channels in staggered structures (also referred to as top contact structures) are located on the opposite sides of a semiconductor film that includes drain and source contacts. The next section briefly discusses the basic working of OTFT.

8.3.1 Working principle of OTFT

In a given OTFT, under the influence of the applied voltage between the drain and source, there will be a current flowing between the terminal of source and drain in device and this current can be modulated by changing the voltage applied at the gate, as shown in Figure 8.3(a). It is observed that the gate dielectrics are used for capacitive separation of gate electrode with the semiconductor. If no voltage is applied at the gate, the charge carrier's density is very low in the semiconductor layer. Therefore, due to low carrier density, developed channel conductance between the drain and source is low. However, if the potential between gate-source is applied (if the p-channel is operated by OTFT using DNTT OSC), additional carriers (holes) are triggered and collected in the interface between semiconductor and insulator, so the creation of a carrier channel takes place. Consequently, the drain and source contacts are electrically connected with conductive channel, causing the flow of large electrical currents.

8.3.2 OTFT parameter

The transfer curve as shown in Figure 8.3(b) provides an estimation of the electrical parameters to analyze the device performance.

(A) Threshold voltage (V_T)—It corresponds to V_{GS} that accumulates large amounts of charge near the interface between dielectric and semiconductor. With the help of linear extrapolation of I_{DS}–V_{GS} graph at low V_{DS}, we can find this parameter.

(B) On–off ratio—On–off ratio is the ratio of peak value of I_{DS} to minimum I_{DS}. It is obvious that the driving capability will be better if higher "on" currents, while lower "off" currents result in lower leakage currents. Therefore, a higher ratio of on-off current is preferred.

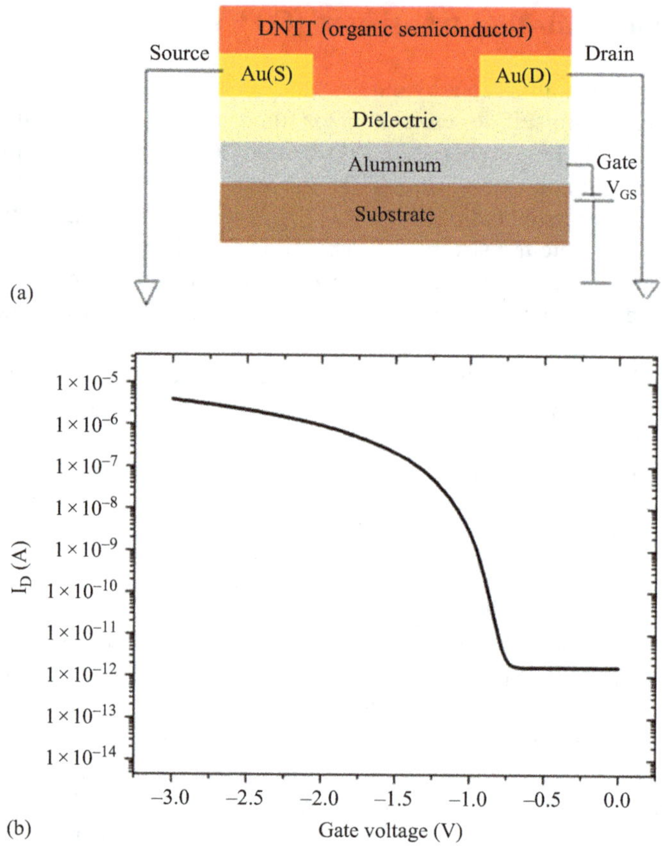

Figure 8.3 (a) Device operation of an organic field-effect transistor and (b) transfer characteristics of an OTFT [14]

(C) Subthreshold swing (*S*)—This electrical parameter represents that V_{GS} needed to increase I_{DS} by one decade, in subthreshold regime. It is defined as V/decade:

$$S = \left(\frac{d \log (I_{DS})}{dV_{GS}} \bigg| \max \right)^{-1}$$

For low power consumption with high speed, the value of *S* should be less.

(D) Mobility—Regarding mobility (μ), it directly affects the switching speed and peak value of drain current. There are various ways to find μ, the important of which are given below:

Effective mobility (μ_{eff})—It is calculated using the following formula and also provides the most accurate results:

$$\mu_{eff} = \frac{g_{DS}}{C_i \frac{W}{L} (V_{GS} - V_T)}$$

Field-effect mobility (μ_{FE})—It is calculated by the transconductance at low voltage:

$$\mu_{FE} = \frac{g_m}{C_i \frac{W}{L} V_{DS}}$$

8.4 Modeling and simulation of DNTT-based OTFT

8.4.1 Configurations of DNTT-based OTFT

Four types of configurations of DNTT-based OTFT are available, which are shown in Figures 8.4 and 8.5. In Figure 8.4, top gate configurations are given which is further classified as top gate top contact (TGTC) and top gate bottom contact TGBC. In Figure 8.5, bottom gate configurations are given which are further classified as bottom gate top contact (BGTC) and bottom gate bottom contact (BGBC).

Energy band diagram of a DNTT-based metal insulator semiconductor (MIS) structure is given as in Figure 8.6. Maximum valence band energy (E_V) and minima of conduction band energy (E_C) of the inorganic semiconductor are substantially

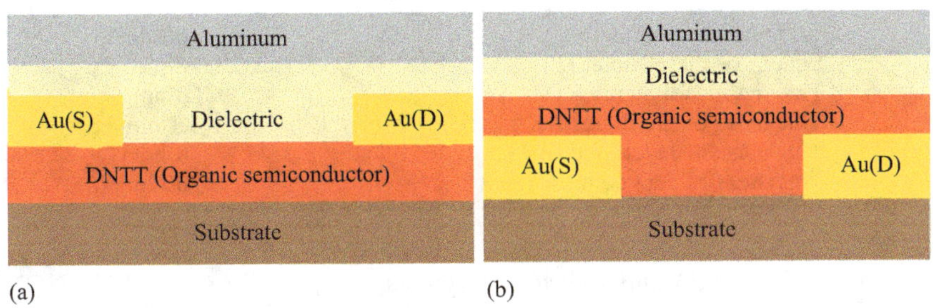

Figure 8.4 Structure of top gate configuration: (a) top gate top contact (TGTC) and (b) top gate bottom contact (TGBC)

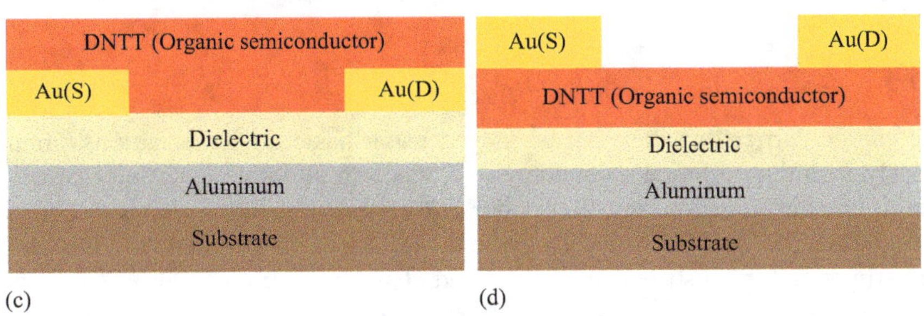

Figure 8.5 Structure of bottom gate configuration (c) Bottom gate bottom contact (BGBC) and (d) bottom gate top contact (BGTC)

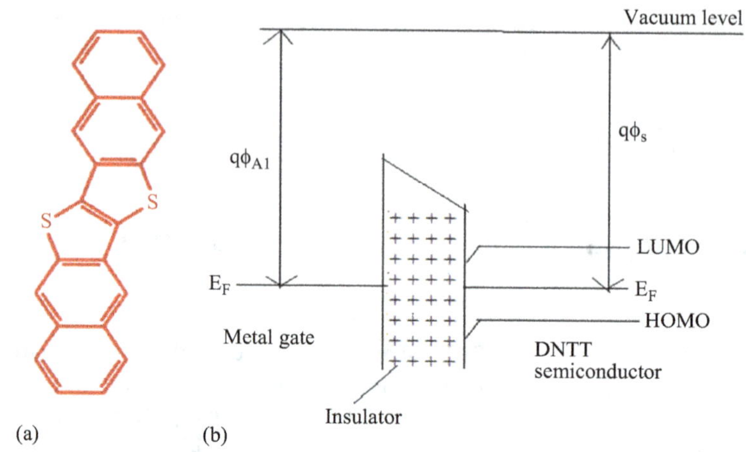

Figure 8.6 (a) Structure of DNTT OSC and (b) energy diagram of MIS structure [14]

Table 8.1 Dimensions used for device simulation

Dimensions used in the device	Value
Width of the channel (W)	100 μm
Length of the channel (L)	10 μm
Thickness of the gate, t_g	20 nm
Thickness of the dielectric, t_{ox}	5.7 nm
Thickness of layer of the OSC, t_{osc}	30 nm
Thickness of S/D contact, $t_{S/D}$	30 nm

similar to OSC HOMO and LUMO energy gap. Especially for DNTT, HOMO energy level is approximately −5.19 eV and energy level of LUMO is almost −1.81 eV [13,14]. This introduces a large enough 3.38 eV HOMO–LUMO energy gap, which is sufficient for transistor operation. Numerical simulation of the bottom gate and top gate configuration is completed using the parameter given in Table 8.1.

8.4.2 Device physical modeling

The charge carrier densities can be solved using basic device equations simultaneously including Poisson's equation [15], electron and hole continuity equation, charge transfer equation and defect density of states equation. Poisson equation finds the electric field intensity in a given device based on the internal movement of the carriers and the distribution of the fixed charges given by (8.1):

$$\nabla.\mathrm{E} = \frac{\rho(x,y)}{\epsilon} \tag{8.1}$$

where ∈ is the permittivity of the region, and $\rho(x, y)$ is the charge density given as

$$\rho(x,y) = q[p(x,y) - n(x,y) + N_D^+(x,y) - N_A^+(x,y)] \tag{8.2}$$

where $p(x, y)$ = density of holes, $n(x, y)$ = density of electrons, $N_D^+(x,y)$ ionization donor density, and $N_A^+(x,y)$ the ionization acceptor density.

Poisson's equation is modified by adding trapped charge (Q_T), representing as

$$\rho(x,y) = q[p(x,y) - n(x,y) + N_D^+(x,y) - N_A^+(x,y)] - Q_T \tag{8.3}$$

Due to charge accumulation, a potential is generated, which affects the intensity of electric field distribution and current.

As shown in (8.4) and (8.5), the continuity equation describes the process of distribution of charge carrier over time. In the following equations, q is the magnitude of electronic charge, n is the carrier density of electron, carrier density of hole represented by p, J is the corresponding current density, G is the corresponding rate of charge generation, and R is the corresponding rate of charge recombination:

$$\frac{\partial \mathrm{n}}{\partial \mathrm{t}} = \frac{1}{\mathrm{q}} \nabla . J_n + G_n - R_n \tag{8.4}$$

$$\frac{\partial \mathrm{p}}{\partial \mathrm{t}} = -\frac{1}{\mathrm{q}} \nabla . J_p + G_p - R_p \tag{8.5}$$

For OTFTs, there is no optical absorption, so the term is simplified and the properties of the material are described by the minority carrier recombination lifetime. Since MOSFETs are majority carrier devices, the behavior of carrier recombination and generation are relatively unimportant. Physical properties of the layer of OSCs depend on the movement and generation of polarons [16].

Third important set of equations for explaining device physics for charge carrier is given by

$$J_p = qn\mu_p E - qD_p \nabla_p \tag{8.6}$$

$$J_n = qn\mu_p E + qD_n \tag{8.7}$$

It contains drift and diffusion parts. These equations determine the current density based on carrier mobility (μ), electric field (E), the carrier density (n, p) and the diffusion coefficient of the carrier (D). Diffusion coefficient operators are related to Einstein's mobility relationship:

$$D_n = \frac{kT}{q} \mu_n \tag{8.8}$$

$$D_p = \frac{kT}{q} \mu_p \tag{8.9}$$

In conclusion, Poisson equations, continuity equations, and current density equations [17,18] are solved by the Silvaco ATLAS software in a two-dimensional grid of the given device structure simultaneously with itself and is subject to boundary conditions (including those applied at the contacts).

8.4.2.1 Density of states model

The assumed total density (DOS) $g(E)$ consists of four bands: two tail (analogous to acceptor-like conduction band and donor-like valence band) and there will be two deep energy bands (one donor-like and other acceptor-like), and all are modeled with the help of Gaussian distribution [19]:

$$g(E) = g_{TA}(E) + g_{TD}(E) + g_{GA}(E) + g_{GD}(E) \tag{8.10}$$

Trap energy is given by E, conduction energy is E_C, valence energy E_V and the subscripts (G, A, D, T) stand for, Gaussian (deep level), acceptor, donor, tail states, respectively:

$$g_{TA}(E) = NTA_{\exp}\left[\frac{E - E_C}{WTA}\right] \tag{8.11}$$

$$g_{TD}(E) = NTD_{\exp}\left[\frac{E - E_V}{WTD}\right] \tag{8.12}$$

$$g_{GA}(E) = NGA_{\exp}\left[-\left[\frac{EGA - E}{WGA}\right]^2\right] \tag{8.13}$$

$$g_{GD}(E) = NGD_{\exp}\left[-\left[\frac{E - EGD}{WGD}\right]^2\right] \tag{8.14}$$

DOS for the exponential tail is defined by its conduction and valence band edge intercept density (NTA and NTD) and its attenuation energy (WTD and WTA). DOS for Gaussian distributions is given by the total state density (NGD and NGA), its attenuation energy (WGD and WGA) and peak distribution (EGD and EGA).

8.4.2.2 Trapped carrier density

Ionized densities of donor and acceptor states are given by (8.15) and (8.16):

$$n_T = n_{TD} + n_{GD} \tag{8.15}$$

$$p_T = p_{TA} + p_{GA} \tag{8.16}$$

where p_{TA}, p_{GA}, n_{TD}, and n_{GD} are given below.

$$p_{TA} = \int_{E_V}^{E_C} g_{TA}(E) . f_{t_{TA}}(E, n, p) dE \tag{8.17}$$

$$p_{GA} = \int_{E_V}^{E_C} g_{GA}(E).f_{tGA}(E,n,p)dE \tag{8.18}$$

$$n_{TD} = \int_{E_V}^{E_C} g_{TD}(E).f_{tTD}(E,n,p)dE \tag{8.19}$$

$$n_{GD} = \int_{E_V}^{E_C} g_{GD}(E).f_{tGD}(E,n,p)dE \tag{8.20}$$

$f_{tGA(E,n,p)}$ and $f_{tTA(E,n,p)}$ are the ionization probabilities for the Gaussian acceptor and tail DOS, and $f_{tTD(E,n,p)}$ and $f_{tGD(E,n,p)}$ are defined as the probability of occupation of a trap level at energy E for the Gaussian and tail acceptor and donor states in steady state are given by following (8.21)–(8.24):

$$f_{tTA}(E,n,p) = \frac{v_n SIGTAE.n + v_p SIGTAH.n_i \exp\left[\frac{E_i - E}{kT}\right]}{v_n SIGTAE\left(n + n_i \exp\left[\frac{E_i - E}{kT}\right]\right) + v_p SIGTAH\left(p + n_i \exp\left[\frac{E_i - E}{kT}\right]\right)} \tag{8.21}$$

$$f_{tGA}(E,n,p) = \frac{v_n SIGGAE.n + v_p SIGGAH.n_i \exp\left[\frac{E_i - E}{kT}\right]}{v_n SIGGAE\left(n + n_i \exp\left[\frac{E_i - E}{kT}\right]\right) + v_p SIGGAH\left(p + n_i \exp\left[\frac{E_i - E}{kT}\right]\right)} \tag{8.22}$$

$$f_{tTD}(E,n,p) = \frac{v_p SIGTDH.p + v_n SIGTDE.n_i \exp\left[\frac{E - E_i}{kT}\right]}{v_n SIGTDE\left(n + n_i \exp\left[\frac{E - E_i}{kT}\right]\right) + v_p SIGTDH\left(p + n_i \exp\left[\frac{E_i - E}{kT}\right]\right)} \tag{8.23}$$

$$f_{tGD}(E,n,p) = \frac{v_p SIGGDH.p + v_n SIGGDE.n_i \exp\left[\frac{E - E_i}{kT}\right]}{v_n SIGGDE\left(n + n_i \exp\left[\frac{E - E_i}{kT}\right]\right) + v_p SIGGDH\left(p + n_i \exp\left[\frac{E_i - E}{kT}\right]\right)} \tag{8.24}$$

v_n is defined as the thermal velocity of electron, v_p is defined as the thermal velocity of hole, and intrinsic carrier concentration is given as n_i. SIGGAE and SIGTAE present the electron capture cross-sections subject to the Gaussian states and main tail, respectively. SIGGAH and SIGTAH present the hole trap cross-sections of the Gaussian states and acceptor tail, respectively, SIGGDE, SIGTDE SIGTDH, and SIGGDH are the equivalent for donor states [19].

8.4.2.3 Poole–Frenkel mobility model

The model which helps to find the carrier mobility is given by Poole–Frenkel mobility model [20]:

$$\mu_{n_{PF}}(E) = \mu_{n0}\exp\left(-\frac{DELTAEN.PFMOB}{kT_{neff}} + \left(\frac{BETAN.PFMOB}{kT_{neff}} - \text{GAMMAN.PFMOB} \right)\sqrt{|E|} \right) \quad (8.25)$$

$$\mu_{p_{PF}}(E) = \mu_{p0}\exp\left(-\frac{DELTAEP.PFMOB}{kT_{peff}} + \left(\frac{BETAP.PFMOB}{kT_{peff}} - \text{GAMMAP.PFMOB} \right)\sqrt{|E|} \right) \quad (8.26)$$

where $\mu_{pPF}(E)$ and $\mu_{nPF}(E)$ are the Poole–Frenkel mobility's for holes and electrons, respectively, μ_{n_0} and μ_{p_0} are defined as the zero-field motilities and E is the electric field. The activation energies at zero electric field for holes and electrons are presented by DELTAEP.PFMOB and DELTAEN.PFMOB, respectively. BETAN.PFMOB, BETAP.PFMOB is the Poole–Frenkel factor. T_{neff}, T_{peff} are the effective temperature.

8.4.3 Simulation results of DNTT-based OTFT

Numerical simulation of the devices and electrical characteristics are measured using ATLAS TCAD by Silvaco International software (Table 8.2).

Figure 8.7(a) shows the transfer characteristics obtained from the TCAD simulation of DNTT-based organic thin-film transistor, and Figure 8.7(b) shows the output characteristic of DNTT-based OTFT.

Table 8.2 Simulation parameters of material of the OTFT [14]

Material simulation parameters	**Value**
DNTT energy bandgap (eV)	3.38 eV
Molecular orbital Of DNTT (highest)	−5.19 eV
Molecular orbital Of DNTT (lowest)	−1.81 eV
Intrinsic p-type doping in DNTT	3×8^{17} cm^{-3}
Aluminum gate work function	4.1 eV
Work function of Au contact	5.0 eV
Dielectric constant	4.5

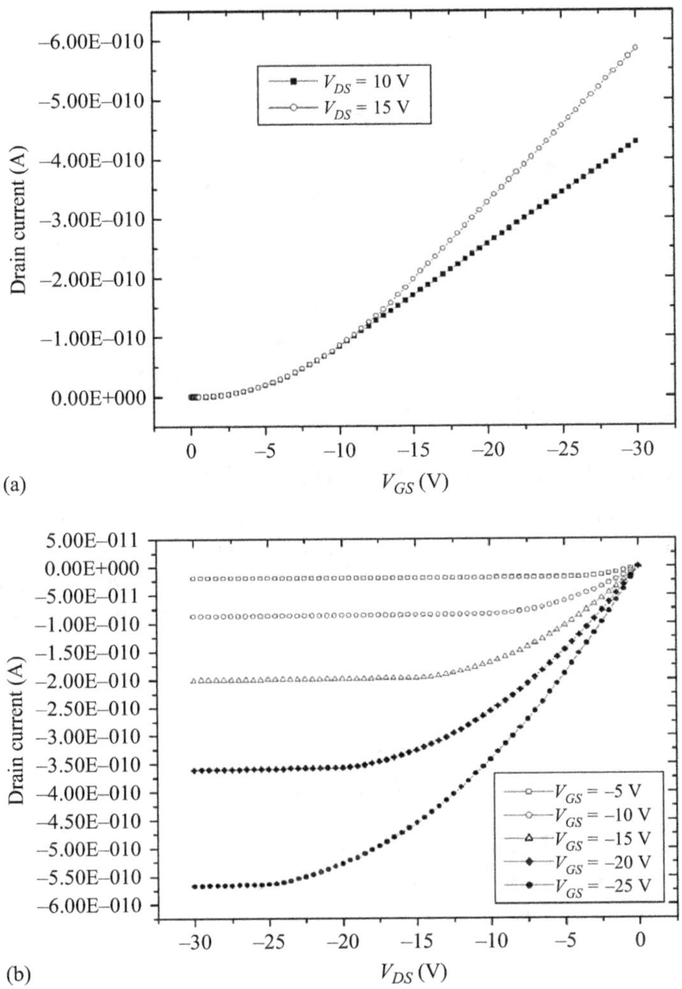

(a) (b)

Figure 8.7 (a) Transfer characteristics of the DNTT-based OTFT simulated results and (b) output characteristics of the DNTT-based OTFT simulated results

8.5 Applications of OTFT

8.5.1 Organic light-emitting diodes (OLEDs)

Display on the flexible substrate like plastic or glass can be made using OTFTs. Liquid crystal display can be made by dot array AM OLED which is available in commercial market, however, due to non-uniformity in the brightness of AMOLED, it is less popular. On the other side, because OTFTs only require a sufficient on/off current ratio, they can be used to manufacture excellent displays

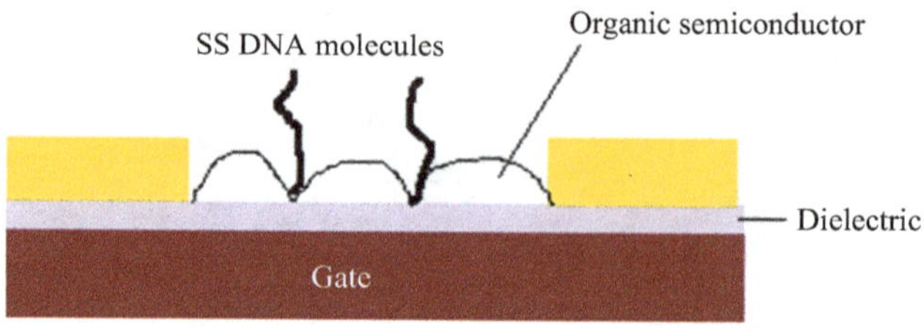

Figure 8.8 DNA sensors using OTFT [8]

for electronic paper or LCDs. OLEDs [21–23] are used in televisions, mobile phones, and various display systems.

8.5.2 Radio frequency identification (RFID) tags

Organic electronic-based devices have a wide range of applications in flexible RFIDs. In order to develop a very low-cost single RFID tag [24], item-level RFID tags have been developed by using OTFTs and printed electronics technologies. Silicon-based RFID tags cannot be used in water and also in the environment which is metal-contaminated.

8.5.3 DNA sensors

The current use of fluorescence scanning and imaging techniques to analyze the microarrays of DNA has less spatial resolution, portability, and sensitivity. These techniques are complicated, expensive, and time-consuming. Using organic TFTs (OTFTs), various DNA sensors have been made. OTFTs fabricated on insulating substrates are very flexible and low cost. OTFTs have various advantages in comparison to silicon wafer-based devices, particularly in disposable application. They can be used directly for sensing applications, as shown in Figure 8.8 [22].

8.6 Conclusion

In this chapter, novel technology and modeling of DNTT-based OTFT is presented by considering field-dependent mobility model and density of states model. Finite element method (FEM)-based device simulation model is able to produce output characteristics in linear and saturation region and transfer characteristics below and above the threshold region of a DNTT-based OTFT. Some important applications, advantages, and disadvantages of OTFTs are also discussed briefly.

Acknowledgments

The authors are thankful to the SERB, DST Government of India, for the financial support of Project no. ECR/2017/000179.

References

[1] M. Mizukami, N. Hirohata, T. Iseki, *et al.* "Flexible AM OLED panel driven by bottom-contact OTFTs" IEEE Electron Device Letters 27, 249, 2006

[2] M. Takamiya, T. Sekitani, Y. Kato, H. Kawaguchi, T. Someya, and T. Sakurai "An organic FET SRAM with back gate to increase static noise margin and its application to Braille sheet display" IEEE Journal of Solid State Circuits 42, 93, 2007

[3] E. Cantatore, T.C.T. Geuns, G.H. Gelinck, *et al.* "A 13.56 MHz RFID system based on organic transponders" IEEE Journal of Solid-State Circuits 42, 84, 2007

[4] D. Brianda, A. Opreab, J. Courbata, and N. Barsanb "Mater making environmental sensors on plastic foils" Today 14, 416, 2011

[5] A. Lodha, and R. Singh "Prospects of manufacturing organic semiconductor-based integrated circuits" IEEE Transactions on Semiconductor Manufacturing 14(3), 281–296, September 2001

[6] L. Petti, N. Münzenrieder, C. Vogt, *et al.* "Metal oxide semiconductor thin-film transistors for flexible electronics," Applied Physics Reviews, 3 021303, 2016

[7] T. Yamamoto, and K. Takimiya "Facile synthesis of highly π-extended heteroarenes, dinaphtho[2,3-b:2′,3′f]chalcogenopheno[3,2-b]chalcogenophenes, and their application to field-effect transistors." Journal of the American Chemical Society 129, 22242225, 2007

[8] B. K. Kaushik, B. Kumar, S. Prajapati, and P. Mittal "Organic Thin-Film Transistor Applications, Materials to Circuits," Boca Raton: Taylor & Francis, CRC Press, 2017.

[9] P. G. LeComber, W. E. Spear, and A. Ghaith "Amorphous-silicon field-effect device and possible application," Electronic Letters 15, 179–181, 1979

[10] A. Tsumura, H. Koezuka, and T. Ando "Macromolecular electronic device: Field effect transistor with a polythiophene thin film," Applied Physics Letters 49, 128–1212, 1986

[11] K. Kudo, M. Yamashina, and T. Moriizumi "Field effect measurement of organic dye films," Japanese Journal of Applied Physics 23, 130–130, 1984

[12] F. Ebisawa, T. Kurokawa, and S. Nara "Electrical properties of polyacetylene/polysiloxane interface," Journal of Applied Physics 54, 3255-1–3255-6, 1983

[13] T. Yamamoto, and K. Takimiya, "Facile synthesis of highly π-extended heteroarenes, dinaphtho[2,3-b:2′,3′f]chalcogenopheno[3,2-b]chalcogenophenes, and their application to field-effect transistors." Journal of the American Chemical Society 129, 22242225, 2007

[14] T. Zaki, S. Scheinert, I. Hörselmann, *et al.* "Accurate capacitance modeling and characterization of organic thin-film transistors" IEEE Transactions on Electronic Devices 61, 98, 2014

[15] A. Buonomo, and C. di Bello "On solving Poisson's equation in two-dimensional semiconductor devices" Electronics Letters 20, 4 1984

[16] D. Hertel and H. Bässler "Photoconduction in amorphous organic solids" ChemPhysChem, 9, 666, 2008

[17] W. van Roosbroeck, "Theory of the flow of electrons and holes in germanium and other semiconductors view document" Bell System Technical Journal 29, 560, 1950

[18] A. Juengel "Drift-diffusion equations," (Springer, 2009) Chap. 5, pp. 99–127.

[19] SILVACO. ATLAS User's Manual – device simulation software (USA, 2008).

[20] C. H. Shim, F. Maruoka, and R. Hattori "Structural analysis on organic thin-film transistor with device simulation" IEEE Transactions on Electronic Devices 57, 195, 2008

[21] D. Li, and J. Guo "Organic thin film transistors and polymer light emitting diodes patterned by polymer inking and stamping." Journal of Physics D, Applied 41, 85115, 2008

[22] Q. T. Zhang and V. Subramanian "DNA hybridization detection with organic thin film transistors: toward fast and disposable DNA microarray chips." Biosensors and Bioelectronics 22 3182–3187, 2007

[23] F. Yan, M. Man, Y. Sheung, Jinjiang, H. L. W. Chan, and Y. Mo "Label-free DNA sensor based on organic thin film transistors" Biosensors and Bioelectronics. 24, 1241–1245, 2009.

[24] P. F. Baude, D. A. Ender, T. W. Kelley, M. A. Haase, D. V. Muyres, and S. D. Theiss "Organic semiconductor RFID transponders." IEEE International Electron Devices Meeting Technical Digest 8.1.1–8.1.4, 2003

Chapter 9

Doping-free tunnelling transistors – technology and modelling

Chitrakant Sahu[1] and Avinash Lahgere[2]

As device dimensions are continuously down-scaling into sub-10 nm regimes, complementary metal-oxide semiconductor technology is facing severe challenges such as increased static power consumption, poor gate controllability, enhanced short-channel effects (SCEs), random dopant fluctuations (RDFs) and process-voltage temperature (PVT) variation, hence, conventional metal-oxide-semiconductor field-effect transistors (MOSFETs) have failed to be a worthy candidate for nano-electronics and micro-electronics regime. Recently, the tunnel FETs (TFET) gain tremendous attention because of their low standby power consumption and scalable subthreshold swing (SS). The TFET is a gated P-I-N diode, where ON-state current would be due to band-to-band tunnelling (BTBT) instead of thermionic emission, and they exhibit very low OFF-state current of the order of fA/μm, which makes them a potential candidate for low power consumption. The heavily doped nature of TFETs causes certain problems, and to address them, the concept of dynamically configurable doping-free (DF) TFETs was recently proposed. In this chapter, a detailed analysis of dynamically configurable TFETs such as the working principle, *I–V* characteristics, fabrication flow and the effect of process and temperature variation is presented.

9.1 Introduction

As we are moving towards sub-20 nm regime, the tunnel field-effect transistors (TFETs) are captivating tremendous attention as the semiconductor switches. To know why TFETs can be used as the semiconductor switch, it is essential to comprehend what are the problems with conventional metal-oxide-semiconductor field-effect transistors (MOSFETs). In order to understand that, we will begin with the scaling of conventional MOSFETs and its merits and demerits.

According to Dennard's scaling rules, the doping of source (S) and drain (D) regions should increase by a factor of *S*, while keeping the electric filed inside the

[1]Department of Electronics and Communication Engineering, Malaviya National Institute of Technology, Jaipur, India
[2]Department of Electrical Engineering, Indian Institute of Technology Delhi, Delhi, India

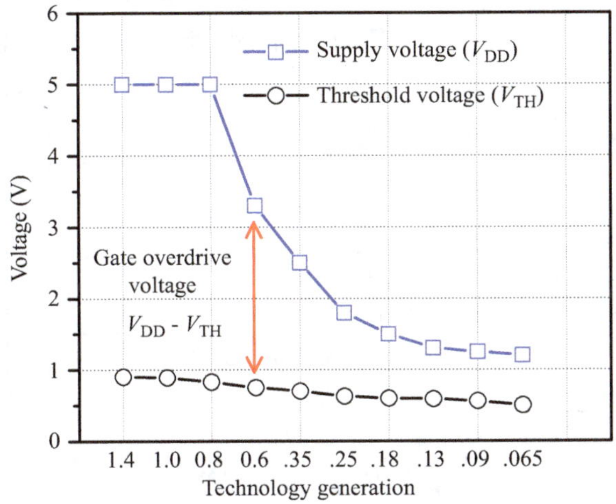

Figure 9.1 The trend of supply voltage and threshold voltage scaling vs. technology generations [1]

device unchanged [1]. In addition, all the device dimensions and applied voltages should be scaled by 1/*S*. These rules have been roughly followed ever since, until rather recently. The reason why Dennard's scaling rules are no longer to be followed for the conventional MOSFETs scaling as they were followed in the past because of the threshold voltage (V_{TH}) did not follow the natural rule of Dennard scaling. One can observe from Figure 9.1, as we are scaling the technology node from the 1.4 um to 65 nm node, the supply voltage (V_{DD}) decreased to about 78% of its original value, while the V_{TH} only went down to approximately half of its starting value, consequently, gate overdrive voltage is also cut down [2].

The most serious consequence of gate overdrive voltage reduction causes decrement in ON-state current (I_{ON}). In addition, decrement in I_{ON} directly affects the conventional MOSFET performance, the I_{ON}/I_{OFF} ratio, and dynamic speed (C_G*V_{DD}/I_{ON}). So, in order to acquire high gate overdrive voltage for keeping the conventional MOSFET performance good, there are two possible solutions, which are as follows:

- Scaling of threshold voltage, and.
- Need of slow supply voltage (V_{DD}) scaling.

9.1.1 Scaling of threshold voltage

In conventional MOSFETs, carriers flow from S to D regions over the potential barrier using thermionic emission mechanism, as shown in Figure 9.2. Since thermionic emission is a thermal process, its SS is constrained by thermal limit, i.e. kT/q mV, which is given in (9.1), where k is the Boltzmann constant, T is the temperature in Kelvin and q is the carrier charge. This thermal constraint corresponds to 60 mV/dec at room temperature for the conventional MOSFETs [3]:

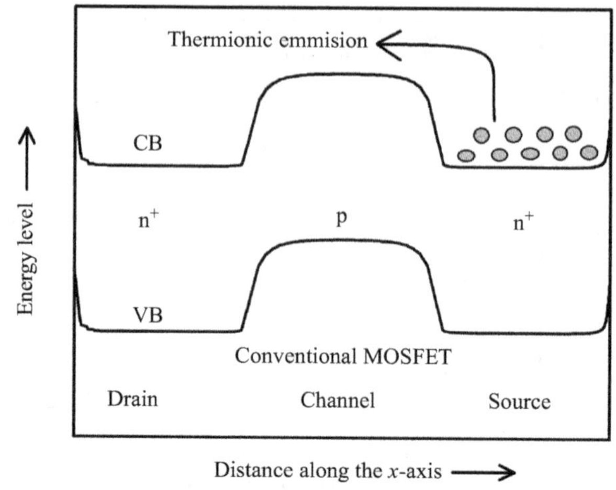

Figure 9.2 Energy band diagram of a MOSFET in which current takes place by thermionic emission of carriers

$$SS = \frac{dVg}{dlogId} = 2.33\frac{mkT}{q} \tag{9.1}$$

where m is the body-effect coefficient, whose value is close to 1 for a well optimized device. To be more precise, $m = 1 + C_{DM}/C_{OX}$ [4], where C_{DM} is the bulk depletion capacitance at threshold, when $\phi_s = 2\phi_F$. A MOSFET with very good gate control leads to COX/CDM $\ll$ 1, then m can be approximated to 1. Then (9.1) becomes $S = \ln(10) \times 26$ mV, or about 60 mV/decade. Then this limit in SS has impeded the scaling of the threshold voltage as can be seen in Figure 9.1.

The only way to scale the threshold voltage further is to reposition the I_{DS}–V_{GS} characteristic of the conventional MOSFET horizontally left on the x-axis, as shown in Figure 9.3(a). We can infer that if we want to shift V_{TH} by 60 mV, then we have to compensate an increase of one decade of OFF-state current (I_{OFF}), and in turn of passive power dissipation.

9.1.2 Need of slow supply voltage (V_{DD}) scaling

In order to maintain the performance of a conventional MOSFET that is I_{ON}/I_{OFF} ratio, and dynamic speed ($C_{G}*V_{DD}/I_{ON}$) acceptable, supply voltage scaling has been slowed down drastically. If supply voltage does not scale, then active and passive power densities increase drastically. For the conventional MOSFET, the active and passive power densities can be expressed as [5]:

$$P_{active} = fC_L V_{DD}^2 \tag{9.2}$$

where f is the frequency, and C_D is the total switched capacitive load, and

$$P_{Passive} = I_{leak} V_{DD} \tag{9.3}$$

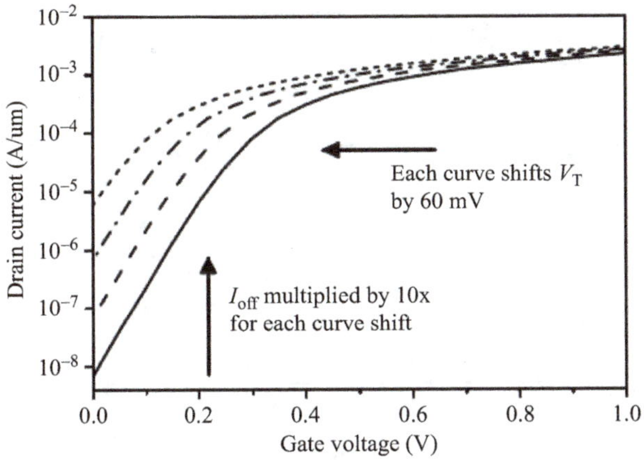

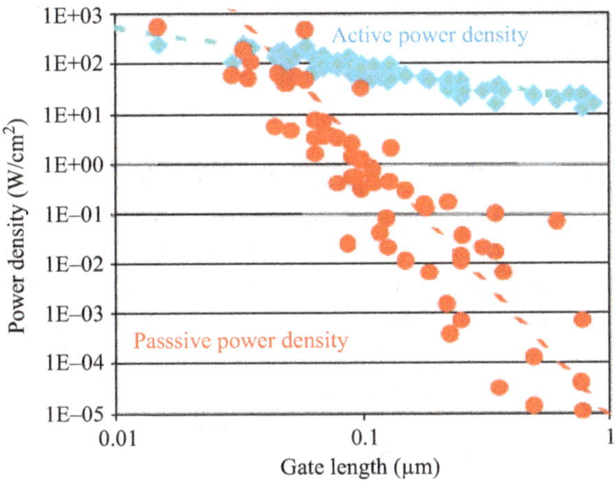

Figure 9.3 (a) I_{DS}–V_{GS} characteristics of the conventional MOSFET [4], and (b) active and passive power densities plotted against gate lengths [5]

where I_{leak} is the sum of leakage currents in the device when the conventional MOSFET is in OFF-state. One can observe from Figure 9.3(b), the active as well as the passive power densities are increasing continuously with scaling of gate length. This is becoming a major concern in semiconductor devices.

Why power consumption is such a problem? There are some fundamental problems, some of which will be mentioned here. The first reason is increasing cost of powering and cooling the servers. On a more personal level, mobile devices suffer from shorter battery life due to increased power dissipation. In addition, Figure 9.4 illustrates the power dissipation and heatsink size (volume) of various Intel processors [6]. One can depict that, as we are moving from Intel 386 to Core 2 Duo, the power and heatsink volume have increased by approximately 92.34%. Since, we would like our appliances, computers and gadgets to stay the same size or shrink, not get larger in order to accommodate a large heatsink required by the power-hungry chip inside.

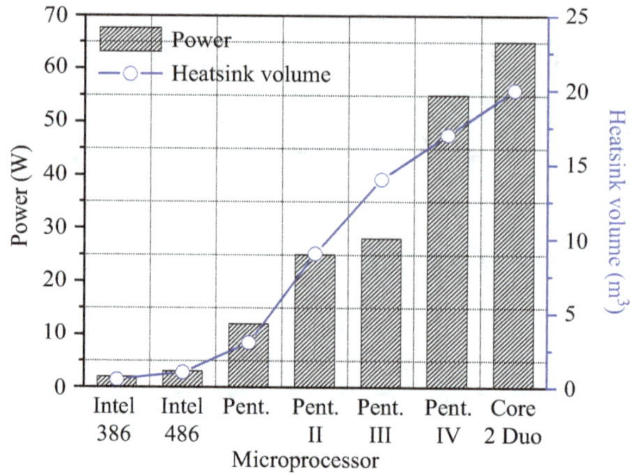

Figure 9.4 Computer chip power trends, along with the accompanying increase in heatsink volume [6]

9.1.3 Possible solution to the power consumption

For obtaining low power consumption, an alternate current transport mechanism is highly desirable that overcomes the 60 mV/dec limit of SS in conventional MOSFETs [7–9]. In other words, a new current mechanism should be exploited that is not based on semi-classical current transport phenomena. That is, carriers do not have to travel over a potential barrier by using thermionic emission. To alter the current transport mechanism many researchers have studied that as transistors have gotten smaller, the distances between different transistor regions have decreased. So, electronic barriers that were once thick enough to restrict current, are now so thin that electrons can barrel right through them. Since we cannot stop electrons from tunnelling through thin barriers, it is better we can turn this phenomenon to our advantage. In the last few years, the BTBT mechanism is a promising candidate for the same. It is basically a quantum mechanical phenomenon where carriers travel through the barrier instead of over from the potential barrier [10]. The TFETs are the ones which are based on BTBT current transport mechanism [11], and they will now be presented in great detail in the next section.

9.2 Tunnel field-effect transistor

9.2.1 Operating principle of TFET

BTBT of electrons across the tunnelling junction is the prominent injection mechanism in the TFETs. The TFET is a gated reverse-biased P-I-N diode and its device architecture is shown in Figure 9.5. In a TFET, the tunnelling junction is created at the source-channel junction by applying a gate bias. As shown in the energy band profile of Figure 9.6(a), in the OFF-state ($V_{GS} = 0$ and $V_{DS} = V_{DD}$), due to the wider tunnel barrier width at the source-channel junction, the OFF-state current

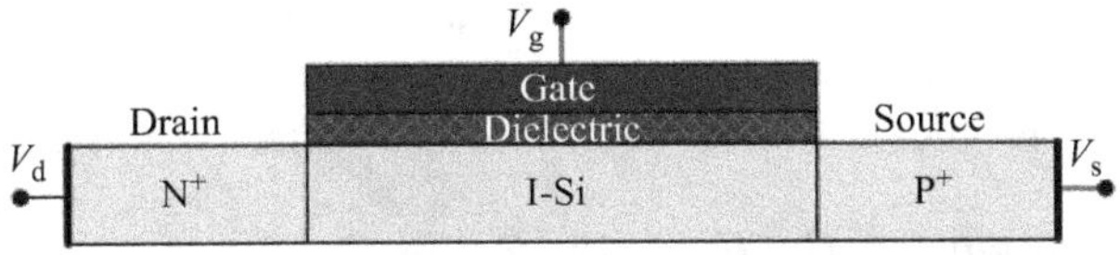

Figure 9.5 Standard n-type tunnel FET structure

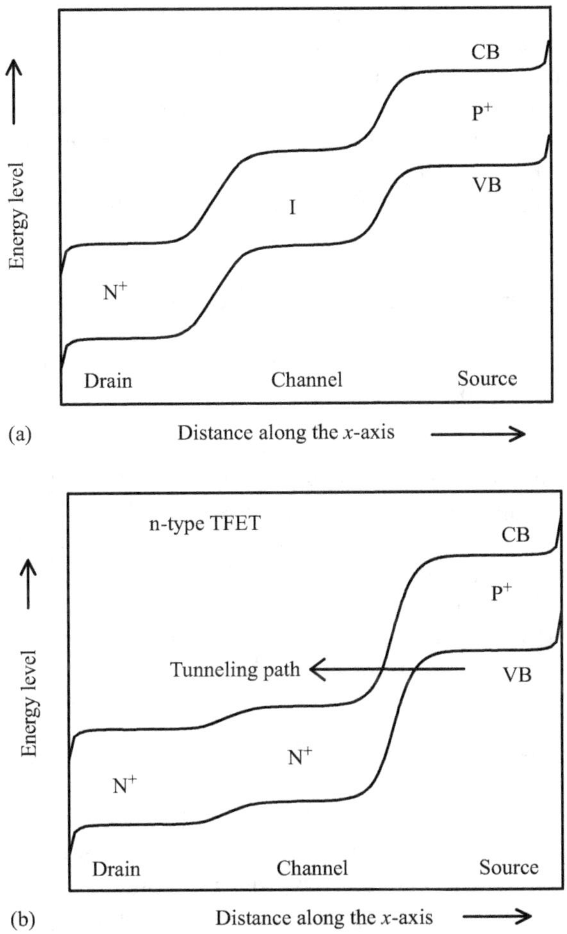

Figure 9.6 Energy band diagram of (a) OFF-state and (b) ON-state for n-type TFET in which current carries by BTBT

is extremely low and is equivalent to the reverse leakage current of the P-I-N diode. By modulating the channel potential, the gate (i) opens the tunnelling window by aligning the minimum of conduction band of the channel with the maximum of valance band of the source, as shown in Figure 9.6(b) and (ii) modulates the effective tunnel barrier width at the source-channel junction, which determines the probability of tunnelling and thus the level of injection. This role of the gate in controlling the tunnelling current enables the TFET to operate as an effective switch.

9.2.2 The conventional TFET limitations

Apart from having so many advantages over conventional MOSFET, the tunnel field transistor poses severe challenges such as:

- The random variability in transistor performance due to RDFs can become significant. The effects of RDF, such as an unacceptably large increase in the OFF-state current and threshold voltage variation.
- The presence of doped source and drain regions in TFETs also necessitates a complex thermal budget due to the need for ion implantation and expensive thermal annealing techniques.
- Abrupt junctions are essential for efficient tunnelling in TFETs. However, creating abrupt junctions using high-temperature processes is not easy due to the diffusion of the dopant atoms from the source/drain regions into the channel.
- PVT-induced variations, such as RDF, thickness of silicon thin film and gate oxide with temperature, enormously affect the device electrical parameters. This variation significantly increases the chances of failure and cuts down the reliability margins of the devices.

From the above discussion, it is clear that the TFETs also have major limitations, and these limitations must be addressed before exploiting their full potential for future nano-complementary metal-oxide semiconductor (CMOS) computing applications. It can be concluded that a major source of these limitations may be high doping concentration requirements for TFET. Intuitively, if we reduce the doping concentration in the TFET, there will be chances to get rid of the PVT variability due to RDFs. However, reduction in doping concentration in TFET will reduce the drive (ON-state) current significantly.

9.3 DF dynamically configurable TFET

In TFETs, PVT-induced variations, such as RDF, thickness of silicon thin film and gate oxide with temperature, enormously affect the device electrical parameters [12,13]. The major issue of variability is from RDF due to the presence of highly doped junctions which may be mitigated by considering a lightly doped silicon film from the source to the drain regions. As in TFET, it is equally important to have high doping concentration throughout from source, channel, and drain regions to maintain sufficient ON-state current. Therefore, to meet this requirement in DF dynamically configurable TFET, with the gate metal electrodes at source, gate, and drain regions are attached over a thin intrinsic (undoped/DF) silicon nanowire of the same work functions. In the proposed device, DF (or lightly doped) Si is used and formation of drain (D)-channel-source (S) is achieved with the application of appropriate bias at polarity gates (PGs) (i.e. electrically-doped) yield P^+ and N^+

S/D regions, hence, device becomes dynamically configurable to switch between n- and p-type TFETs. These unique features eliminate the requirements of abrupt doping profile at junctions, higher thermal budget and ion-implantation process for device fabrications. The combination of junctionless and doping mechanisms in the proposed device yields simpler fabrication process and very good electrostatic control over the channel.

9.3.1 Device structure and simulation parameter

Figure 9.7(a) and (b) shows the 2-D schematic view of the conventional TFET and proposed DF-TFET, respectively. The simulation parameters for both devices are indicated in Table 9.1 [14]. The source and drain regions (either P^+ or N^+) are electrically induced on ultrathin lightly doped silicon body by applying an external bias to the PGs of same metal such as TiN, as control gate (CG). Further, source and drain contacts are made up of nickel silicide (NiSi) as it features near mid-gap work function with barrier height of 0.45 eV. The configurable device structure

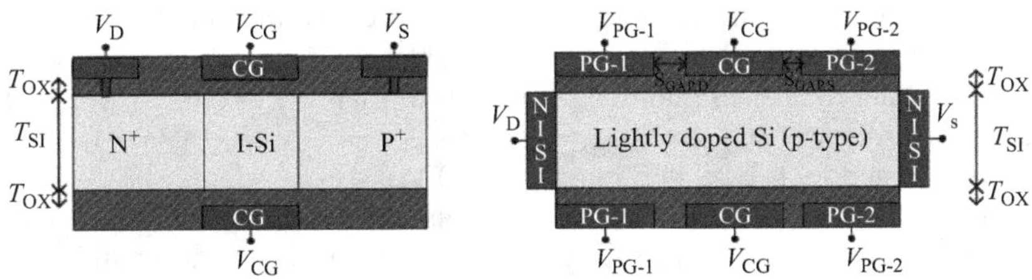

Figure 9.7 Cross-sectional view of (a) conventional TFET [8] and (b) proposed DF-TFET

Table 9.1 Simulation parameters for conventional and DF-TFETs

Parameters	**Values**
N_D	5×10^{18} cm^{-3}
N_A	1×10^{20} cm^{-3}
Channel doping	1×10^{17} cm^{-3}
Lightly doped Si	1×10^{15} cm^{-3}
W.F. of CG, PG-1, and PG-2	4.5 eV
T_{SI}	10 nm
T_{OX}	0.8 nm
L_{CG}, $L_{PG\text{-}1}$ and $L_{PG\text{-}2}$	50 nm
$S_{GAP;S}$	5 nm
$S_{GAP;D}$	20 nm

shown in Figure 9.7, is simulated using 2D ATLAS TCAD tool [15]. The working mechanism of the proposed device is illustrated by the electrical characteristics and functionalities supported by TCAD simulator after incorporating different models. In addition, the potential benefits of polarity control concepts are as follow:

- It allows the degree of freedom to dynamically switch between n- and p-type DFTFETs. In other words, we can change polarity during operation by setting the PGs bias and obtained a different functionality from the same circuit.
- It is easy to fabricate and could be less sensitive for SCE than the conventional TFETs, since the device does not require any exotic materials or advanced techniques, such as strain technology, etc.
- It does not require thermal annealing and ion implantation process, and using polarity control concept we can approach large-scale integration.

Moreover, the conventional TFET and proposed DF TFET are analysed using 2-D ATLAS Silvaco simulator. To consider the BTBT, the non-local BTBT model is incorporated in the device simulation. Along with this trap-assisted tunnelling (TAT) by Schenk is also incorporated in device simulation.

The dynamic configurable TFET can be achieved by applying the appropriate biasing on PGs. To show the dynamic configurable TFET initially, the energy band diagrams for different configurations is shown in Figure 9.8(a) and (b). The bias condition for n-type TFET are $V_{CG} = 0$ V, $V_{PG\text{-}1} = 1.2$ V, $V_{PG\text{-}2} = -1.2$ V, and $V_{DS} = 50$ mV and for p-type TFET are $V_{CG} = 0$ V, $V_{PG\text{-}1} = -1.2$ V, $V_{PG\text{-}1} = 1.2$ V, and $V_{DS} = -50$ mV TFETs in the OFF-state can be seen in Figure 9.8(a). One can observe that both devices (TFETs) in OFF-state offer sufficient barrier and obstruct the flow of carriers. Similarly, on the application of control signal (i.e. ON-state), there is an occurrence of appropriate band-bending for n- and p-type TFET, as shown in Figure 9.8(b).

To validate the dynamic configurability of the proposed device, Figure 9.9 shows the carrier concentration contour plots for the proposed device. These plots help us in understanding how the electron and hole carrier concentrations vary across the device under OFF-state and ON-state conditions. On the application of suitable bias over PGs, the configurable FET shows perfectly doped S/D regions under different configurations. For example, an n-TFET in OFF-state ($V_{CG} = 0$ V, $V_{PG\text{-}1} = 1.2$ V, $V_{PG\text{-}2} = -1.2$ V, and $V_{DS} = 50$ mV) perfectly exhibits the N^+-I-P^+ structure, however, on the application of control signal (i.e. ON-state), the structure becomes N^+-N^+-P^+. Hence, carrier concentration in the ON and OFF-states of configurable TFET under different configurations perfectly matches with static conventionally doped TFET.

9.3.2 Proposed fabrication process flow

Fabrication of the device require top to down approach. The nano-channels/wires in the device are formed by deep reactive ion etching Bosch process step [11].

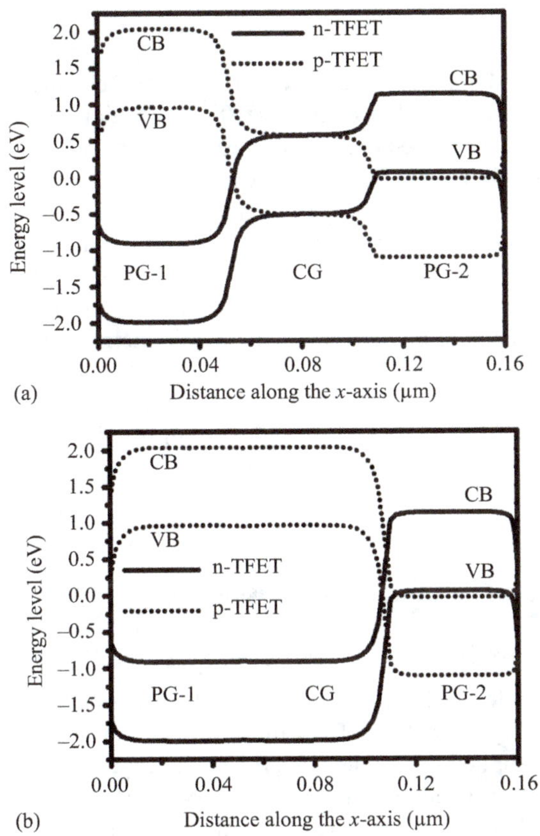

Figure 9.8 Band diagrams of a configurable n- and p-TFET under (a) OFF-state and (b) ON-state condition

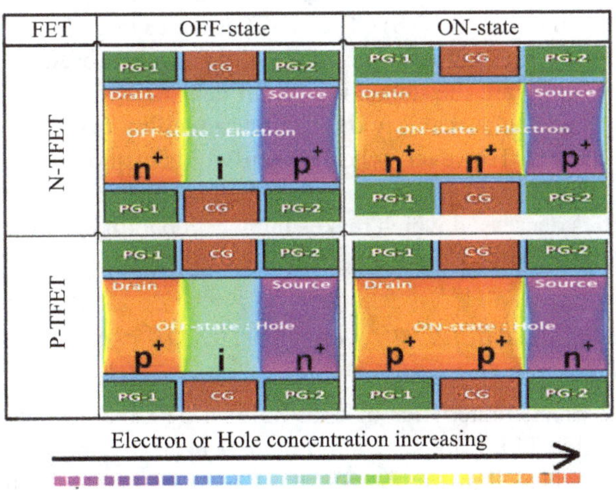

Figure 9.9 Carrier concentration of a configurable n- and p-TFET under (a) OFF-state and (b) ON-state condition

SOI substrate is used and a vertical stack of four nano channels/wires of 350 nm lengths on a p-type lightly doped (nearly 10^{15} atoms/cm^{-3}) silicon on insulator substrate. In order to form HSQ patterns of 60 nm wide and 350 nm long patterns, e-beam lithography is done to obtain nano channels/wires through single Bosch DRIE process [16]. Then self-limiting oxidation followed by poly-Si deposition is required to form gate oxide. To form PG all around the nano channels/wires, PG material with mid-gap work function need to be deposited and then it can be patterned through e-beam lithography [17]. To form the CG, another oxidation and poly-Si deposition need to be performed and then patterning is required. Two PGs and CG is isolated by deposition and patterning by forming spacer of Si_3N_4. Then, to form the source and drain contacts, Ni is deposited through e-beam evaporation and need to anneal around 200 °C–400 °C to form silicide (NiSi). Further wet etching through hot Piranha is required to strip /remove/wipe out the Nickel which is unreacted.

9.4 Simulation results and discussion

9.4.1 Carrier concentration and energy band diagram

The electron and hole carrier concentration in the OFF-state and ON-state conditions is shown in Figure 9.10(a). To realize a P-I-N kind device, the appropriate voltage biases are applied across the PGs. As a result, n-type and p-type doping can be acquired in different regions. It can be observed that the carrier concentration near the source and the drain is falling due to the NiSi contact, however, the surface concentration is similar to the conventional TFET. Moreover, when $V_{CG} = 1.2$ V, the electron concentration in intrinsic (I) region becomes of the order of 10^{19} cm^{-3} likely to the N^+ region and which ensures an abrupt p–n junction for tunnelling of electron from valence band (VB) of P^+ to conduction band (CB) of I region.

Furthermore, in Figure 9.10(b) CB and VB energies in OFF- and ON-states are shown for analysing the tunnelling mechanism in the proposed device. It can be seen that in the OFF-state the tunnelling width near the CG and PG-2 junction is very large, therefore, the probability of electron tunnelling from VB of P^+ to CB of I is also very minimal. However, when a positive bias is applied to CG (i.e. ON-state), CB and VB in I region get aligned with CB and VB of N^+ region, as a result, tunnelling width is decreased which makes higher probability of electron tunnelling from VB of P^+ into CB of I region.

9.4.2 Transfer characteristics comparison of conventional and DF-TFET

The I_D–V_G transfer characteristics of conventional TFET and proposed DF dynamically configurable TFET are shown in Figure 9.11(a). It can be observed that the proposed device is also showing the SS below 60 mv/dec, similar to conventional TFET. Moreover, the proposed device exhibits very low OFF-state current of the order of 10^{-19} A/mm. This happens due to lightly doped Si film on SOI and it does

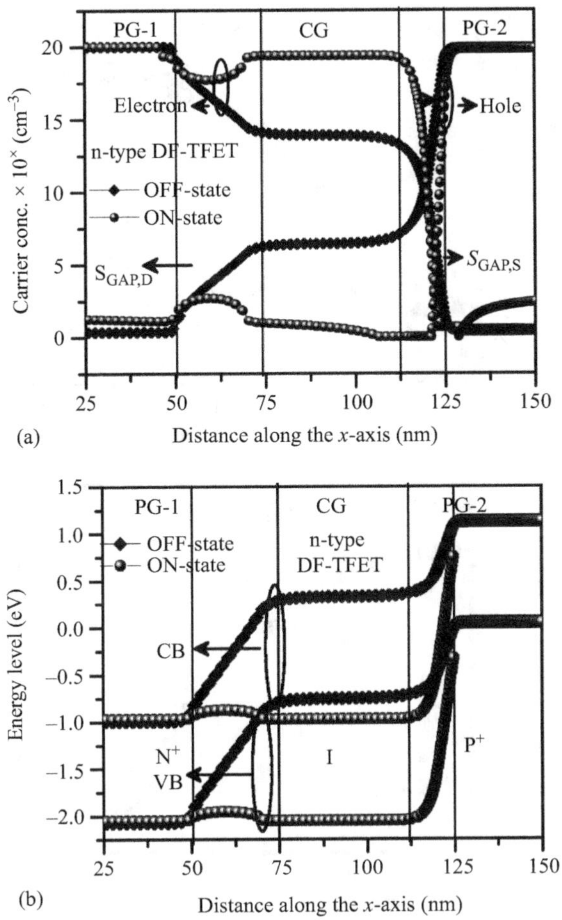

Figure 9.10 *(a) Electron/hole concentration, and (b) energy band diagram of n-type DF-TFET ($V_{DS} = 1.0$ V, $V_{PG\text{-}1} = 1.2$ V, and $V_{PG\text{-}2} = -1.2$ V) in OFF- ($V_{CG} = 0$ V) and ON-state ($V_{CG} = 1.2$ V)*

not have any p-n junctions in thermal equilibrium. Further, to guarantee the dynamic configurability and symmetric I_D–V_G characteristics of the proposed device, we reversed the bias across PGs that represents p-type DF dynamically configurable TEFT and observed I_D–V_G characteristics, as shown in Figure 9.11(b).

However, the ON-state current of the proposed device is smaller than the conventional TFET as shown in Figure 9.11(a). It is because the resistance offered by the proposed device in the ON-state is higher than the conventional TFET, which is similar to the past reported results [14]. Furthermore, for conventional TFET, the ON-state resistive components are channel resistance (R_{ch}) and

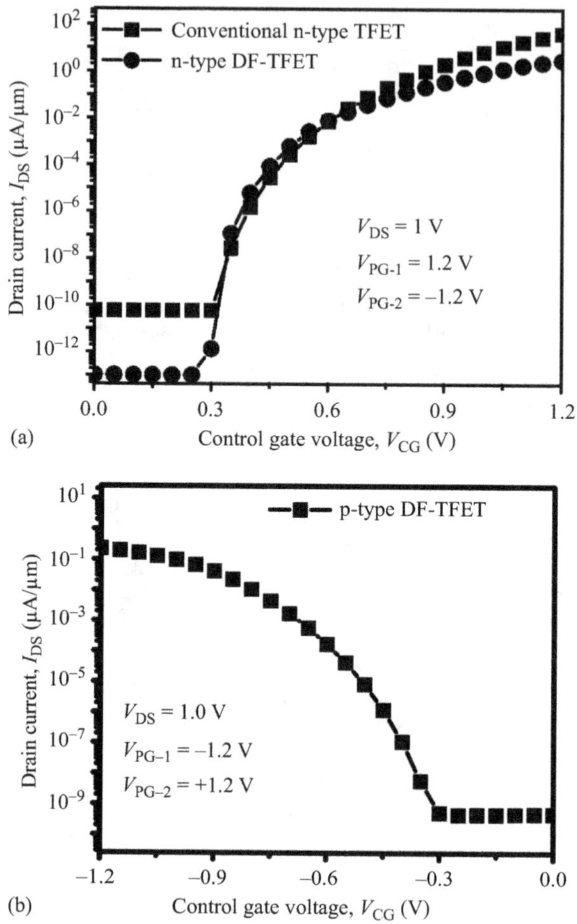

Figure 9.11 *(a) ID-VG characteristics of conventional and proposed n-type TFETs for $V_{DS} = 1.0$ V and (b) I_D–V_G characteristics for p-type DF-TFET*

tunnelling width resistance (R_{tunnel}) whereas, for DF-TFET there are two more additional resistive components, i.e. due to source side spacer ($R_{SGAP;S}$) and drain side spacer ($R_{SGAP;S}$), as shown in Figure 9.12(a) and (b).

9.4.3 Output characteristics of conventional and DF-TFET

In Figure 9.13(a) and (b), we have shown the output characteristics of DF dynamically configurable TFET. We can see from Figure 9.13(a) and (b) that the proposed device clearly demonstrates the triode and saturation region of operations.

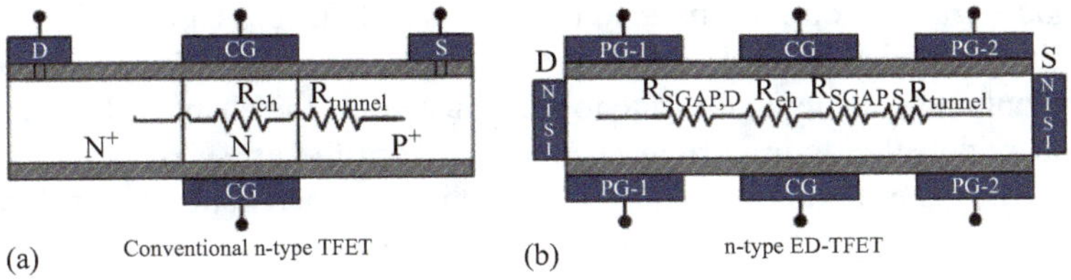

Figure 9.12 ON-state resistive components of (a) conventional TFET and (b) DF-TFET

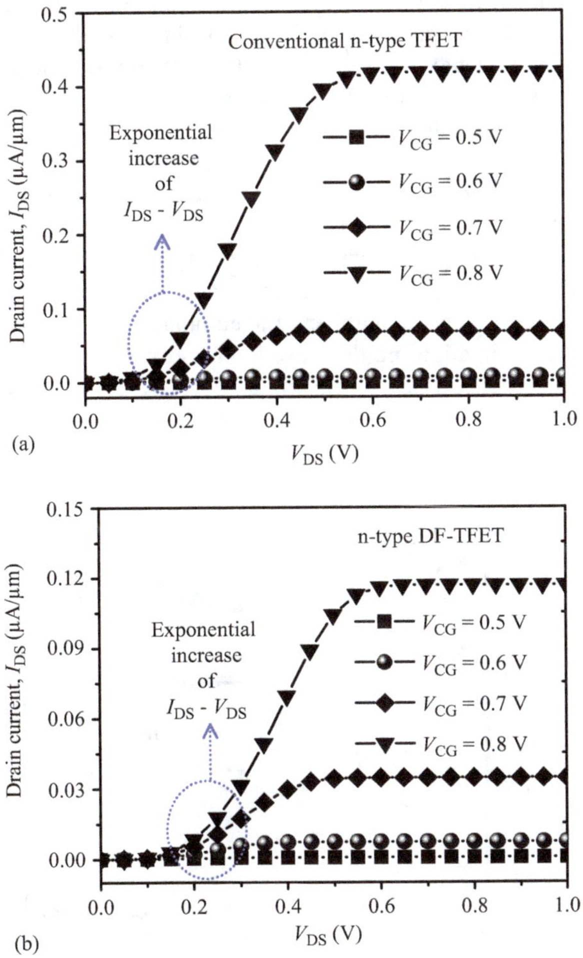

Figure 9.13 Output characteristics of (a) conventional TFET and (b) n-type DF-TFET for different control-gate (CG) voltages

The triode region in output characteristic can be defined as $V_{DS} < V_{GS} - V_{TH}$. Further, in the triode region, I_D and V_{DS} follow a linear relationship, and saturation region in output characteristic is due to the pinched off the channel. Further, in the saturation region the drain current (I_D) is free from the effect of drain to source voltage (V_{DS}). The saturation region in the output characteristic can be defined as $V_{DS} \geq V_{GS} - V_{TH}$.

9.4.4 Impact of supply voltage and PG bias scaling on DF-TFET

In addition, we have also investigated the impact of voltage supply and PG bias scaling on device performance. The impact of supply voltage (V_{DS}) scaling from 0.5 V to 0.3 V reduces I_{ON} approximately 88%, as shown in Figure 9.14, which is happening because of the decrement in the lateral direction electric field, however, it has insignificant effect on leakage current. Furthermore, the proposed device does not require ion-implantation and high-temperature annealing process [18] that makes it less sensitive to RDF.

Figure 9.15(a) and (b) shows the impact of PG bias on transfer characteristics and OFF-state band diagrams of the proposed device. As the PG bias reducing the OFF-state current is increasing, this is similar to the previously published results. It happens due to reduction in the energy barrier at the drain terminal, which allows carrier to go from drain region to the channel region. In other words, reduced PG bias results in decrement in the number of accumulation carriers due to less longitudinal electric field. In addition, decrease in carrier concentration lower the bend edges in drain side that can be seen in Figure 9.15(b), consequently higher OFF-state current.

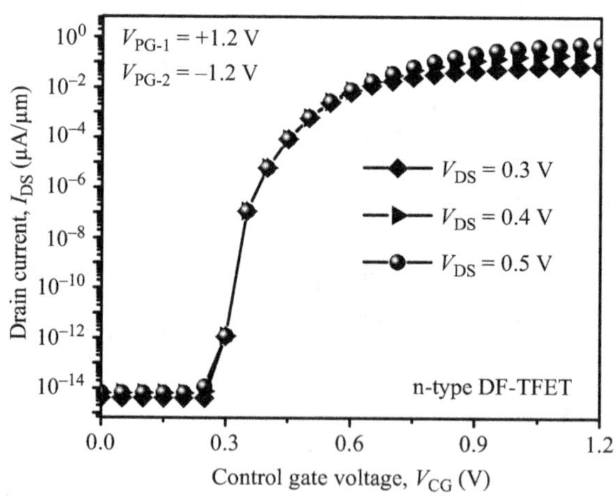

Figure 9.14 I_D–V_G characteristics of n-type DF-TFET for different V_{DS}

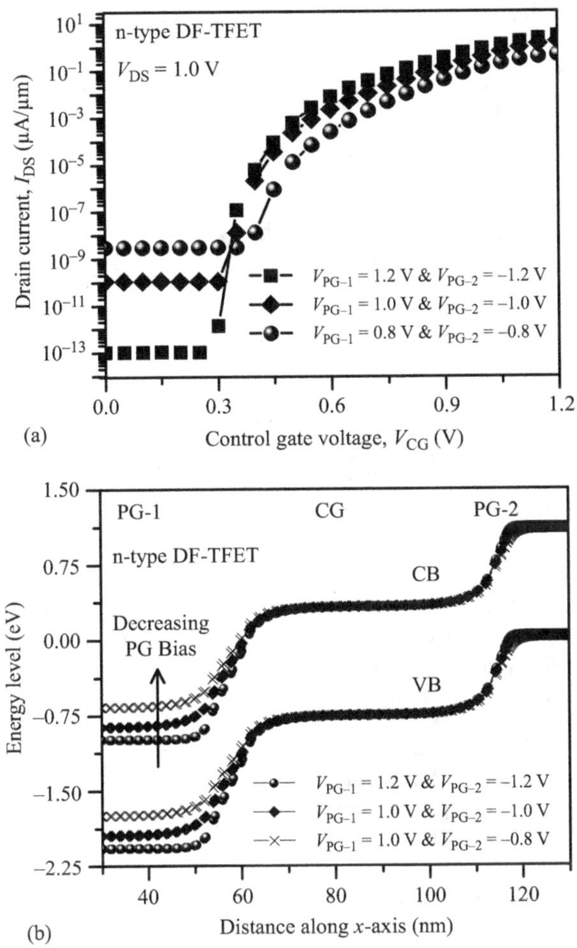

Figure 9.15 Impact of PG bias on (a) transfer characteristics and (b) OFF-state band diagrams

9.4.5 Impact of control gate voltage on tunnelling rate and energy barrier width

In addition, for obtaining a large ON-state current and steep SS in TFET, it is indeed to have the tunnelling width small and the tunnelling rate higher. Therefore, Figure 9.16 shows the effect of V_{CG} over energy barrier width and electron tunnelling rate for the proposed device. The energy barrier width is exponentially reduced from 20 nm to 10 nm as we are increasing V_{CG} from 0 V to 0.4 V, however, further increase in V_{CG} till 1.0 V exhibits a very less effect on the energy barrier width. Also, as V_{CG} increases, the energy in the intrinsic region (I) gets down, as a result the tunnelling width reduces and the tunnelling rate increases.

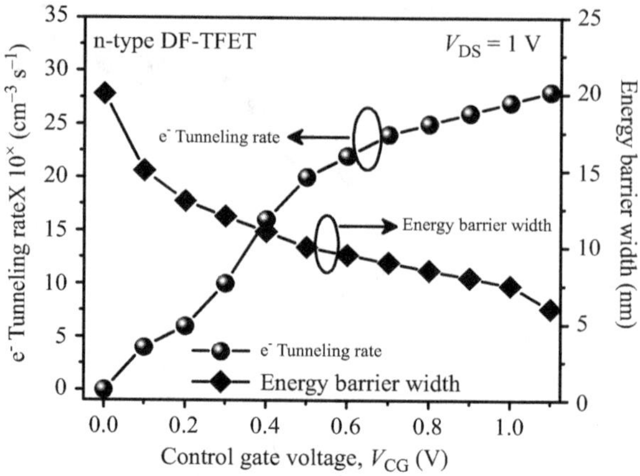

Figure 9.16 Effect of V_{CG} over energy barrier width and electron tunnelling rate for n-type DF-TFET

Since energy barrier width is directly related to the electron tunnelling rate, in other words, lower the energy barrier width higher the electron tunnelling rate and consequently higher will be I_{ON}. One can justify the same with the (9.4), which is based on the WKB approximation:

$$T_{WKB} = \exp\left(\frac{-4\lambda\sqrt{E_g{}^3}\sqrt{2m^*}}{2qh(E_g + \Delta\phi)}\right) \tag{9.4}$$

where m^* is the effective carrier mass, E_g is the bandgap, Δf is the energy range over which tunnelling can take place, and λ can be given as:

$$\lambda = \sqrt{\frac{\in_{SI}}{\in_{OX}} T_{OX} T_{SI}} \tag{9.5}$$

where T_{OX} , T_{SI} , $\in_{SI}$ and $\in_{OX}$ are the silicon and oxide film thickness and dielectric constant, respectively.

9.4.6 *Impact of source spacer thickness*

In the proposed device, energy barrier width and tunnelling rate can also be influenced by source side spacer thickness ($S_{GAP;S}$), consequently I_{ON} and SS are shown as functions of $S_{GAP;S}$ in Figure 9.17. An increment in $S_{GAP;S}$ causes a reduction in I_{ON} due to increased tunnelling width and decreased tunnelling of electrons which is coherent with the past reported results [8]. Therefore, $S_{GAP;S}$ should be as small as possible to get better tunnelling efficiency, as a result, very good I_{ON} and SS can be achieved.

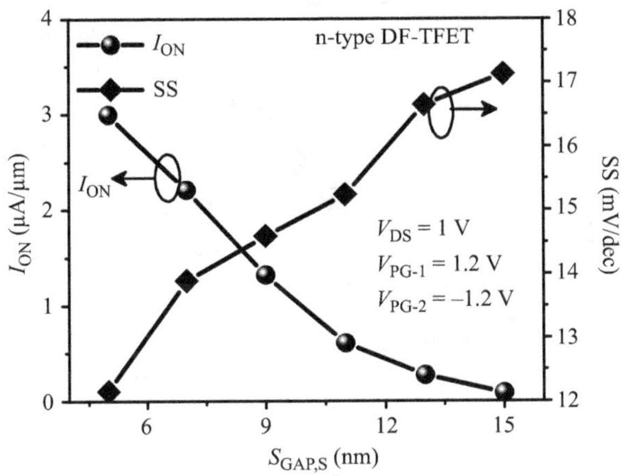

Figure 9.17 ON-state current (I_{ON}) and SS of n-type DF-TFET for different source side spacer thickness ($S_{GAP;S}$)

9.4.7 Sensitivity towards control gate length scaling

In Figure 9.18(a) and (b), the effect of CG length scaling on transfer characteristics of conventional TFET and the proposed DF dynamically configurable TFET is analysed by keeping the spacer thickness and device length from source to the drain same. One can see that as the CG length is reducing, the I_{OFF} current for both devices is increasing. It is because the tunnelling barrier width at source junction is reducing, which is similar to the past literature [14]. It can also be observed that the variation in I_{OFF} for the proposed device is much smaller than the conventional TFET. This happens because the CG length scaling results in higher reduction in effective channel length in conventional doped devices as compared to the dopingless FETs [19,20]. Furthermore, as the CG length scales from 50 nm to 20 nm, the I_{ON}/I_{OFF} ratio for the proposed device is almost same of 10^{12}, whereas the conventional TFET shows variation from ~10^{11} to 10^{8}. Therefore, the proposed device can easily be scaled down without affecting the I_{ON}/I_{OFF} ratio.

9.4.8 Sensitivity towards temperature

The environmental temperature has very influences on the device electrical parameters, therefore, in Figure 9.19(a)–(d), the temperature dependence on I_{ON} and V_{TH} (keeping other device parameters constant) is observed for both devices. It can be seen that as the temperature is increasing the I_{ON} is increasing and the V_{TH} is decreasing, this happens due to bandgap narrowing. Moreover, the simulation results show that the conventional TFET is much more sensitive to the temperature than the proposed DF dynamically configurable TFET. For example, at $T_{SI} = 6$ nm, V_{TH} increases 0.36 mV/K and I_{ON} changes 0.2026 μA/K, while in the case of the proposed device, these changes remain 0.10 mV/K and 0.076 μA/K, respectively, for per K change in temperature when scaled from 500 K to 100 K. Similarly, for

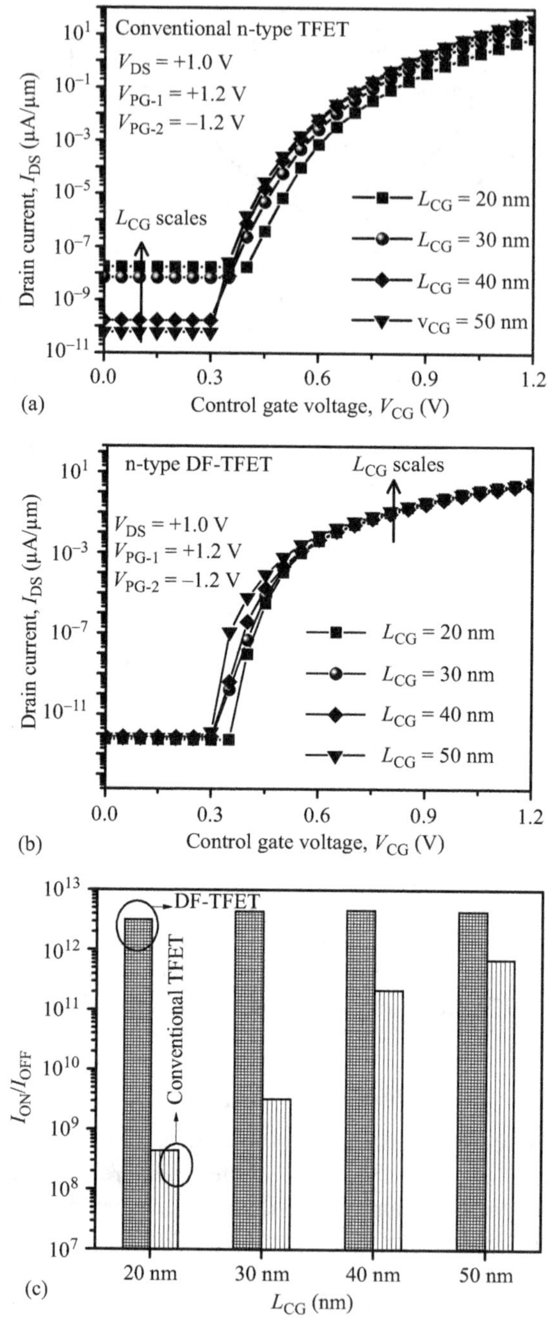

Figure 9.18 Impact of CG length scaling on transfer characteristics (a) conventional TFET, (b) DF-TFET, and (c) I_{ON}/I_{OFF} ratio

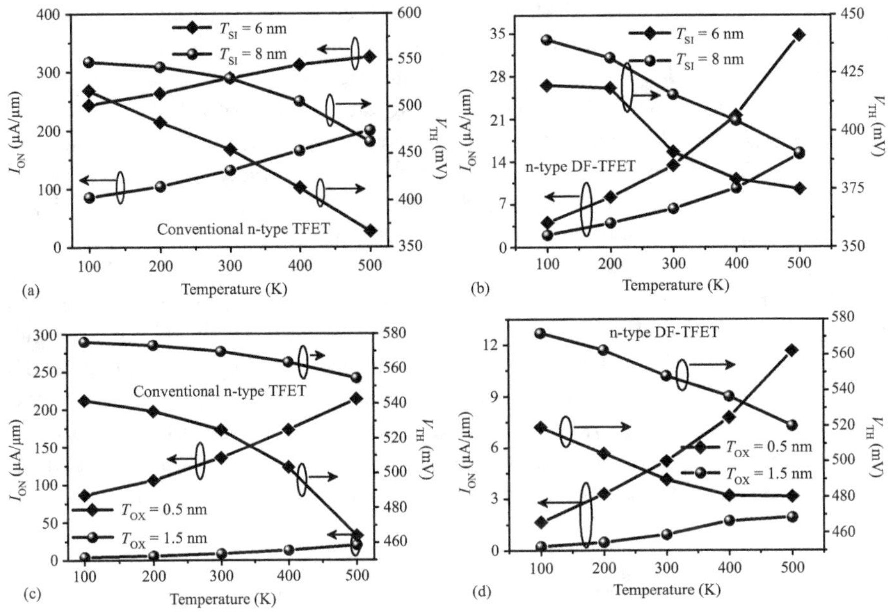

Figure 9.19 I_{ON} and V_{TH} fluctuation with variation in temperature for different silicon thickness (a) conventional TFET, (b) DF-TFET and for different oxide thickness, (c) conventional TFET and (d) DF-TFET

Table 9.2 Comparison of temperature variation for conventional and DF-TFETs

Device	**Conventional TFET**	**DF-TFET**	**Conventional TFET**	**DF-TFET**
Parameters	$T_{SI} = 6$ nm		$T_{OX} = 0.5$ nm	
I_{ON} (μA/K)	0.2026	0.076	0.3174	0.0248
V_{TH} (mV/K)	0.36	0.1	0.1949	0.0981

$T_{OX} = 0.5$ nm and same temperature change, yields approximately 0.1949 mV/K change in V_{TH} and 0.3174 μA/K change in I_{ON}, while in the case of the proposed device, these changes remain 0.0981 μA/K and 0.0248 μA/K, respectively, that are very small (about 1/10) in comparison to conventional TFET. These results are also summarized in Table 9.2.

9.4.9 Sensitivity towards oxide thickness

The most important parameter of MOSFETs is its oxide thickness (T_{OX}) because it helps in determining the gate capacitance (C_{OX}). The ON current of conventional MOSFETs is inversely proportional to the oxide thickness whereas for TFETs I_{ON}

is specifically dependent on BTBT and also on the gate oxide capacitance [8]. For different T_{OX} Figure 9.20 depicts the sensitivity of both conventional and proposed device in context of I_{ON} and SS. Due to enhancement in the gate electric field, it is observed that reduction in T_{OX} enhances I_{ON} and SS in both devices and even reduces SCEs [21]. Furthermore, when comparing conventional TFET and the proposed DF-TFET, conventional TFET shows a large variation in SS and I_{ON}. To improve the short-channel characteristics in DF-TFET, there is no necessity to aggressively scale T_{OX} as in conventional TFETs, hence DF-TFET device exhibits better control over channel. The SS and I_{ON} change by 11.2805 mV/dec and 34.867 mA/mm, respectively, for per nm change in T_{OX} (scaling from 2.5 to 0.5 nm) of conventional TFET, while SS and I_{ON} remain only 4.0963 mV/dec and 2.59 mA/mm, respectively, for DF-TFET. From the results mentioned in Table 9.3, we can infer that conventional TFET exhibits very high sensitivity (about 10×) towards the device parameters. The RDF in the devices is induced due to the presence of doped S/D and channel regions under the influence of vertical electrical field, it even results in significant differences in device electrical characteristics [12].

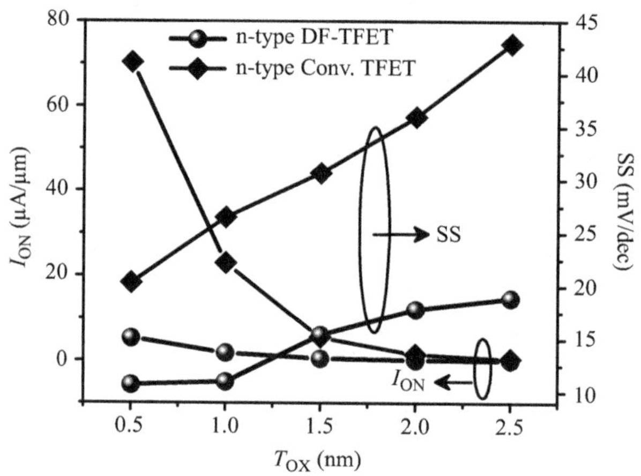

Figure 9.20 Oxide thickness sensitivity analysis of conventional TFET and DF-TFET at $V_{DS} = 1.0$ V

Table 9.3 Comparison of performance variation with per nm change in T_{OX} and T_{SI} for conventional and DF-TFETs

Device	Conventional TFET	DF-TFET	Conventional TFET	DF-TFET
Parameters	T_{OX}		T_{SI}	
SS (mV/dec)	11.2805	4.0963	2.68	0.588
I_{ON} (μA/μm)	34.867	2.59	77.42	5.35

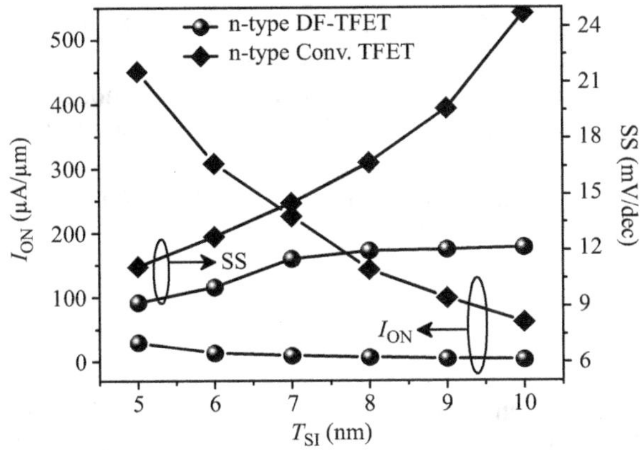

Figure 9.21 Silicon thickness sensitivity analysis of conventional TFET and DF-TFET at V_{DS} = 1.0 V

9.4.10 *Sensitivity towards silicon thickness*

In today's era, most of the advanced devices are built on thin silicon film. Therefore, it is very necessary to analyse the silicon thickness variation on the electrical parameters such as I_{ON} and SS (keeping other device parameters constant) in Figure 9.21. One can see that the small thickness allows better electrostatic control over the tunnelling width. Furthermore, when comparing conventional TFET and the proposed device shows small variation in SS and I_{ON}. To improve the short-channel characteristics in the proposed device there is no necessity to aggressively scale T_{SI} as in conventional TFETs hence DF dynamically configurable TFET device exhibits better control over channel. The SS and I_{ON} change by 2.68 mV/dec and 77.92 μA/μm, respectively, for per nm change in T_{SI} of conventional TFET, while SS and I_{ON} remain only 0.588 mV/dec and 5.35 μA/μm, respectively, for the proposed device. These results are also summarized in Table 9.3. Thus, proposed device demonstrates very less sensitivity towards device parameter. This happens due to less reliant RDF.

9.5 Summary

The concept of DF-TFET with PGs for dynamically configurable applications is affirmed using 2D TCAD simulations. The P+ and N+ are induced on a lightly doped silicon film via PG biasing. The conduction mechanism in DF-TFET can be controlled by tunnelling barrier width similar to that of a conventional TFET. This chapter also highlights that the proposed DF-TFET does not require ion-implantation and thermal annealing process. From simulation results, it is observed that the DF-TFET shows excellent control of SCEs and less sensitivity towards I_{ON}, even in presence of fluctuation in process parameters, especially T_{SI} and T_{OX} which are the most sensitive among other device parameters. This idea can be extended to an undoped silicon nanowire with surround gate electrodes making it compatible with the future nanowire-based CMOS technology since polarity-based silicon nanowire MOSFET has already been experimentally demonstrated. The chapter also highlighted the advantages of DF-TFET towards immunity to PVT, RDFs and SCEs. Hence, proposed results may provide incentives and guidelines for further research and experimental exploration of the DF-TFET for the circuit and system level.

References

[1] Dennard RH, Gaensslen FH, Yu HN *et al.* Design of ion-implanted MOSFET's with very small physical dimensions. IEEE Solid-State Circuits Society Newsletter. 2007; 12(1):38–50.

[2] Packan P. Short Course, IEDM, 2012.

[3] International Technology Roadmaps for Semiconductor, ITRS, 2012.

[4] Taur Y, and Ning TH. Fundamentals of modern VLSI devices. New York, NY, USA: Cambridge University Press; 1998.

[5] Nilsson P. Arithmetic reduction of the static power consumption in nanoscale CMOS. In: 2006 13th IEEE International Conference on Electronics, Circuits and Systems; 2006. p. 656–59.

[6] Dang B, Bakir MS, Sekar DC *et al.* Integrated microfluidic cooling and interconnects for 2D and 3D chips. IEEE Transactions on Advanced Packaging. 2010; 33(1):79–87.

[7] Choi WY, Park BG, Lee JD *et al.* Tunneling field-effect transistors (TFETs) with subthreshold swing (SS) less than 60 mV/dec. IEEE Electron Device Letters. 2007; 28(8):743–45.

[8] Boucart K, and Ionescu AM. Double-gate tunnel FET with high-954; gate dielectric. IEEE Transactions on Electron Devices. 2007; 54(7):1725–33.

[9] Wang PF, Hilsenbeck K, Nirschl T, *et al.* Complementary tunneling transistor for low power application. Solid-State Electronics. 2004; 48(12):2281–86.

[10] Bhuwalka KK, Schulze J, and Eisele I. Scaling the vertical tunnel FET with tunnel bandgap modulation and gate work function engineering. IEEE Transactions on Electron Devices. 2005; 52(5):909–17.

[11] Saurabh S, and Kumar MJ. Impact of strain on drain current and threshold voltage of nanoscale double gate tunnel field effect transistor: theoretical investigation and analysis. Japanese Journal of Applied Physics. 2009; 48(6R): 064503.

[12] Chiang MH, Lin JN, Kim K *et al.* Random dopant fluctuation in limited-width finFET technologies. IEEE Transactions on Electron Devices. 2007; 54(8):2055–60.

[13] Jhaveri R, Nagavarapu V, and Woo JCS. Effect of pocket doping and annealing schemes on the source-pocket tunnel field-effect transistor. IEEE Transactions on Electron Devices. 2011; 58(1):80–86.

[14] Lahgere A, Sahu C, and Singh J. PVT-aware design of dopingless dynamically configurable tunnel FET. IEEE Transactions on Electron Devices. 2015; 62(8):2404–09.
[15] Silvaco, Version 5.19.20.R. (2014) [Online]. Available: http://www.silvaco.com.
[16] Marchi MD, Sacchetto D, Frache S *et al.* Polarity control in double-gate, gate-all-around vertically stacked silicon nanowire FETs. In: 2012 International Electron Devices Meeting; 2012. pp. 8.4.1–8.4.4.
[17] Marchi MD, Zhang J, Frache S *et al.* Configurable logic gates using polarity-controlled silicon nanowire gate-all-around FETs. IEEE Electron Device Letters. 2014; 35(8):880–82.
[18] Lahgere A. Electrically doped dynamically configurable field-effect transistor for low-power and high-performance applications. Electronics Letters. 2015; 51(2):1284–86.
[19] Sahu C, and Singh J. Charge-plasma based process variation immune junctionless transistor. IEEE Electron Device Letters. 2014; 35(3):411–13.
[20] Sahu C. Junction and DF transistors for future computing. Institution of Engineering and Technology; 2016.
[21] Sahu C, and Singh J. Potential benefits and sensitivity analysis of dopingless transistor for low power applications. IEEE Transactions on Electron Devices. 2015; 62(3):729–35.

Chapter 10

Tunnel junctions to tunnel field-effect transistors—technologies, current transport models, and integration

Ashok Srivastava[1] *and Muhammad Shamiul Fahad*[1]

10.1 Introduction—band-to-band tunneling graphene nanoribbon tunnel FETs

Graphene nanoribbon field-effect transistors (GNR-FETs) have been fabricated and characterized which demonstrated promising performance. Wang *et al.* [1] first observed an on/off current ratio of 10^6 in a GNR FET operating at 2,000 μA/μm on-state drain current for a channel width of ~2 nm. However, the channel length is 236 nm and the subthreshold slope is 210 mV/decade which clearly makes such GNR FET unsuitable for current high-performance CMOS-based integrated circuit (IC) design. Therefore, the constant demand for GNR transistors operation at low supply voltage at CMOS compatible channel length still exists.

Graphene nanoribbon (GNR) is promising for tunneling FETs due to its symmetric band structure, low band-gap, light effective mass, and monolayer-thin body. Zhang *et al.* [2] in 2008 first reported the conceptual theoretical study of a GNR tunnel field-effect transistor. The idealistic theoretical model predicted a GNR TFET performance of 800 μA/μm drain current at 0.1 V supply voltage for a GNR width of 5 nm and a channel length of 20 nm. The computed subthreshold slope was 0.19 mV/decade, which was neither validated from numerical simulations nor from experiments.

Further, band-to-band tunneling in GNR tunnel FET has been studied by numerical simulations for graphene homojunctions [3], heterojunctions [4], single and bilayers [5,6], and dissipative transport through rough edges for the understanding of current transport [7]. Majority of these numerical simulations are obtained by solving 3D Poisson's equation coupled with Schrodinger's equation for a nearest-neighbor tight binding (NNTB) Hamiltonian for a finite width and specific edge type GNR. Compared to numerical simulations, physics-based compact analytical models of such novel emerging devices not only allow a better

[1]Department of Electrical and Computer Engineering, Louisiana State University, Baton Rouge, LA, USA

understanding of the transistor operation but also enable their potential for circuit-level synthesis.

In this chapter, an analytical current transport model of a p-i-n n-type armchair GNR TFET is developed which is compared with numerical simulation. Two separate current transport models are derived analytically from semi-classical and semi-quantum modeling approaches. Non-equilibrium Green function (NEGF)-based numerical simulation study is also carried out. Results obtained from these two methods are compared with the numerical simulation to establish analytical models. The analytical model in the work of Zhang *et al.* [2] is revisited and results are also compared with the analytical and numerically simulated results in this work. Furthermore, GNR TFET's performance is studied for varying GNR width using semi-classical, semi-quantum and NEGF simulation-based current transport models. Finally, complementary GNR TFET inverter for digital circuit design is demonstrated through the computation of voltage transfer characteristic from all three modeling approaches.

10.2 Device structure and operation of GNR TFET

Schematic of a GNR TFET is shown in Figure 10.1(a) and (b) where GNR is placed on top of SiO_2 substrate. For nanoscale GNR of 4.9 nm width and 20 nm length, lithographically patterned GNR is highly desirable. Silicon dioxide of 1 nm is considered as a top gate oxide. Length of gate dielectric is 20 nm as shown in both Figure 10.1(a) and (b), for a GNR channel width of 2.3 nm and 0.52 eV energy band-gap. Cr/Au or Ti/Cu contacts are typically used.

By solving 3D Poisson's equation coupled with 1D Schrodinger's equation within the NEGF formalism, three-dimensional potential is obtained for GNR TFET in all three (*X*, *Y*, and *Z*) spatial directions. Corresponding energy band diagrams along the channel in the *Z*-direction are plotted from the *Z*-component of this potential. Figure 10.1(c) and (d) shows the energy band diagram during an off/on condition for n-i-p (p-type) and p-i-n (n-type) TFET, respectively. E_C^S, E_C^C, and E_C^D are source, channel, and drain conduction bands, respectively, whereas E_V^S, E_V^C, and E_V^D are the source, channel, and drain valence bands, respectively. The solid and dashed lines show the off- and on-states of TFET, respectively. Note for both types of transistors, off-state is defined for $|V_{DS}| = 0.1$ V and $|V_{GS}| = 0$ V and on-state is defined for $|V_{DS}| = 0.1$ V and $|V_{GS}| = 0.1$ V. Throughout this chapter, positive bias of V_{GS} and V_{DS} is considered assuming n-type GNR TFET operation. Junction electric field of 3.85×10^6 V/cm is taken into account which is identical with the estimated electric field in [2]. For the p-i-n n-type GNR TFET, source and drain are assumed to be p- and n-type where Fermi levels are assumed to coincide with the valence band and conduction bands, respectively.

In thermal equilibrium, Fermi levels in the source, channel, and drain regions are aligned together. During the off-state, there is a difference of $|V_{DS}|$ between E_C^S and E_V^D in p-type TFET and E_V^S and E_C^D in n-type TFET. However, source and channel Fermi levels remain aligned together. Therefore, no tunneling of carriers

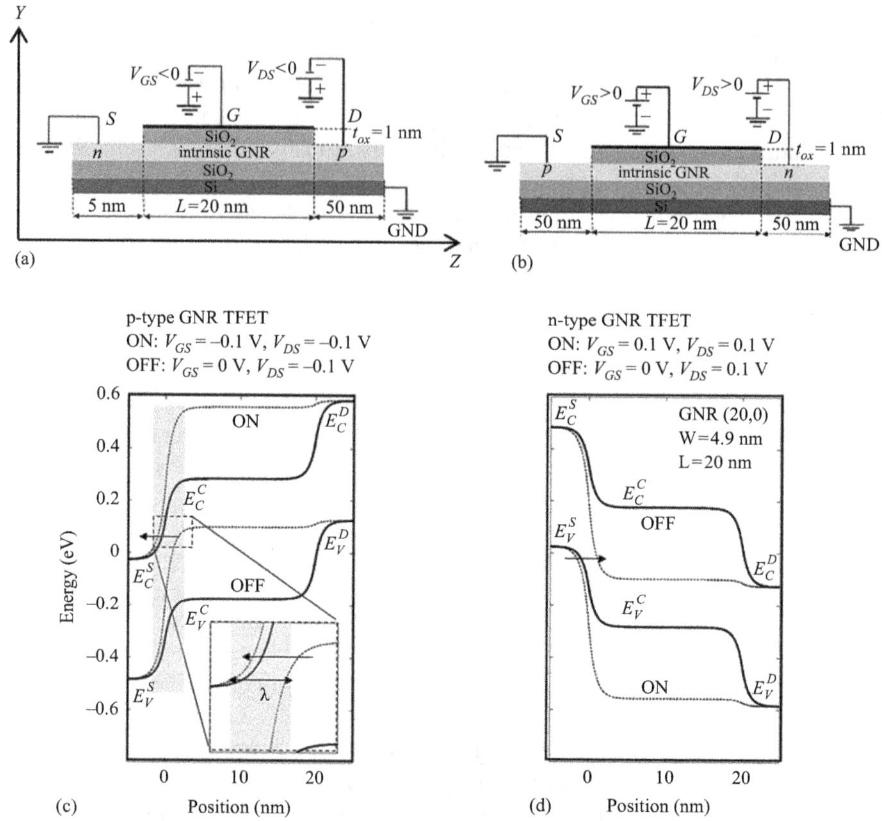

Figure 10.1 Schematic of GNR TFET. (a) Vertical cross-section of p-type GNR TFET with 1 nm SiO_2 *top gate dielectric. Channel length is 20 nm with 5 nm of source and drain extension making the total length of GNR 30 nm, (b) n-type GNR TFET, (c) energy band diagram of n-i-p GNR TFET (p-type GNR TFET where both* V_{GS} *and* V_{DS} *are "−" ve) and (d) energy band diagram of p-i-n GNR TFET (n-type GNR TFET where both* V_{GS} *and* V_{DS} *are "+" ve). Note: In both (c) and (d), solid line is for off-state whereas dashed line is for on-state. The off-state is defined as* $|V_{DS}| = 0.1$ *V and* $|V_{GS}| = 0$ *V and on-state is defined as* $|V_{DS}| = 0.1$ *V and* $|V_{GS}| = 0.1$ *V. Semiconducting GNR (20,0) has a bandgap of 0.52 eV for its corresponding 2.3 nm width. Inset: Enlarged view of potential variation*

occurs through source-channel tunnel junction. Further, for $V_{GS} > 0$ (in n-type TFET) and $V_{GS} < 0$ (in p-type TFET), a tunneling window opens and initiates band-to-band tunneling. Direction of arrows shows flow of carriers due to tunneling between source and channel. GNR TFET is less sensitive to channel mobility since band-to-band tunneling dominates over the scattering in channel. Both source and channel are of same material assuming momentum conservation in both conduction and valence bands. The inset and the shaded area in Figure 10.1(c) show the relevant length scale for potential variation (λ), which is usually dependent on the device geometry. For 1D geometry of GNR, λ is determined from $\lambda = \sqrt{(\varepsilon_{GNR}/\varepsilon_{ox})t_{GNR}t_{ox}}$, where ε_{GNR} and ε_{ox} are the GNR and oxide dielectric constants, respectively, and $t_{GNR} = 0.35$ nm is the thickness of the GNR. In this work, we consider λ to be significantly lower than the channel length L. For $L \gg \lambda$, it has been found that the drain induced barrier lowering (DIBL) is significantly suppressed thereby yielding an ideal turn-off characteristic [8].

10.3 Current transport model

In the following subsections, three types of current transport models are presented and compared.

10.3.1 Semi-classical analytical model

In the conventional inversion mode MOSFETs, threshold voltage is well defined. However, definition of threshold voltage in TFET is not so well defined rather varies depending upon the geometry and the channel material. The definition of threshold voltage proposed by Boucart and Ionescu for Si p-i-n TFET [9] considers threshold voltage as the voltage where the I_D–V_{GS} characteristic makes a transition between quasi-exponential and linear dependence of the drain current. It is termed as either gate threshold voltage or the drain threshold voltage depending on its reference point and depends strongly on the tunnel junction design and gate geometry. Recently, Ortiz-Conde *et al.* [10] proposed an extrapolated threshold extraction method for the bulk semiconductor and compared with experimental fin type TFETs. The method however considers strong conduction modeling scheme and does not explain transition type threshold voltage for the weak conduction region.

For TFETs having GNR as the channel material, contact materials play a crucial role. The graphene gets doped by adsorption on metal substrates based on studies from the density functional theory. Graphene establishes a weak bond with metal atoms while preserving its electronic structure. A significant shift of the Fermi level with respect to the conical point by ~0.5 eV is observed [11]. In contrast to graphene, GNR has inherent non-zero and direct bandgap. Nevertheless, there is still a high probability of GNR to get doped by adsorption on metal. Hence, for GNR TEFT to operate in its actual bias condition, such inherent contact potential needs to be overcome. Hence, their contribution toward calculating GNR TFET threshold voltage comes into existence.

Here, we consider a simple expression of threshold voltage (V_{TH}) for a-GNR TFET similar to a MOSFET threshold voltage. However, unlike in MOSFET, this expression is assumed to be dominated by contact potentials. In the absence of dangling bonds, mobile charges and fixed ions, V_{TH} can be expressed as follows:

$$V_{TH} = \varphi_{BI} + \varphi_S + \varphi_G + \varphi_{ox} \tag{10.1}$$

where φ_G and φ_S are contact potentials due to gate and source contacts. The built-in potential φ_{BI} is defined as follows [12]:

$$\varphi_{BI} = \frac{E_G}{2} - V_T \ln\left(\frac{N}{n_i}\right) \tag{10.2}$$

where E_G is the GNR bandgap (0.289 eV), V_T is the thermal voltage (0.0259 V at 300 K), N is the doping density ($\sim 5 \times 10^{11}/\mathrm{cm}^2$), and n_i is the intrinsic carrier density ($9 \times 10^{10}/\mathrm{cm}^2$) [13]. φ_{OX} is the potential drop due to gate oxide over the channel. Corresponding change in GNR bandgap due to additional intermediate energy states from edge roughness can be considered through (10.2). Potential drop through the gate oxide is defined as follows:

$$\varphi_{ox} = \frac{Q_o}{Cox} \tag{10.3}$$

In (10.3), $Q_0 = n_s q$ is the total charge, where n_s is the induced surface charge density through gate oxide and is calculated as follows [14]:

$$n_s = \frac{\varepsilon_o \varepsilon_{ox}(V_{GS} - V_{TH})}{q t_{ox}} \tag{10.4}$$

Here, V_{GS} is input gate-source voltage. For 1 nm SiO_2 gate oxide (relative permittivity 3.9) and 0.1 V gate-source input voltage, calculated n_s is $2.16 \times 10^{12}\ \mathrm{cm}^{-2}$ [15]. Oxide capacitance is defined as, $C_{ox} = \varepsilon_o\varepsilon_{ox}/t_{ox}$. Substituting values of C_{OX} in (10.3) and replacing φ_{OX} in (10.4), V_{TH} can be calculated as a function of both dielectric permittivity and oxide thickness.

Integrating product of charge flux and tunneling probability from 0 to energy window of $\Delta\varphi$, 1D Zener tunneling current is calculated as follows [12]:

$$I_T = \int_0^{\Delta\varphi} q V_g \rho_{GNR}(k)\, [f_S(E) - f_D(E)] T_{WKB}\, dk \tag{10.5}$$

In (10.5), I_D is the tunneling drain current, V_g is the group velocity $(1/\hbar\ (dE/dk))$; $\rho_{GNR}(k)$ is the 1D density of states of graphene in k-plane $(1/\pi)$ [2] and $f_D(E)$ is the Fermi level position at drain (qV_{DS}) and $f_S(E)$ is the Fermi level position at source (0). T_{WKB} is the tunneling probability in a semiconducting p-n junction GNR and is expressed as follows [16]:

$$T_{WKB} = \exp\left(-\frac{\pi E_G^2}{4q\hbar v_F \xi}\right) \tag{10.6}$$

Here, $v_F \sim 10^8$ cm/s is the Fermi velocity, E_G is GNR bandgap, $\hbar$ is the reduced Plank's constant and ξ is the electric field at the source-channel tunnel junction. Based on the universal analytic model for TFET proposed by Lu *et al.* [17], electric field at the tunnel junction is linearly dependent on the junction built-in electric field, V_{GS} and V_{DS}. This is expressed as follows:

$$\xi = \xi_0(1 + \gamma_1 V_{GS} + \gamma_2 V_{DS}) \tag{10.7}$$

where ξ_0 is the built-in electric field at the source-channel tunnel junction when $V_{GS} = V_{DS} = 0$ V. Parameters γ_1 and γ_2 are the linear coefficients in unit of inverse of volt (V^{-1}). An increase in gate bias enhances the electric field at the tunnel junction by narrowing the tunneling barrier whereas an increase in drain bias also does the same with a lesser degree as the drain field is screened by the gate electrode. The limit considered in this work for γ_1 ranges from 1 to 5 whereas for γ_2 from 5 to 10, which is higher than that proposed in [17]. The model derived in [17] describes the parameters with respect to bulk three-dimensional heterojunction material. It is to be noted that the electrical properties and energy band structure of GNR are significantly different from such materials. Built-in electric field is dependent on both the built-in potential and the length of potential screening at the source-channel tunnel junction as follows:

$$\xi_0 = \varphi_{BI}/\lambda \tag{10.8}$$

The Fermi levels at the drain and source are expressed as follows:

$$f_D(E) = \frac{1}{1 + e^{\left(E - E_f^D\right)/kT}} \tag{10.9}$$

$$f_S(E) = \frac{1}{1 + e^{\left(E - E_f^S\right)/kT}} \tag{10.10}$$

Here, E is the energy of electron with a unit in electron-volt (eV) during the operation of band-to-band tunneling occurs. During the off-state, source Fermi level is at 0 V and drain Fermi level is at V_{DS} with reference to source. Considering proper limits of integration from 0 to $\Delta\varphi = V_{GS} - V_{TH}$, (10.5) can be expressed as follows:

$$I_T = \int_0^{\Delta\varphi} q\frac{1}{\hbar}\frac{dE}{dk}\frac{1}{\pi}\left[\frac{1}{1 + e^{\left(E - E_f^D\right)/kT}} - \frac{1}{1 + e^{\left(E - E_f^S\right)/kT}}\right] T_{WKB}\, dk \tag{10.11}$$

$$I_T = q\frac{1}{\hbar}\frac{1}{\pi} T_{WKB} \int_0^{\Delta\varphi} \left[\left(1 - \frac{e^{(E - qV_{DS})/qV_T}}{1 + e^{(E - qV_{DS})/qV_T}}\right) - \left(1 - \frac{e^{(E)/qV_T}}{1 + e^{(E)/qV_T}}\right)\right] dE \tag{10.12}$$

We obtain:

$$I_T = \frac{14q^2}{2\hbar\pi} V_T T_{WKB} \left[\begin{array}{l} -\ln\left(1 + \exp\left(\frac{V_{GS} - V_{TH} - V_{DS}}{V_T}\right)\right) + \ln\left(1 + \exp\left(\frac{V_{GS} - V_{TH}}{V_T}\right)\right) \\ + \ln\left(1 + \exp\left(-\frac{V_{DS}}{V_T}\right)\right) - \ln(2) \end{array}\right] \tag{10.13}$$

The term $4q^2/2\pi\hbar$ in (10.13) can be termed as the minimum conductivity of graphene (σ). Following Drude model, minimum conductivity in graphene can be expressed in terms of mobility and charge density as follows:

$$\sigma = 4q^2/2\pi\hbar = \mu_n(n_s)q \tag{10.14}$$

where μ_n is carrier mobility. Combining (10.4), (10.13) and (10.14), tunneling current equation for GNR TFET is expressed as follows:

$$I_T = \frac{\mu_n\varepsilon_0\varepsilon_{ox}(V_{GS}-V_{TH})}{t_{ox}} V_T\, T_{WKB}\left[-\ln\left(1+\exp\left(\frac{V_{GS}-V_{TH}-V_{DS}}{V_T}\right)\right)\right.$$
$$\left.+\ln\left(1+\exp\left(\frac{V_{GS}-V_{TH}}{V_T}\right)\right)+\ln\left(1+\exp\left(-\frac{V_{DS}}{V_T}\right)\right)-\ln(2)\right] \tag{10.15}$$

Considering built-in potential and thermal voltage, leakage current for GNR TFET can be defined as follows [2]:

$$I_L = \frac{q^2}{\pi\hbar} V_T \exp\left(-\frac{\varphi_{BI}}{V_T}\right) \tag{10.16}$$

Combining (10.15) and (10.16), drain current for GNR TFET can be expressed as follows:

$$I_D = I_T + I_L \tag{10.17}$$

$$I_D = \frac{\mu_n\varepsilon_0\varepsilon_{ox}(V_{GS}-V_{TH})}{t_{ox}} V_T\, T_{WKB}\left[-\ln\left(1+\exp\left(\frac{V_{GS}-V_{TH}-V_{DS}}{V_T}\right)\right)\right.$$
$$\left.+\ln\left(1+\exp\left(\frac{V_{GS}-V_{TH}}{V_T}\right)\right)+\ln\left(1+\exp\left(-\frac{V_{DS}}{V_T}\right)\right)-\ln(2)\right]$$
$$+\frac{q^2}{\pi\hbar} V_T \exp\left(-\frac{\varphi_{BI}}{V_T}\right) \tag{10.18}$$

Equation (10.18) has been derived for semi-classical current transport model for the n-type GNR TFET. Since the minimum conductivity of graphene of $4q^2/2\pi\hbar$ is maintained at a charge density corresponding to (10.3), mobility in (10.18) is estimated as 223.6 cm^2/V-s. Such a small value of mobility has little or no effect on tunneling phenomena as tunneling dominates over the scattering in TFETs [8]. The current transport model as described in [2] does not account for any leakage current effect on drain current which may lead to an erroneous result.

10.3.2 Semi-quantum analytical model

Compared to the semi-classical analytical model, a semi-quantum "mode"-based analytical model is developed for GNR TFET and performance is compared with both semi-classical analytical model and numerical simulations. Considering transverse "mode" of current transport and transmission coefficient for the channel to conduct charge carriers from source to drain, conductance of the channel defined according to Landauer expression is as follows [18]:

$$G(E) = \frac{2q^2}{\hbar} M(E) T_{WKB}(E) \tag{10.19}$$

where:

$$M(E) = W \frac{2|E_{TM}|}{\pi(\hbar v_F)} \tag{10.20}$$

W is the width of GNR and $|E_{TM}|$ is the energy of electron in transverse mode. In this work, E_{TM} is described in terms of gate-source voltage and is applied to control energy window through which number of modes are calculated. The number of conducting channels at energy E_{TM} is proportional to the width of the conductor in two-dimensional and to the cross-sectional area in three-dimensional geometry. Band structure of the conducting channel also affects the total number of modes. Expression of $M(E)$ in (10.20) is specific to graphene which is different from the expression of mode usually adopted for a parabolic band structure [19]. In ballistic transport, transmission coefficient $T_{WKB}(E)$ is assumed as 1. However, in order to apply the similar concept for a tunneling transistor, transmission coefficient is assumed to be equal to tunneling probability as described by (10.6) in [20]. Considering source and drain Fermi-Dirac statistics and channel conductance expressed in Landauer formalism, current can be calculated as follows:

$$I = \int dE G(E)(f_S(E) - f_D(E)) \tag{10.21}$$

where drain Fermi function $f_D\ (E)$ and source Fermi function $f_S(E)$ are described in (10.9) and (10.10), respectively, and can be rewritten for $|E_{TM}|$ instead of E. Combining (10.9), (10.10), (10.19) and (10.21), drain current is expressed as follows:

$$I_D = \int dE \frac{2q^2}{\hbar} M(E) T_{WKB}(E)(f_S(E) - f_D(E)) \tag{10.22}$$

Substituting expression of $T_{WKB}(E)$ from (3.6) and $M(E)$ from (10.20), (10.22) becomes:

$$I_D = \int dE \frac{2q^2}{\hbar} W \frac{2|E_{TM}|}{\pi(\hbar v_F)} \exp\left(-\frac{\pi E_G^2}{4q\hbar v_F \xi}\right) \left(\frac{1}{1 + e^{\left(E_{TM} - E_F^S\right)/kT}} - \frac{1}{1 + e^{\left(E_{TM} - E_F^D\right)/kT}}\right) \tag{10.23}$$

$$I_D = \frac{2q^2}{\hbar} \exp\left(-\frac{\pi E_G^2}{4q\hbar v_F \xi}\right) \frac{2W}{\pi(\hbar v_F)} \int \left(\frac{|E_{TM}|}{1 + e^{\left(E_{TM} - E_F^S\right)/kT}} - \frac{|E_{TM}|}{1 + e^{\left(E_{TM} - E_F^D\right)/kT}}\right) dE \tag{10.24}$$

$$I_D = \frac{4q^3 W}{\pi(\hbar^2 v_F)} \exp\left(-\frac{\pi E_G^2}{4q\hbar v_F \xi}\right)$$

$$\left[-V_T(V_{GS}-V_{TH})\left\{\ln\left(1+\exp\left(\frac{V_{GS}-V_{TH}}{V_T}\right)\right)\right.\right.$$

$$\left.\left. -\ln\left(1+\exp\left(\frac{V_{GS}-V_{TH}-V_{DS}}{V_T}\right)\right)\right\} - \frac{(\pi V_T)^2}{12}\right] \qquad (10.25)$$

In (10.25), in order to obtain a closed form of solution, complex polylog expression is avoided. For $V_{GS} \gg k_B T/q$ polylog terms become insignificant compared to other terms. Only the nonvanishing term remains after the integration in (10.25).

10.3.3 NEGF-based numerical model: simulation method and approach

In this section, we model the GNR TFET with numerical simulation. Device schematic shown in Figure 10.1(b) for n-type TFET is studied through self-consistent solution of the Poisson and Schrödinger equations using NEGF formalism incorporated in open-source device simulation tool NanoTCAD ViDES [20]. The objective of this study is to compare and verify the validity of the previously derived semi-classical and semi-quantum analytical models.

The band structure of armchair GNR of *(20,0)* chirality is modeled using first principles pseudo-potential method by local (spin) density approximation (L(S) DA) in which energy relaxation at the GNR edges is assumed. The Hamiltonian for this calculation is obtained from [21].

The associated three-dimensional potential is obtained by solving self consistently 3D Poisson's equation coupled with Schrodinger equation which is solved for the real space. The carbon to carbon hopping parameter is 2.7 eV. The simulations are performed at room temperature, 300 K, which is also the considered temperature in other two models. The default parameters for *(20,0)* GNR simulations are described as follows: the channel is intrinsic and the doped contacts are considered for better comparison with the analytical TFET models.

The p-type source and n-type drain are doped with a molecular fraction of 2.19×10^{-4}, which is 0.026/nm compared to carbon atom density of 122/nm and is consistent with the considered doping concentration of $5 \times 10^{11}/\text{cm}^2$ used in semi-classical analytical model in (10.2). The SiO_2 layer of thickness 1 nm is used as the gate dielectric at the top of the channel. The length of the nanoribbon is 30 nm with channel length of 20 nm and source drain extension of 5 nm on each side of the channel. With chirality of *(20,0)* GNR width becomes 4.9 nm and the calculated semiconducting bandgap is 0.289 eV. The same GNR bandgap and width considered for the numerical simulations are also used for all three current transport models discussed in this chapter.

10.4 Transfer characteristics of GNR TFET

Performance of GNR TFET obtained from all these three current transport models is discussed in this section. Using the analytical current transport models developed in (10.18) and (10.25), transfer characteristics are plotted in Figure 10.2(a) for the n-type GNR TFET for an idealistic GNR with zero threshold voltage and no defects or edge roughness. The obtained results from the analytical model are compared with the numerical simulation. Results obtained from the model of Zhang *et al.* [2] are also shown in Figure 10.2(a). It is found that semi-classical analytical current transport model gives fairly good agreement with the results obtained from rigorous NEGF simulation. However, the derived semi-quantum analytical model deviates from the NEGF-simulated results to a larger extent.

A supply voltage V_{DS} ($=V_{DD}$) of 0.1 V maintains minimum power consumption. This also ensures the condition $V_{BI} + V_{DS} < 2E_G$ to shut down any ambipolar tunneling characteristics at off-state of the TFET where V_{BI} is the built-in voltage of the p-i-n structure [22]. The on/off current ratio for both semi-classical model and NEGF simulation are calculated as 122 and 116 at $V_{GS} = V_{DS} = 0.1$ V, respectively, which are fairly close within an accepted margin. Drive current for semi-classical model is 6.2×10^{-6} μA/μm, which is also in close agreement with the calculated drive current of 5.95×10^{-6} μA/μm from NEGF simulation. Table 10.1 summarizes the performance

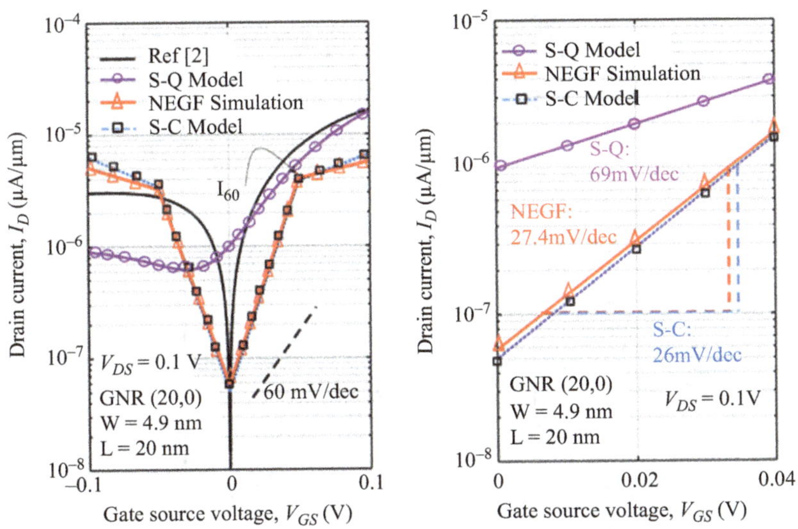

Figure 10.2 (a) Comparison of transfer characteristics of n-type GNR-TFET obtained from three current transport models along with that of [2]. (b) Method of obtaining SS for three current transport models. Note: S-Q stands for semi-quantum and S-C for semi-classical. Values written in Figure 10.2(b) are obtained using (10.27)

Table 10.1 Comparison of n-type GNR TFET performance from different current transport models

Model	V_{DD} (V)	V_{GS} (V)	Channel (nm)	t_{ox} (nm)	Drive current, I_D (μA/μm)	OFF-state leakage current, I_{OFF} (μA/μm)	Leakage power, $V_{DD}I_{OFF}$ (μW/μm)	Dynamic power ½ I_DV_{DD} (μW/μm)	I_{ON}/I_{OFF}	Subthreshold Slope (mV/dec)	I_{60} (μA/μm)
Analytical model [101]	0.1	0.1	$L = 20$ $W = 5$	1	1.51×10^{-5}	1.2×10^{-11}	1.2×10^{-12}	7.55×10^{-7}	1.25×10^{6}	14.15	3.8×10^{-6}
Semi-classical analytical model	0.1	0.1	$L = 20$ $W = 4.9$	1	6.2×10^{-6}	5.05×10^{-8}	5.05×10^{-9}	3.1×10^{-7}	122	26	4.2×10^{-6}
Semi-quantum analytical model	0.1	0.1	$L = 20$ $W = 4.9$	1	1.6×10^{-5}	9.8×10^{-7}	9.8×10^{-8}	8×10^{-7}	16.3	69	Does not provide
NEGF-based simulation	0.1	0.1	$L = 20$ $W = 4.9$	1	5.95×10^{-6}	5.145×10^{-8}	5.145×10^{-9}	2.9×10^{-7}	116	27.4	4.4×10^{-6}

comparison of these three current transport models. Note that the drain current has been normalized along the GNR width.

Though the semi-classical analytical model and numerical simulation for the current transport matches closely, the semi-quantum analytical model differs from both. Before further studies into GNR TFET transfer characteristics; it is to be mentioned that the tunneling probability used in calculating the drain current in semi-quantum model is taken from the semi-classical model which is semi-classical in nature. The transmission coefficient (T_{WKB}) of the Landauer's conductance expression has been considered as the equivalent tunneling probability (T_{WKB}) from semi-classical model following (10.6). A more rigorous calculation considering source and drain contacts and their corresponding self-energy, and Fermi–Dirac distribution between the source and drain and effect from the gate is required to describe "T_{WKB}" properly. Moreover, a self-consistent calculation of the number of "modes" is essential to describe the semi-quantum analytical model completely since the number of modes in on- and off-states differs based on the bias conditions. For these reasons, the semi-quantum analytical model differs in describing the current transport in GNR TFET when compared with semi-classical analytical model and NEGF simulation.

10.5 Subthreshold slope of GNR TFET

For energy-efficient switching technique, subthreshold swing (SS) of TFETs is required to be below the thermionic limit of 60 mV/decade of conventional MOSFETs. In order to verify the suitability of the studied current transport model for digital circuit design, SS of all three models is compared. Figure 10.2(b) shows a decade change of drain current (I_D) from which SS is calculated. This method expresses the conventional SS as, $SS = \log(10)[I_D/(dI_D/dV_{GS})]$. Using this method, the semi-classical model and NEGF simulation give a SS of 26 mV/decade and 27 mV/decade, respectively. SS for semi-quantum model is 71 mV/decade in this case. Moreover, following the method of Seabaugh and Zhang in [23], effective swing is determined as follows:

$$SS_{eff} = (V_{DD}/2)/\log_{10}(I_{TH}/I_{OFF}) \tag{10.26}$$

where I_{TH} is the current at threshold voltage (V_{TH}) and I_{OFF} is the off current determined at $V_{GS} = 0$ V. In [23], V_{TH} is considered as the half of the supply voltage ($V_{TH} = V_{DD}/2$) which returns I_{TH} as I_D at $V_{DD}/2$. Following this notation and after extracting the corresponding value of $V_{DD}/2$ as 0.05 V, SS for all three models is also evaluated from Figure 10.2(b) using (10.26). Here, SS is calculated to be 28 mV/decade for the semi-classical model and 27 mV/decade for the NEGF simulation. Both of these values closely match with previously mentioned values of SS. SS of 68 mV/decade is obtained from this method for semi-quantum model.

Here, we propose a method of estimating average SS using point-slope method which depends on the bias voltage at the gate and corresponding TFET current ratio at that point. This can be written as follows:

$$SS_{avg} \approx (V_{GS})/\log_{10}(I_{D,V_{GS}}/I_{OFF}) \tag{10.27}$$

Note, the above expression is similar to (10.26) with minor changes that make it independent of threshold voltage and applicable to any order magnitude of drain current. Using (10.27), SS for all three models is calculated as specified in Figure 10.2(b). Using V_{GS} of 0.04 V and corresponding I_D at $V_{GS} = 0.04$ V and $V_{GS} = 0$ V from Figure 10.2(b), calculated values of SS from semi-classical, semi-quantum and NEGF simulation are 26 mV/decade, 69 mV/decade and 27.4 mV/decade, respectively. The values of SS mentioned in Figure 10.2(b) and Table 10.1 are obtained using (10.27). Based on the rigorous calculation and comparison of SS for all three models, it is evident that semi-classical analytical model can predict the current transport in GNR TFET very similar to the numerical simulation using NEGF formalism. However, the semi-quantum analytical model lags such proximity due to inherent weakness in calculating SS as discussed earlier. For circuit simulation, the semi-classical analytical model can be fairly adopted for large scale integration.

10.6 Estimation of subthreshold swing point, I_{60}

One of the most important figures of merits for TFET is the highest current where a subthreshold slope of 60 mV/decade is obtained [24]. This parameter is written as "I_{60}" and has the unit of μA/μm. For a TFET to be competitive with MOSFET, I_{60} should be 1–10 μA/μm. However, existing theoretical, experimental and simulated results have shown that I_{60} is still lagging behind this range. Note, current has been normalized along the channel width.

Figure 10.2(a) shows the point for I_{60} estimation where the drain current makes a transition from sub-60 to super-60 with respect to gate bias. Both the semi-classical analytical model and NEGF simulation approximates I_{60} around 4×10^{-6} μA/μm, however, I_{60} remains undeterminable for semi-quantum analytical model. As calculated earlier, average SS for semi-quantum model is 69 mV/decade for which the point slope does not converge to a specific point where SS makes a transition from sub-60 and super-60 region. Compared to the earlier reported I_{60} of 2×10^{-6} μA/μm in [25], 10^{-5} μA/μm in [26] and 3×10^{-5} μA/μm in [27], estimated value of I_{60} falls within an acceptable range.

10.7 Output characteristics of GNR TFET

Figure 10.3 shows the output characteristics (I_D–V_{DS}) of n-type a-GNR TFET using the three current transport models studied in this work for different V_{GS}.

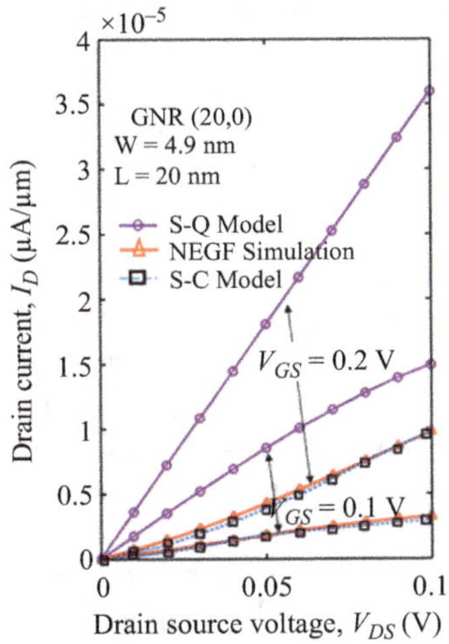

Figure 10.3 I_D–V_{DS} characteristics of n-type GNR TFET for semi-classical analytical model, semi-quantum analytical model and NEGF simulation for $V_{GS} = 0.1$ V and $V_{GS} = 0.2$ V

The semi-classical analytical model shows good agreement with the results obtained from the numerical simulation, however, the semi-quantum model differs largely.

For a fixed V_{GS}, a constant amount of carriers tunnel through the source-channel tunnel junction. For $V_{DS} = 0$ V and $V_{GS} > 0$ V, a small tunneling window is opened at the source-channel tunnel junction which works as the origin of leakage current. From (10.7), maximum electric field at the source-channel tunnel junction has linear dependence on V_{DS} which is used to determine T_{WKB}. It is obvious from (10.7), for a fixed V_{GS}, junction maximum electric field will solely depend on V_{DS}. As a result, tunneling probability depends exponentially on V_{DS}. For a fixed V_{GS} with varying V_{DS}, semi-quantum model is now strongly governed by the difference in source-drain Fermi level. Therefore, any change in drain current calculated by semi-quantum model is also strongly controlled by V_{DS} as opposed to V_{GS} dependence of semi-classical and NEGF simulated current transport models. For this reason a large deviation of semi-quantum model is observed in Figure 10.3 compared to semi-classical analytical model and numerical simulation. Compared to output characteristics of conventional MOSFETs where V_{DS} governs channel electric field and affects pinch-off and velocity saturation, output characteristics in TFET not only depends on V_{GS} but also on V_{DS}. Especially in reduced dimensional materials as in graphene such behavior is often observed. Current transport equations of (10.18) and (10.25) derived from semi-classical and semi-quantum considerations, respectively, can be used also for p-type GNR TFET n-i-p structure shown in Figure 10.1(a) with opposite voltage polarities.

10.8 Width-dependent performance analysis of GNR TFET

In this section, performance of GNR TFET is examined for different number of atoms along the GNR width. As the number of atoms varies along the width, electronic properties of armchair nanoribbon change, based on any one of $3N$, $3N + 1$ and $3N + 2$ configurations along with the associated bandgap of nanoribbon [21]. As a result, GNR TFET performance also changes. To get suitable performance from GNR TFET, appropriate chirality of GNR needs to be selected.

Since the bandgap of GNR is determined using first principles L(S)DA approximation, bandgaps of GNR are nonzero and direct irrespective of width. This can be observed in Table 3.2 for GNR with different width. The major difference between tight binding and first principles based energy gaps calculation is observed for $3N + 2$ configuration. Among the considered chiral armchair nanoribbons, GNR TFET with *(20,0)* and *(11,0)* chiral nanoribbon represents $3N + 2$ configuration. In conventional tight binding method based calculation of GNR bandgap, $3N + 2$ configuration provides metallic GNR whereas first principles method considers $3N + 2$ as semiconducting as well. For this reason a bandgap is observed for *(20,0)* and *(11,0)* chiral nanoribbons. A higher on/off current ratio is seen in Table 10.2 for semi-classical and NEGF simulations. However, on/off current ratio obtained for *(11,0)* GNR in semi-classical model differs largely compared to other two current transport models. For *(11,0)* chiral GNR, the on-state drive current in semi-classical model (5×10^{-7} µA/µm) matches closely with the that obtained from NEGF simulation (4.6×10^{-7} µA/µm), however, off-state leakage current differs by a decade of magnitude. It is important to note that the method of calculating off-state leakage current in these two models is different.

Following (10.2) and (10.16), off-state leakage current in semi-classical analytical model has built-in potential (φ_{BI}) and bandgap (E_G) dependence. A GNR with large bandgap provides a significantly large built-in potential as seen from (10.2). This limits additional thermionic transport over the barrier at off-state and results in low leakage current. Moreover, condition of $V_{BI} + V_{DS} < 2E_G$ to limit additional ambipolar tunneling at the off-state becomes $V_{BI} + V_{DS} \ll 2E_G$ for $V_{DS} \ll 2E_G$. Both of these conditions lower the off-state leakage current for larger GNR bandgap in semi-classical analytical model for which a high on/off current ratio is observed for *(11,0)* GNR. In contrast to compact semi-classical analytical model, NEGF simulation adopts rigorous Newton–Raphson method with a predictor-corrector scheme to calculate the charge density and channel electrostatic potential. The simulation thereby takes into account the deeper detail of current transport mechanism in estimating even the leakage current. This could be one of the limitations of the semi-classical analytical current transport model to differ from NEGF simulation.

However, a better description of off-state leakage current considering quantum confinement and energy states from GNR edges can solve this problem and substantiate the semi-classical analytical model as a reliable tool for circuit simulation.

Table 10.2 Performance comparison of n-type GNR TFET for different GNR width and bandgap

GNR dimer	V_{DS} (V)	V_{GS} (V)	Band-gap (eV)	GNR width (nm)	NEGF simulation			Semi-classical model			Semi-quantum model		
					OFF current (μA/μm)	ON current (μA/μm)	I_{ON}/I_{OFF}	OFF current (μA/μm)	ON current (μA/μm)	I_{ON}/I_{OFF}	OFF current (μA/μm)	ON current (μA/μm)	I_{ON}/I_{OFF}
(7,0)	0.1	0.1	0.13	1.62	9.3×10^{-7}	7.5×10^{-6}	8	1×10^{-6}	7.3×10^{-6}	7.3	4.9×10^{-7}	6.4×10^{-6}	13
(10,0)	0.1	0.1	0.092	2.37	3.8×10^{-6}	7.7×10^{-6}	2	1.9×10^{-6}	3.7×10^{-6}	1.9	5.8×10^{-7}	9.6×10^{-6}	16.6
(11,0)	0.1	0.1	0.52	2.61	3.5×10^{-9}	4.6×10^{-7}	46	2.9×10^{-10}	5×10^{-7}	1,724	2.94×10^{-7}	4.7×10^{-6}	15.98
(12,0)	0.1	0.1	0.313	2.86	1.9×10^{-7}	6.5×10^{-6}	34	2.2×10^{-7}	6.3×10^{-6}	29	5.4×10^{-7}	8.7×10^{-6}	16.1
(15,0)	0.1	0.1	0.252	3.62	9.3×10^{-8}	6.6×10^{-6}	71	8.6×10^{-8}	6.8×10^{-6}	79	7.7×10^{-7}	1.3×10^{-5}	16.9
(20,0)	0.1	0.1	0.289	4.9	5.14×10^{-8}	5.95×10^{-6}	116	5.05×10^{-8}	6.2×10^{-6}	122	9.8×10^{-7}	1.6×10^{-5}	16.3

10.9 Voltage transfer characteristics of GNR TFET complementary inverter

Figure 10.4(a) shows the schematic of a complementary GNR TFET inverter for operation at different supply voltages and is similar to CMOS inverter in design and operation. Characteristics of GNR TFET inverter is plotted from all three current transport models. At input logic level "1" (either 0.1 V or 0.2 V), n-type GNR TFET turns ON, p-type GNR TFET is OFF and output gives logic "0". Similarly, when input is at logic "0" (0 V), p-type GNR TFET turns ON and n-type GNR TFET is OFF, output is at logic "1" (either 0.1 V or 0.2 V for the case in Figure 10.4 (b)). Figure 10.4(b) shows the plot of voltage transfer characteristics (VTC) of the complementary GNR TFET-based inverter of Figure 10.4(a) for GNR for *(20,0)* chirality and $V_{DD} = 0.1$ V and 0.2 V supply voltages. Following the transfer characteristics obtained for all three current transport models, VTC of GNR TFET inverter also shows good agreement between semi-classical analytical model and NEGF simulation. However, semi-quantum analytical model differs from both of these models in this case as well. A decrease in the logic "1" is observed due to inherent leakage current at off-state for both transistors. However, sharp transition between on to off-state is observed at reduced supply voltage. The VTC shown in Figure 10.4(b) confirms the reliable use of semi-classical analytical model for digital circuit simulation with a good agreement with numerical simulation.

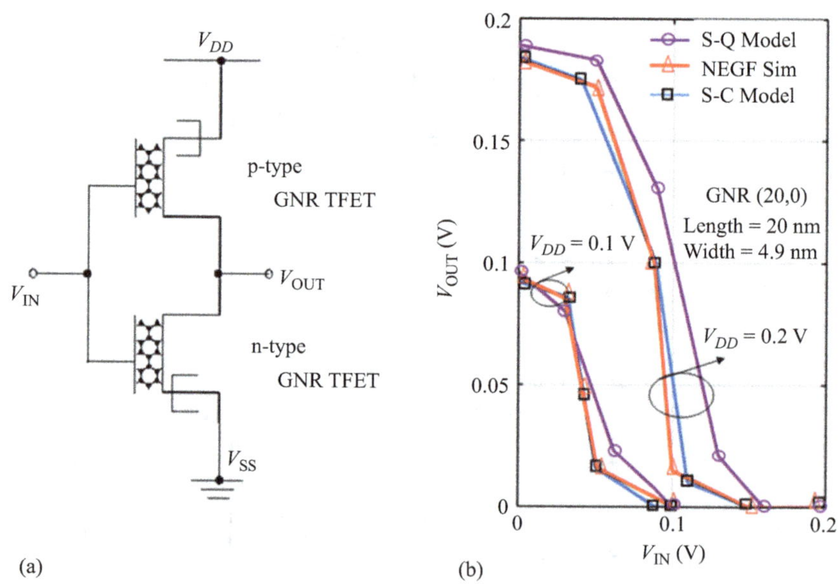

Figure 10.4 (a) A complementary GNR TFET inverter circuit and (b) voltage transfer characteristics of GNR TFET inverter for different supply voltages

10.10 Conclusion

The semi-classical analytical model closely agrees with numerical simulation whereas significant difference between semi-quantum model and NEGF simulation is observed [28]. Performance of n-type GNR TFET is also studied for GNR width variation. The semi-classical analytical current transport model of n-type GNR TFET can be applied to p-type GNR TFETs (n-i-p structure) with opposite voltage polarities. Promise of GNR TFET for digital logic application as a TFET inverter is studied by all three current transport models. Characteristics sharp transition from "on" to "off" condition is observed for lower supply voltage. By comparing the semi-classical analytical model with the numerical simulation, it is evident that the semi-classical analytical model derived can predict near similar performance of GNR TFET for different figure of merits. Readers are suggested to read [15].

However, semi-quantum analytical model differs from simulation due to inherent limitation in calculation and hence it is not yet reliable in its current form. Therefore, we conclude the semi-classical analytical current transport model as a powerful tool for circuit simulation for digital IC design.

References

[1] X. Wang, Y. Ouyang, X. Li, H. Wang, J. Guo, and H. Dai, "Room-temperature all-semiconducting sub-10-nm graphene nanoribbon field-effect transistors," *Phys. Rev. Lett.*, vol. 100, pp. 206803-01–206803-04, 2008.

[2] Q. Zhang, T. Fang, H. Xing, A. Seabaugh, and D. Jena, "Graphene nanoribbon tunnel transistors," *IEEE Elect. Dev. Lett.*, vol. 29, pp. 1344–46, 2008.

[3] P. Zhao, J. Chauhan, and J. Guo, "Computational study of tunneling transistor based on graphene nanoribbon," *Nano Letters*, vol. 9, pp. 684–88, 2009.

[4] G. Fiori, A. Betti, S. Bruzzone, and G. Iannaccone, "Lateral graphene–hBCN heterostructures as a platform for fully two-dimensional transistors," *ACS Nano*, vol. 6, pp. 2642–48, 2012.

[5] R. K. Ghosh and S. Mahapatra, "Proposal for graphene boron nitride heterobilayer-based tunnel FET," *IEEE Trans. Nanotech.*, vol. 12, pp. 665–67, 2013.

[6] G. Fiori and G. Iannaccone, "Ultralow-voltage bilayer graphene tunnel FET," *IEEE Elect. Dev. Lett.*, vol. 30, pp. 1096–98, 2009.

[7] Y. Yoon and S. Salahuddin, "Dissipative transport in rough edge graphene nanoribbon tunnel transistors," *Appl. Phys. Lett.*, vol. 101, p. 263501, 2012.

[8] J. Knoch and J. Appenzeller, "Tunneling phenomena in carbon nanotube field-effect transistors," *Physica Status Solidi (A)*, vol. 205, pp. 679–94, 2008.

[9] K. Boucart and A. M. Ionescu, "A new definition of threshold voltage in Tunnel FETs," *Solid State Elect.*, vol. 52, pp. 1318–23, 2008.

[10] A. Ortiz-Conde, F. J. García-Sánchez, J. Muci *et al.* "Threshold voltage extraction in tunnel FETs," *Solid State Elect.*, vol. 93, pp. 49–55, 2014.

[11] G. Giovannetti, P. A. Khomyakov, G. Brocks, V. M. Karpan, J. van den Brink, and P. J. Kelly, "Doping graphene with metal contacts," *Phys. Rev. Lett.*, vol. 101, pp. 026803-1–026803-4, 2008.

[12] S. M. Sze and K. K. Ng, "Tunnel devices," in *Physics of Semiconductor Devices*, ed: New York: John Wiley & Sons, Inc., 2006, pp. 415–65.

[13] T. Fang, A. Konar, H. Xing, and D. Jena, "Carrier statistics and quantum capacitance of graphene sheets and ribbons," *Appl. Phys. Lett.*, vol. 91, p. 092109, 2007.

[14] K. S. Novoselov, A. K. Geim, S. V. Morozov *et al.* "Electric field effect in atomically thin carbon films," *Science*, vol. 306, pp. 666–69, 2004.

[15] Md. S. Fahad, A. Srivastava, A. K. Sharma, and C. Mayberry, "Current transport in graphene tunnel field effect transistor for RF integrated circuits," *Proc. IEEE MTT-S International Wireless Symposium (IWS)*, pp. 1–4, (13–18 April, 2013, Beijing, China).

[16] D. Jena, T. Fang, Q. Zhang, and H. Xing, "Zener tunneling in semiconducting nanotube and graphene nanoribbon p-n junctions," *Appl. Phys. Lett.*, vol. 93, p. 112106, 2008.

[17] H. Lu, D. Esseni, and A. Seabaugh, "Universal analytic model for tunnel FET circuit simulation," *Solid State Elect.*, vol. 108, pp. 110–17, 2015.

[18] S. Datta, *Quantum Transport: Atom to Transistor*. United Kingdom: Cambridge University Press, 2005.

[19] M. Lundstrom and C. Jeong, *Lessons from Nanoscience: A Lecture Note Series Near-Equilibrium Transport*. New Jersey: World Scientific, 2012.

[20] G. Fiori and G. Iannaccone. (2013). *NanoTCAD ViDES*. Available: www.nanotcad.com/vides

[21] Y. W. Son, M. L. Cohen, and S. G. Louie, "Energy gaps in graphene nanoribbons," *Phys. Rev. Lett.*, vol. 97, pp. 216803-1–216803-4, 2006.

[22] M. Luisier and G. Klimeck, "Performance analysis of statistical samples of graphene nanoribbon tunneling transistors with line edge roughness," *Appl. Phys. Lett.*, vol. 94, p. 223505, 2009.

[23] A. C. Seabaugh and Q. Zhang, "Low-voltage tunnel transistors for beyond CMOS logic," *Proc. of the IEEE*, vol. 98, pp. 2095–110, 2010.

[24] W. G. Vandenberghe, A. S. Verhulst, B. Sorée *et al.* "Figure of merit for and identification of sub-60 mV/decade devices," *Appl. Phys. Lett.*, vol. 102, pp. 013510-1–013510-4, 2013.

[25] R. Gandhi, C. Zhixian, N. Singh, K. Banarjee, and L. Sungjoo, "Vertical Si-nanowire n-type tunneling FETs with low subthreshold swing (<50 mV/decade) at room temperature," *IEEE Elect. Dev. Lett.*, vol. 32, pp. 437–39, 2011.

[26] B. Ganjipour, A. W. Dey, B. M. Borg *et al.* "High current density Esaki tunnel diodes based on GaSb-InAsSb heterostructure nanowires," *Nano Letters*, vol. 11, pp. 4222–26, 2011.

[27] K. Sung Hwan, H. Kam, C. Hu, and T. J. K. Liu, "Germanium-source tunnel field effect transistors with record high I_{ON}/I_{OFF}," in *Symposium on VLSI Technology*, pp. 178–79, 2009.

[28] Md. S. Fahad, A. Srivastava, A.K. Sharma and C. Mayberry, "Analytical current transport modeling of graphene nanoribbon tunnel field-effect transistor for digital circuit design," *IEEE Trans. Nanotechnology*, vol. 15, pp. 39–50, 2016.

Chapter 11

Low-dimension materials-based interlayer tunnel field-effect transistors: technologies, current transport models, and integration

Muhammad Shamiul Fahad[1] and Ashok Srivastava[1]

11.1 Introduction

Atomically thin two-dimensional graphene has emerged as a potential candidate for the next generation electronics due to its unique electronic properties. However, single-layer graphene is a zero-bandgap semiconductor. Scientifically it is well known that bandgap engineering is required for obtaining a bandgap in graphene. Undoubtedly, this makes the fabrication process complicated. In the absence of a bandgap, graphene field-effect transistor suffers from poor on/off current ratio with high off-state leakage current. Following the ITRS requirement [1] of energy-efficient circuit design, a minimum on/off current ratio of 10^4 is required for a supply voltage below 0.7 V for digital applications. Performance of the existing graphene-based MOS-type transistors is still lagging behind unless graphene is lithographically patterned to GNR or chemically synthesized.

In order to resolve the issue of on/off current ratio, vertical heterostructure consisting of multilayer stacks of graphene and other atomically thin two-dimensional materials such as boron nitride and the transition metal dichalcogenides have been proposed for different transistor structures [2–11]. Basically, interlayer tunneling technique is employed in these types of transistors. Majority of these transistors contain two graphene conducting layers separated by a thin tunneling barrier. These transistors are commonly known as interlayer tunnel field-effect transistors (iTFETs). Schematic of a conventional *n*-channel MOSFET, a p-i-n band-to-band tunnel field-effect transistor (TFET) and an iTFET are shown in Figure 11.1(a), (b), and (c), respectively, for distinction. High on/off current ratio, sharp resonant tunneling characteristic, and suitability for flexible and transparent electronics are some of the reported key features of this graphene iTFET. However, these transistors lag the potential for digital integrated circuit design considering the requirements of steep subthreshold slope and high drive current. Due to fundamental physical

[1]Department of Electrical and Computer Engineering, Louisiana State University, Baton Rouge, LA, USA

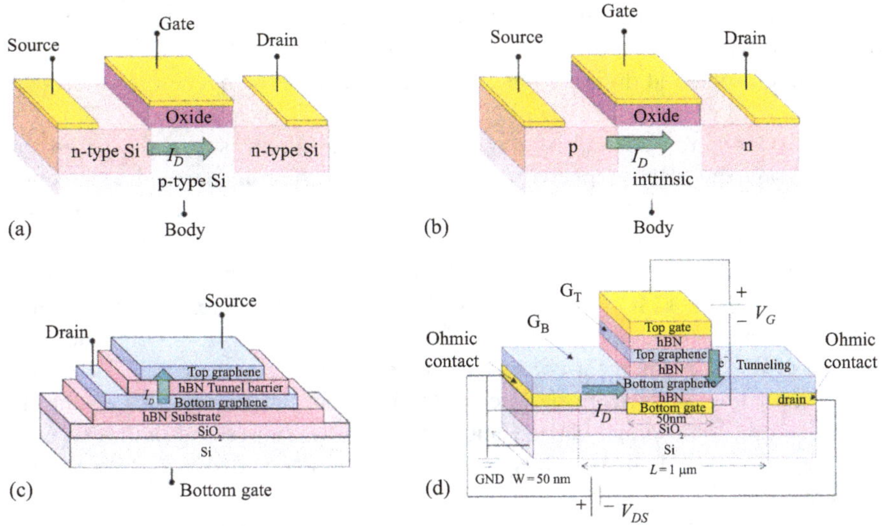

Figure 11.1 *(a) Schematic of a conventional n-channel depletion type MOSFET, (b) schematic of conventional p-i-n band-to-band TFET, (c) schematic of iTFET and (d) schematic of proposed graphene JTET. Note, G_B refers to the bottom graphene layer and G_T refers to the top graphene layer. The arrow in each of the device structures shows the direction of current flow with the second arrow in Figure 11.1(d) denotes the vertical interlayer tunneling of the electron from the top to bottom of the graphene layer*

limit, subthreshold slope of such iTFET cannot go below the thermionic limit of 60 mV/decade. Moreover, high supply voltage (>2 V) is also required for operation in some of the reported iTFETs. Studies of some of these devices have been carried out at cryogenic temperatures with poor performance at room temperature. Therefore, an improved current transport mechanism in a novel device structure is essential for making such iTFETs competitive for next generation more than the Moore's era.

In iTFET, source and drain contacts are placed at the two opposite conducting layers as seen in Figure 11.1(c) contrary to the contacts in conventional four-terminal MOSFET shown in Figure 11.1(a) and TFET, as shown in Figure 11.1(b), In this way, a bias between drain and source (V_{DS}) controls the vertical interlayer tunneling of carriers between the two conducting materials separated by a tunneling barrier. However, V_{DS} overshadows the actual control of channel electrostatic potential by the gate voltage [7]. For this reason, linear resistive behavior is obtained as opposed to the current saturation at different gate biases of output characteristics [8]. This impedes iTFETs' prospects in digital logic circuits. Apart from this, observed negative differential resistance (NDR) also undermines the scope of iTFETs. Therefore, a graphene switching transistor meeting the ITRS requirement with high on/off current ratio, steep subthreshold slope, non-resonant high drain current, and drain current saturation at sub-0.5 V operation is necessary for graphene to be suitable for digital integrated circuit design.

In this chapter, modeling of a graphene switching transistor is discussed considering graphene-hBN-graphene vertical heterostructure and named as junctionless tunnel effect transistor (JTET). JTET is one of the types of iTFET. Schematic of graphene JTET is shown in Figure 11.1(d), which is significantly different from the generic iTFET, as shown in Figure 11.1(c). However, graphene JTET ensembles similarity with a MOSFET, as shown in Figure 11.1(a) in terms of the location of source and drain. Compared to MOSFET, TFET and iTFET, graphene JTET adopts a different method for controlling the channel barrier height. JTET utilizes vertical tunneling of electrons between the top and bottom graphene layers through hBN to control the channel barrier height between the source and drain that eventually regulates the ballistic transport between source and drain at the bottom graphene layer.

Compared to planar MOSFET where a gate bias fully depletes the channel by "field effect" and inverts the channel's majority carrier type, JTET operates based on the gate-induced "tunneling effect". In addition to that, JTET does not require any doping in source, channel or drain regions and inherently remains junctionless for which it is termed as "junctionless tunnel effect transistor (JTET)". Compared to planar TFET, JTET is also free from any depletion region originating from high doping concentration and thus becomes suitable for both channel length scaling and vertical integration. For transport mechanism in JTET, analytical compact current transport model has been derived in this chapter for understanding the device physics of JTET. Further, the performance of graphene JTET is compared with ITRS-projected 2020 nMOSFET as well. Similar to a CMOS inverter, a complementary graphene JTET (p-type JTET and n-type JTET) inverter is designed and voltage transfer characteristics studied.

11.2 Device structure and operation

Figure 11.1(d) shows the schematic of the graphene JTET based on graphene-hBN-graphene. Over the Si/SiO_2 substrate, a bottom gate contact is placed followed by the multilayer boron nitride deposition as the gate dielectric. Thermal evaporation or sputtering technique can be employed for the formation of contacts. First principle density functional theory (DFT) has shown that graphene doped by adsorption on metal substrates still preserves its unique electronic properties. A small shift in the Fermi level at the graphene Dirac point by ~0.5 eV is observed [12]. For simplicity, in our current transport model [13], we have assumed zero shift in graphene Fermi level due to the metal contact. Multilayer hBN can be deposited by micromechanical cleavage technique from boron nitride crystal. The buried layers of hBN work as the bottom gate dielectric for the gate contact and a substrate for the bottom graphene layer.

Boron nitride substrate preserves graphene's electronic properties compared to SiO_2 substrate for which hBN is considered as both top and bottom gate dielectrics [14]. Moreover, hBN graphene lattice mismatch is 1.7% for which hBN is suitable as an interlayer tunneling barrier [15]. We have assumed ohmic contacts in source and drain. For the top and bottom gates, a metal-insulator-graphene tunneling junction is formed through metal-hBN-graphene heterostructure. This provides a low differential contact resistance because of hBN-graphene very low lattice mismatch. The graphene layer on top of the buried hBN is referred to as bottom graphene layer (G_B). Source and drain contacts are placed at the two ends of G_B as seen in Figure 11.1(d). Atomically thin multilayer hBN are then deposited on the top of G_B followed by a second layer deposition of graphene. This layer is defined as the top graphene layer (G_T).

Finally, multiple layers of hBN are further deposited on G_T as the top gate dielectric followed by the metal contact deposition. The top metal contact is termed as the top gate contact. The graphene JTET discussed in this chapter considers an effective channel area of 0.05 μm^2, with a channel length of 1 μm and a width of 50 nm. It has been observed experimentally that electrons can propagate without scattering, a distance in micrometer range in graphene [16] for which we have assumed an idealistic scattering free graphene channel of 1 μm. Moreover, such a channel length simplifies the current transport model from the complexity arising from short channel effects and reduces the probability of direct source-drain tunneling effect. The drain current is also less affected by the drain-induced barrier lowering for the considered channel length in graphene JTET.

It is found that quantum-confined graphene in its nanoribbon shape (length >> width) demonstrates an observable bandgap depending on its edge type [16]. The bandgap of graphene nanoribbon increases as the width of nanoribbon reduces for armchair graphene nanoribbon. Therefore, a graphene channel of 50 nm width ensures a zero-bandgap semiconductor. It is to be noted that the channel width << 50 nm (1–10 nm) will open up a bandgap which will change the current transport mechanism in graphene JTET whereas channel width > 50 nm will have potentially no additional effect on the current transport. Therefore, the assumption of 50 nm channel width in this section provides a good approximation between graphene and graphene nanoribbon. Following the work of Britnell *et al.* [2], graphene JTET considers top and bottom hBN gate dielectrics of 20 nm thickness each. The thickness of the interlayer tunneling barrier is 1.02 nm for three hBN layers. Sciambi *et al.* [17] have studied that two graphene layers separated by a nanometer-scale tunneling barrier, preserves not only the coherent length of tunneling but also conserves the out of plane momentum of carriers. The coherent length of tunneling drastically degrades as the tunneling barrier thickness increases [17]. Therefore, we have considered 1.02 nm of hBN as the thickness of the tunneling barrier of three layers of hBN. It is to be noted that a single layer of hBN is 0.34 nm thick [18]. Nevertheless, single or bilayer of hBN can also be adopted which are more susceptible to etching in such vertical heterostructures.

In the off-state, the Fermi levels of the top and bottom graphene layers remain in equilibrium. We assume at equilibrium the Fermi level coincides with the channel Dirac point and no intermediate energy states exist due to roughness or defects. Gate voltage (V_G) is defined as the difference between the bottom (V_{GB}) and top gate voltages (V_{GT}). To turn-on the transistor, V_G ($V_G = V_{GB} - V_{GT}$) is applied between G_T and G_B. V_{DS} is applied between source and drain located at G_B. Device off-state is defined for $|V_G| = 0$ V, $|V_{DS}| = 0.1$ V and on-state for $|V_G| \neq 0$ V, $|V_{DS}| = 0.1$ V.

For low power dissipation, the supply voltage of any switching transistor needs to be compliable with one of the ITRS requirements. Existing silicon and III-V material-based TFETs operate at sub-0.5 V supply voltage for which it is essential for the switching transistor to operate at equal or low supply voltage. Moreover, it is found that the graphene-based transistors can be operated at low supply voltages for which the assumption of 0.1 V operation of graphene JTET is in accordance with the existing TFET performance and ITRS requirement.

Figure 11.2(a) shows the off-state of graphene JTET. E_F^S, E_F^C and E_F^D are the source, channel, and drain Fermi levels, respectively. As V_G is applied, interlayer tunneling of carrier occurs between the top and bottom graphene layers. The carrier concentration (N) due to tunneling shifts E_F^C from the Dirac point of the channel graphene layer by an amount of ΔE_F, as shown in Figure 11.2(b) [19].

This shift in Fermi level results in the change of barrier height between E_F^S and E_F^D which controls the current transport between source and drain due to V_{DS}. In this way, drain current becomes a function of vertical tunneling of carriers between the top and bottom graphene layers. It should be noted that the bottom graphene layer is also the channel graphene layer.

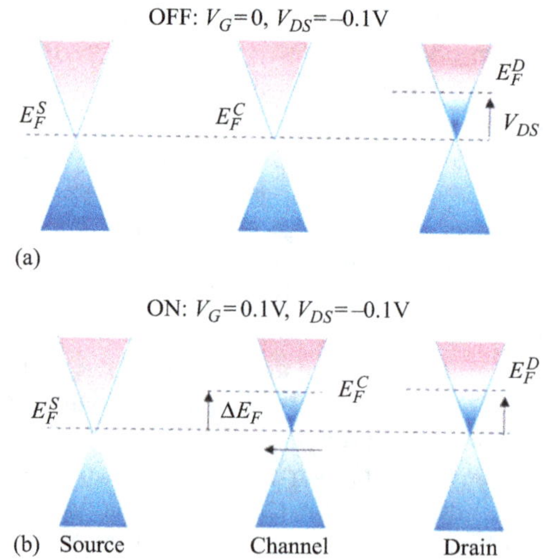

Figure 11.2 (a) Energy band diagram of graphene JTET in the off-state for $V_G = 0$ V and $V_{DS} = -0.1$ V and (b) on-state for $V_G = 0.1$ V and $V_{DS} = -0.1$ V

Based on the experimental study in [4], it is found that a positive gate bias shifts the Fermi level above the Dirac point whereas a negative gate bias shifts the Fermi level below the Dirac point. Therefore, $V_G > 0$ provides $\Delta E_F < 0$ and $V_G < 0$ provides $\Delta E_F > 0$. The channel barrier height is controlled by the vertical interlayer tunneling between two graphene layers. It is important to note that in conventional iTFET, the interlayer tunneling bias results in the tunnel drain current whereas in graphene JTET, the interlayer tunneling bias changes the channel barrier height which regulates the source-drain ballistic transport. Conventional iTFET does not discuss any source-drain ballistic transport mechanism.

As the Dirac point at the top and bottom graphene layers is misaligned, an interlayer tunneling carrier crosses the tunneling barrier. Electrons having the energy halfway between the Dirac points contribute toward this flow [4]. A change in such tunneling of carriers due to gate voltage is also confirmed by the phenomena of wave function extension of one graphene layer to the other and a corresponding overlap at the bottom graphene layer. For both positive and negative gate voltages, the wave function extension is observed [17]. In this way, the out-of-plane momentum is conserved for a longer coherent length for tunneling, preferably in a nanometer range [8]. We assume that the source and drain wave functions do not result in any interference with the wave function extended from the top graphene layer to the bottom graphene layer. The vertical interlayer tunneling, therefore, only contributes toward the barrier control of the channel electrostatic potential.

11.3 Current transport model

11.3.1 Estimation of tunneling probability

The change of the effective barrier height via the shift in Fermi level of graphene is dominated by the height and shape of the barrier [4]. It has been observed that using wide bandgap monolayer of two-dimensional semiconductor, the changes in Fermi level of the graphene due to external bias are near to or more than the height of the tunneling barrier. However, in the case of wide bandgap insulator, such changes in the Fermi level of graphene are insignificant. Wide bandgap insulator such as hBN (bandgap > 5 eV) helps in this regard.

In this section, tunneling probability for a specific tunneling energy barrier height (Δ) and thickness (d) are considered in determining the tunneling probability of carriers from the top graphene layer to the bottom graphene layer and vice versa. Tunneling probability (T_{WKB}) is calculated from the well-known WKB approximation and is expressed as follows [4]:

$$T_{WKB}(E) = \exp\left(-2d\frac{\sqrt{2m^*\Delta}}{\hbar}\right) \tag{11.1}$$

In (11.1), d is the thickness of the tunneling barrier material, Δ is the energy gap between either graphene valence band to hBN valence band for holes or graphene conduction band to hBN conduction band for electrons, and m^* is the effective mass of electron in the tunneling barrier material.

The separation between the graphene Dirac point and the top of the valence band of hBN (Δ) is 1.5 eV whereas this value is >4 eV in case of hBN conduction band [19]. Following the work of Britnell *et al.* [2], we have chosen Δ as 1.5 eV. This yields an effective tunneling mass of holes, $m^* = 0.5m_o$ (m_o is the free electron mass) which is also the effective mass for holes in hBN.

It has been observed that a barrier separating two graphene layers where Fermi surface in one side is electron like and is hole like on the other side demonstrates that electrons incident normally at one side continue to propagate as holes with 100% efficiency at the other side [20]. For this reason, the choice of Δ as 1.5 eV for hole conduction remains consistent. For relativistic carriers, a perfect tunneling probability of 1 can be obtained. However, for nonrelativistic electrons, this is not the case for which the tunneling probability is always less than 1. With a negligible inter-valley scatterings and very low lattice mismatch, a potential barrier shows no reflections for the electron's incident normal to the potential barrier [21]. In graphene JTET, it is assumed that electrons incident normal to the hBN barrier where graphene and hBN has a lattice mismatch of only 1.7%.

11.3.2 Estimation of charge density

When a bias is applied between the top and bottom graphene layers, a corresponding potential difference between the two Fermi levels is observed. Considering the potential difference between the top and bottom graphene layers as $\Delta\varphi$, carriers tunneling from the top to bottom graphene layers are described as follows [22]:

$$N_1 = \int_0^{\Delta\varphi} D(E)T_{WKB}(E)f_T(E)dE \tag{11.2}$$

Similarly, the carriers tunneling from the bottom to top graphene layers can also be expressed as follows:

$$N_2 = \int_0^{\Delta\varphi} D(E)T_{WKB}(E)f_B(E)dE \tag{11.3}$$

Net carriers tunneling from top to bottom graphene layers can be written as follows:

$$N = N_1 - N_2 = \int_0^{\Delta\varphi} D(E)T_{WKB}(E)f_T(E)dE - \int_0^{\Delta\varphi} D(E)T_{WKB}(E)f_B(E)dE \tag{11.4}$$

$$N = \int_0^{\Delta\varphi} D(E)T_{WKB}(E)(f_T(E) - f_B(E)dE \tag{11.5}$$

Here, $D(E)$ is the density of states of graphene, $f_T(E)$ is Fermi function for the top graphene layer, $f_B(E)$ is Fermi function for the bottom graphene layer, $T_{WKB}(E)$

is the tunneling probability obtained from (11.1). $\Delta\varphi$ is the limit of integration. In this case, it is the total energy window between the top and bottom graphene layers through which the tunneling occurs. Density of states in graphene layer is defined as follows [21]:

$$D(E) = \frac{g_s g_v E}{2\pi(\hbar v_F)^2} \tag{11.6}$$

where E is the energy of electron tunneling. For the proposed current transport model of graphene JTET, energy range E is limited between 0 and $\Delta\varphi$; g_s and g_v are spin and valley degeneracy, respectively. For graphene, $g_s = 2$ and $g_v = 2$ [21]. Fermi functions in the top and bottom graphene layers are defined as follows:

$$f_T(E) = \frac{1}{1 + e^{\left(E - E_f^T\right)/k_B T}} \tag{11.7}$$

$$f_B(E) = \frac{1}{1 + e^{\left(E - E_f^B\right)/k_B T}} \tag{11.8}$$

In (11.7) and (11.8), E_f^T and E_f^B are the positions of the Fermi levels at the top and bottom graphene layers, respectively. E is the energy of the electron during tunneling. Fermi level in the top graphene layer is at $E_f^T = qV_G$ and the Fermi level in the bottom graphene layer is at $E_f^B = 0$. Combining (11.5)–(11.8),

$$N = \int_0^{\Delta\varphi} \frac{g_s g_v E}{2\pi(\hbar v_F)^2} T_{WKB}(E) \left(\frac{1}{1 + e^{\left(E - E_f^T\right)/k_B T}} - \frac{1}{1 + e^{\left(E - E_f^B\right)/k_B T}} \right) dE \tag{11.9}$$

Replacing the values of E_f^T and E_f^B by qV_G and 0, (11.9) can be expressed as follows:

$$N = \int_0^{\Delta\varphi} \frac{2E}{\pi(\hbar v_F)^2} T_{WKB}(E) \left(\frac{1}{1 + e^{(E - V_G)/kT}} - \frac{1}{1 + e^{(E)/kT}} \right) dE \tag{11.10}$$

The energy window for tunneling ($\Delta\varphi$) from the top to bottom graphene layers is assumed as $\Delta\varphi = E_f^T - E_f^B = qV_G - 0 = qV_G$. Now integrating (11.10) from $E = 0$ to $E = \Delta\varphi = qV_G$, closed form of Fermi-Dirac integration becomes:

$$N = \frac{g_s g_v}{2\pi(\hbar v_F)^2} T_{WKB}(E) \left(\frac{V_G{}^2}{12} - (V_G) k_B T \ln\left[1 + \exp(V_G/k_B T)\right] - \frac{(\pi k_B T)^2}{12} \right.$$
$$\left. - (k_B T)^2 Poly\log(2, -\exp(V_G/k_B T)) \right) \tag{11.11}$$

Now for any $qV_G >> k_BT$, it is found that the first few terms dominate over the later parts of (11.11) for which the higher energy terms in (11.11) can be simplified as follows:

$$\frac{V_G^2}{2} - (V_G)k_BT\ln[1+\exp(V_G/k_BT)] \gg \frac{(\pi k_BT)^2}{12} - (k_BT)^2 Polylog(2,-\exp(V_G/k_BT))$$

Therefore, the closed-form solution of (11.11) can be expressed as follows:

$$N = \frac{2}{\pi(\hbar v_F)^2} T_{WKB}(E)(\frac{V_G^2}{12} - (V_G)k_BT\ln[1+\exp(V_G/k_BT)] \tag{11.12}$$

Equation (11.12) expresses the doping density which is the net number of carriers tunneling from top to bottom graphene layers due to applied voltage, V_G as shown in Figure 11.3(a). Following the work of Georgiou *et al.* [4], a positive bias generates electron tunneling whereas the negative bias generates hole tunneling. The electron tunneling is shown in blue curve and hole tunneling is shown in red curve in Figure 11.3(a) for positive and negative biases, respectively. The induced doping density through interlayer tunneling (N), calculated using (11.12), has a square root dependence on Fermi level of the bottom graphene layer which is expressed as follows [4]:

$$\Delta E_F = \pm\hbar v_F\sqrt{\pi|N|} \tag{11.13}$$

The sign of the Fermi level shift (positive or negative) is determined from the polarity of the gate voltage [20]. A positive bias shifts the Fermi level upward which is shown in Figure 11.2(b). Figure 11.3(b) shows the change in the amount of shift in Fermi level (ΔE_F) due to induced carrier concentration (N) at the bottom graphene layer. The red and blue lines in Figure 11.3(b) represent the change of Fermi level based on the polarity of V_G.

11.3.3 Estimation of drain current

Based on "mode" (M)-based modeling approach of nanoscale transistor, drain current in graphene JTET can be calculated considering channel conductivity and transmission coefficients. Considering the change of Fermi level at the bottom graphene layer due to vertical tunneling of carriers between the top and bottom graphene layers due to V_G and the source-drain lateral transport due to V_{DS}, drain current in graphene JTET can be expressed using Landauer's expression as follows [23]:

$$I = \int dE[(G(E)(f_S(E) - f_D(E)))] \tag{11.14}$$

Here, $G(E)$ is channel conductance; $f_S(E)$ and $f_D(E)$ are source and drain Fermi functions, respectively, which can be expressed similar to (11.7) and (11.8). Based on Landauer expression, conductance ($G(E)$) can be expressed as follows [23]:

$$G(E) = (2q^2/\hbar)M(E)T_B(E) \tag{11.15}$$

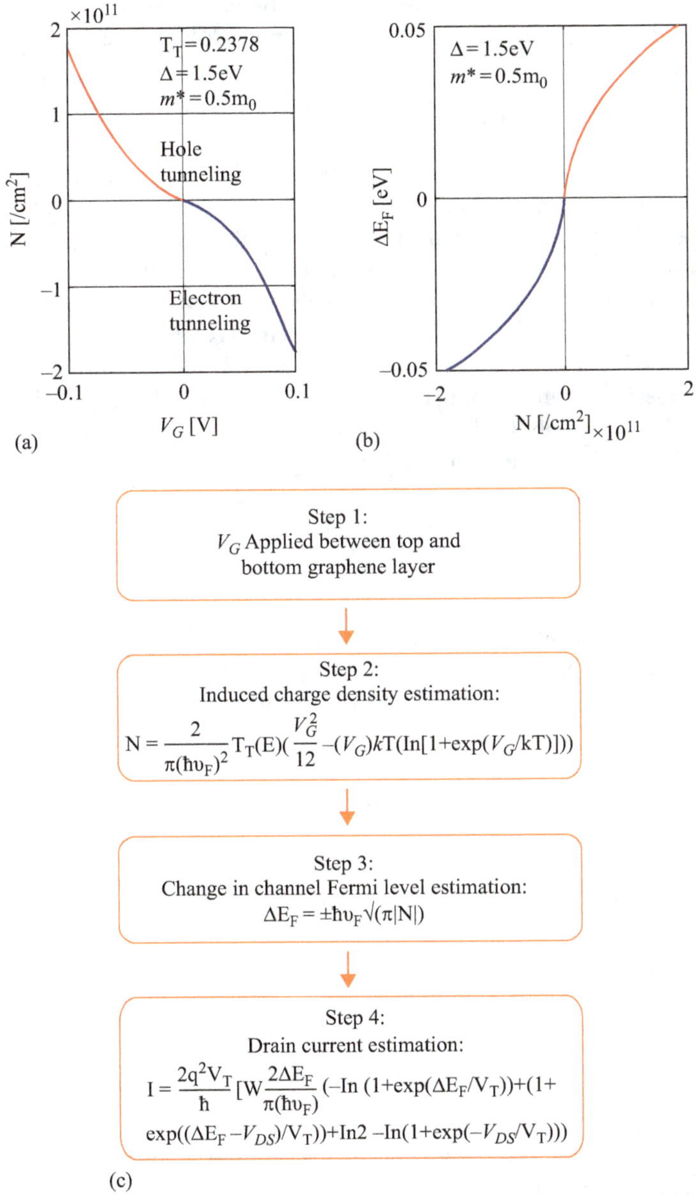

Figure 11.3 (a) Carrier concentration (N) versus V_G and (b) change of Fermi level (ΔE_F) with N and (c) flow chart showing the operation of graphene JTET

Here, $T_B(E)$ is transmission coefficient in ballistic transport and $M(E)$ is the number of modes in graphene. The number of modes (M) in graphene is expressed as follows [24]:

$$M(E) = 2W|E_{TM}|/\pi(\hbar v_F) \tag{11.16}$$

In (11.16), W is the width of the channel and $|E_{TM}|$ is the energy range for calculating transverse mode. In this work, E_{TM} is considered as the amount of shift in Fermi level in the channel (ΔE_F) which controls the number of modes in the channel between source and drain. The number of conducting channels at energy E_{TM} is proportional to the width of the conductor in two-dimensional and to the cross-sectional area in three-dimensional geometry. Total number of modes is also affected by the band structure of the channel material [24]. Expression of $M(E)$ in (11.16) is specific to the graphene which differs from the expression of mode usually used for a parabolic band structure. v_F is the Fermi velocity. Combining from (11.14) to (11.16), drain current can be written as follows:

$$I = \frac{2q^2}{\hbar}\int dE[(M(E)T_B(E)(f_S(E) - f_D(E)))] \tag{11.17}$$

Considering a scattering free source-drain ballistic transport in the channel, we have assumed the transmission coefficient, $T_B(E)$ as 1 in (11.17). Now, combining the energy window for ballistic transport from 0 to qV_{DS} and the change in channel barrier height from 0 to ΔE_F, (11.17) can be written as follows:

$$I = \frac{2q^2}{\hbar}\left[\int_0^{\Delta E_F} W\frac{2|E_{TM}|}{\pi(\hbar v_F)}\left(\frac{1}{1+e^{\left(E_{TM}-E_F{}^S\right)/kT}} - \frac{1}{1+e^{\left(E_{TM}-E_F{}^D\right)/kT}}\right)dE\right] \tag{11.18}$$

The closed-form analytical solution of (11.18) is as follows:

$$I = \frac{2q^2V_T}{\hbar}\left[W\frac{2\Delta E_F}{\pi(\hbar v_F)}(-\ln(1+\exp(\Delta E_F/V_T)) + \ln(1+\exp((\Delta E_F - V_{DS})/V_T)) + \ln(2) - \ln(1+\exp(-V_{DS}/V_T)))\right] \tag{11.19}$$

In (11.19), k_BT is replaced by the thermal voltage qV_T, value of which is defined as 0.0259 eV at 300 K. We consider this as the equation of drain current in graphene JTET which is applicable for both the electronic conduction (n-type behavior) and hole conduction (p-type behavior), provided appropriate bias is considered. Figure 11.3(c) provides the flow chart of the operation of graphene JTET with necessary current transport equations. Mobility is an important parameter in graphene JTET. Considering Drude model for conductivity ($\sigma = \mu_n Nq$, where σ is conductivity, μ_n is the carrier mobility, and q is the charge on electron) and graphene minimum conductivity ($\sigma = 4q^2/h$ where h is Planck's constant), we have calculated the mobility of the graphene JTET as 5468 cm^2/V-s. The doping density through tunneling (N) of $1.76 \times 10^{11}/\text{cm}^2$ at 0.1 V gate bias is considered for the mobility extraction. Graphene band structure is symmetric around the Dirac point for which nearly identical value applies for both electron and hole mobility [21].

11.4 Performance analysis of interlayer tunneling-based graphene JTET

Using (11.19), drain current is calculated which has both T_{WKB} and N dependence. The plotted transfer characteristic in Figure 11.4(a) considers a fixed tunneling probability ($T_{WKB} = 0.2378$) for different V_{DS}. For $V_{GS} = 0.1$ V and $V_{DS} = 0.1$ V, on-current density of 88 μA/μm^2 is obtained for the effective channel area of 0.05 μm^2. With three hBN layers, graphene JTET operating at 0.1 V supply voltage turns on at an average subthreshold slope of 25 mV/decade with 2.45×10^4 on/off current ratio. The off-state leakage current of 3.5 nA/μm^2 gives an off-state static power of 0.35 nW/μm^2. Calculated dynamic power for graphene JTET is 4.4 μW/μm^2 for the drive current of 88 μA/μm^2 at 0.1 V supply voltage. A comparison of the transfer characteristic of graphene JTET with some of the earlier reported iTFETs is presented in Figure 11.4(b) for 0.1 V gate bias. Figure 11.4(b) shows that the earlier reported iTFETs provide low on-current density and high subthreshold slope. Table 11.1 summarizes the comparison obtained from Figure 11.4(b). For the focus on digital circuit, we have avoided the inclusion of similar graphene-insulator-graphene devices showing NDR effects in Table 11.1 and Figure 11.4(b); thus, limited the comparison with non-NDR devices only.

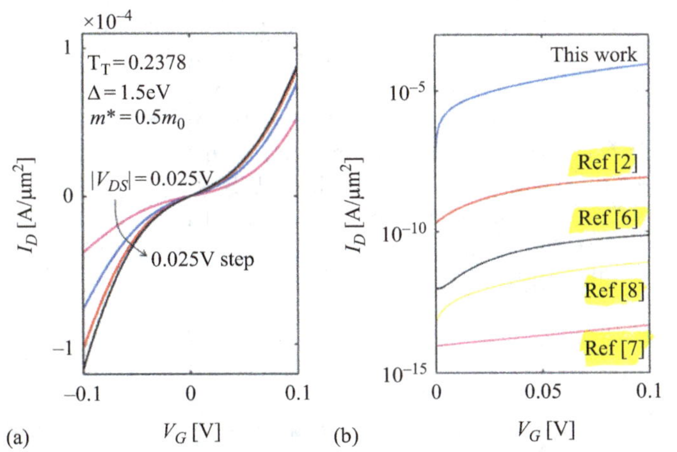

Figure 11.4 Transfer characteristics for the graphene JTET: (a) I_D–V_G curve for different V_{DS} in linear scale with 0.025 V step and (b) comparison of the transfer characteristics of graphene JTET with earlier similar type of iTFETs

Table 11.1 Comparison of graphene JTET performance with similar itfet

Model	$\|V_{DD}\|$ or V_{DS}**	$\|V_G\|$	Tunneling barrier type	I_{ON}/I_{OFF}	Subthreshold slope (mV/decade)
This work	0.1	0.1	hBN, 3 layers	2.45×10^4	25
Ref. [2]	25	0.1	hBN, 4 layers	10 to 10^4	16
Ref. [4]	2	0.1	WS_2, 4 layers	10^6	20
Ref. [7]	0.5	10	hBN, 5 layers	30	300
Ref. [6]	0.8	0	TiO_x/TiO_2, 5 nm*	Unspecified	70

* $x = 0.68$–0.75

** Literature considers both forms of expression for the drain bias.

It is observed from both Figure 11.4(b) and Table 11.1 that graphene JTET performs better than other similar iTFETs. Few explanations are required at this stage for describing the high performance of graphene JTET. We have considered three layers of hBN equivalent to 1.02 nm in thickness as the tunneling barrier, whereas the other listed iTFETs in Figure 11.4(b) and Table 11.1 consider a thicker tunneling barrier. Such a small barrier thickness not only induces a higher charge density at the bottom graphene layer but also energy momentum in vertical direction remains conserved. This is consistent with having a relatively smaller coherence length of tunneling which suppresses the NDR effect [6]. ITRS requires a minimum value of on/off current ratio (I_{ON}/I_{OFF}) as 10^4 at $V_{DD} < 0.7$ V for the next-generation devices for digital applications [7]. From Table 11.1, graphene JTET provides the I_{ON}/I_{OFF} of $2.45\text{x}10^4$ at $V_{DD} = 0.1$ V which meets the ITRS requirement. Although graphene JTET provides low I_{ON}/I_{OFF} compared to some other iTFET, it is still suitable for digital circuit design. It is to be mentioned that Georgiou *et al.* [4] obtained a current ratio of 10^6 at $V_{DD} = 2$ V ($>V_{DD}$ of graphene JTET) range for graphene-WS_2-graphene iTFET, however, subthreshold slope is larger than that obtained for graphene JTET at 0.1 V supply voltage. Moreover, WS_2 is a wide bandgap semiconductor compare to hBN which is a wide bandgap insulator. The electronic properties of graphene-WS_2 superlattice are different from the graphene-hBN superlattice for which I_{ON}/I_{OFF} of graphene JTET differs from the I_{ON}/I_{OFF} in [4]. Using the method of average subthreshold slope, SS can be determined as follows [25,26]:

$$SS = \frac{dV_{GS}}{d(log_{10}I_D)} \tag{11.20}$$

where I_D is the drain current and V_G is the gate bias. For a decade change in drain current in the subthreshold region, required gate bias is calculated which gives the subthreshold slope. Figure 11.5 shows the extraction of subthreshold slope. It is to be mentioned that Figure 11.5 is plotted in log scale compared to the linear scale in Figure 11.4(a). The values of SS mentioned in Table 11.1 are also calculated using Figure 11.5 following the method described in the work of Appenzeller *et al.* [26].

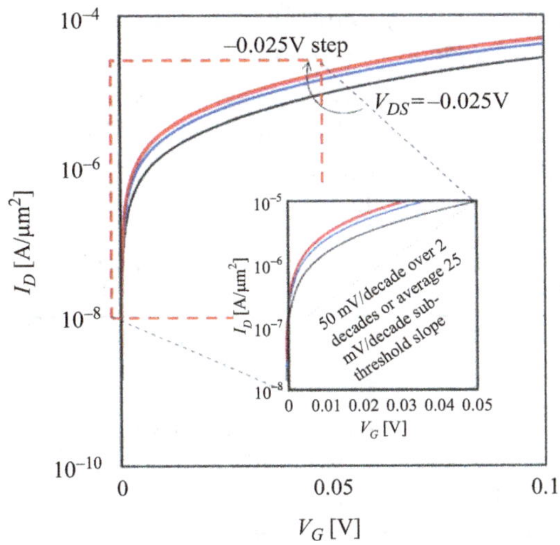

Figure 11.5 Subthreshold slope extraction from I_D–V_G curve of graphene JTET. Inset shows the change in V_G for estimating the average subthreshold slope over three decades of drain current [25]. Note: Drain current is plotted in the log scale compared to the linear scale, as shown in Figure 11.4(a)

For energy-efficient switching technology, it is necessary that a transistor provides subthreshold slope (SS) less than the conventional thermionic limit of 60 mV/decade. Since most iTFETs provide either NDR behavior or linear resistive characteristic, SS of such devices is not always discussed explicitly. The iTFET proposed by Roy *et al.* [6] obtained an SS of 70 mV/decade for the TiO_x/TiO_2 stack for a tunneling barrier ($x = 0.68$–0.75), which is also found to be limited by the gate capacitance. Using the first principles DFT combined with non-equilibrium Green function (NEGF), Fiori *et al.* [7] studied a very large on-current modulation in graphene-hBN-graphene vertical heterobilayer. For a drain-source voltage of 0.5 V, a corresponding SS ~ 300 mV/decade has been obtained. Such performance is observed due to the poor electrostatic control of channel potential by the gate voltage. Ghobadi and Pourfath [8] obtained > 1,000 mV/decade SS for similar iTFETs with three hBN layers. The fundamental physical limitation of such iTFETs in terms of subthreshold slope is also consistent with the high subthreshold slope obtained for similar iTFETs discussed in this chapter. Compared to iTFETs, graphene JTET adopts a mixed-mode mechanism of vertical interlayer tunneling of carriers between two graphene layers and lateral ballistic transport between source and drain for which gate capacitance has little or no effect. Moreover, the shift in Fermi level controlling source-drain ballistic transport provides superior channel electrostatic control. For these reasons, a very steep subthreshold has been obtained for graphene JTET compared to previously reported iTFETs. Table 11.2 enlists the

Table 11.2 Comparison of graphene JTET performance with 2020 nMOSFET projected in the 2012 edition of ITRS

Parameter	**2020 nMOSFET**	**Graphene JTET**	**Unit**
Supply voltage, V_{DD}	0.68	0.1	V
Drive current, I_D	1,942	880	μA/μm
Off-state leakage current, I_{OFF}	100	3.5	nA/μm
Off-leakage power, ~ $I_{OFF}V_{DD}$	68	0.35	μW/μm
Dynamic power, ~$1/2I_DV_{DD}$	660.28	44	μW/μm

performance comparison of graphene JTET with ITRS projected 2020 nMOSFETs. Compared to the on-state drain current of 1942 μA/μm at $V_{DD} = 0.68$ V for 2020 nMOSFET, graphene JTET on-state drain current is calculated as 880 μA/μm at $V_{DD} = 0.1$ V. Calculated off-state leakage current is 3.5 nA/μm compared to 100 nA/μm of 2020 nMOSFET. Therefore, graphene JTET provides 194 times less off-state leakage power and dissipates ~15 times less dynamic power than the 2020 nMOSFET. Note that the current values mentioned in Table 11.2 for graphene JTET have been normalized with the channel length which provides the drain current unit in μA/μm.

Transfer characteristics of graphene JTET are highly dependent on the thickness of the tunneling barrier. Therefore, it is necessary to study the performance of graphene JTET at different tunneling barrier thicknesses. Since graphene JTET is designed as a vertical heterostructure, its tunneling barrier thickness is determined by the number of hBN layers used between the top and bottom graphene layers. Figure 11.6(a) shows the transfer characteristics of graphene JTET for different number of hBN layers. From (11.1), we found that the tunneling probability is exponentially dependent on the thickness of the barrier. Therefore, on-current density of 96.03 μA/μm^2 is observed for the monolayer hBN (0.34 nm thick) as the tunneling barrier, value of which decreases to 0.282 μA/μm^2 for six layers of hBN used.

The ratio between the on-current to the off-current (I_{ON}/I_{OFF}) also changes with the total number of the hBN layers along with subthreshold slopes of graphene JTET. Figure 11.6(b) shows I_{ON}/I_{OFF} and SS for different number of hBN layers. As the tunneling barrier thickness increases with the number of hBN layers, I_{ON}/I_{OFF} decreases. The subthreshold slope of graphene JTET increases with the increase in the number of hBN layers due to reduced tunneling probability. For the monolayer hBN, only 0.9 mV/decade of SS over single decade is estimated which increases to 20.31 mV/decade for six hBN layers. With smaller barrier thickness, precise gate control over the channel is obtained. Moreover, the wave function of the top graphene layer easily extends toward the bottom graphene layer [17]. This provides not only high on-current density but also a reduced off-state leakage current along with the steep subthreshold slope. Therefore, a high I_{ON}/I_{OFF} and low subthreshold slope are observed for a smaller number of hBN layers.

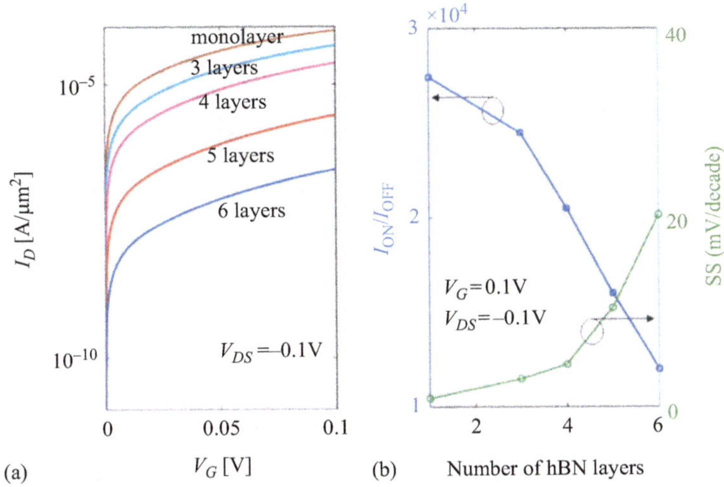

Figure 11.6 (a) Change in the transfer characteristics of graphene JTET for multiple hBN layers as tunneling barrier and (b) change in on/off current ratio (I_{ON}/I_{OFF}) and SS with the number of hBN layers

I_D-V_{DS} characteristics in conventional iTFET suffer large NDR effect. Therefore, their scope in digital circuit design becomes limited. However, the proposed graphene JTET overcomes such limitations and provides NDR free output characteristics with separate n- and p-type behavior. Figure 11.7(a) and (b) depicts the output characteristics (I_D–V_{DS}) of graphene JTET for p-type and n-type graphene JTET for different V_G, respectively. Compared to conventional MOSFETs, n-type electronic transport is obtained for $V_{DS} > 0$ and $V_G < 0$ whereas p-type hole transport is obtained for $V_{DS} < 0$ and $V_G > 0$. Since a positive gate bias induces a negative shift in the Fermi level and a negative gate bias induces a positive shift in Fermi level [27], the sign of notation used in Figure 11.7 is consistent with the overall current transport. Figure 11.7 considers equal tunneling probability (T_{WKB}) in both the p- and n-type transistors. With independently applied bias at the top and bottom graphene layers, a strong Coulomb drag is generated due to the interlayer electron-hole interaction [28]. By applying a positive bias at the gate ($V_G > 0$), electron like Fermi surface is formed at the top graphene layer. Further when a negative bias at drain ($V_{DS} < 0$) is applied, hole-like Fermi surface is formed at the bottom graphene layer. Both of these opposite types of Fermi surfaces are necessary for: (1) scattering free elastic tunneling normal to the barrier and (2) positive Coulomb drag for interlayer electron–hole interaction. Similarly, a negative Coulomb drag with elastic scattering free tunneling is observed when $V_G < 0$ is applied at the top graphene layer and $V_{DS} > 0$ at the bottom graphene layer. Thus, the need for such opposite polarity of biasing for obtaining the output characteristic is understood. Figure 11.8(a) and (b) shows the plot of output characteristics of p-type and n-type graphene JTET at high V_{DS}, respectively. Note that

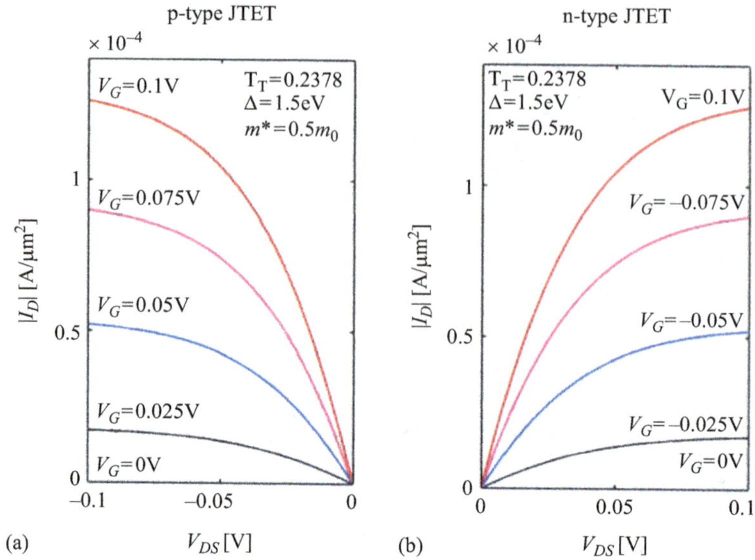

Figure 11.7 Output characteristics for graphene JTET: (a) p-type behavior obtained for $V_G > 0$, $V_{DS} < 0$ and (b) n-type behavior obtained for $V_G < 0$, $V_{DS} > 0$

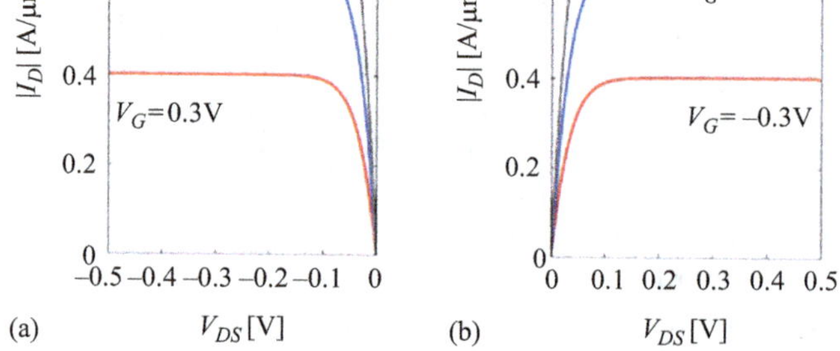

Figure 11.8 Output characteristics of graphene JTET with increasing V_{DS} for varying V_G: (a) p-type graphene JTET and (b) n-type graphene JTET

at higher V_{DS}, drain current saturation is observed. For all three conditions of $V_{DS} < V_G$, $V_{DS} = V_G$ and $V_{DS} > V_G$, graphene JTET provides drain current saturation. This implies that the magnitude of the Coulomb drag originating at higher drain and gate bias provides not only a precise interlayer tunneling but also preserves superior gate control over the channel. For this reason, smooth output characteristics are obtained.

11.5 Voltage transfer characteristics of graphene JTET inverter

The inverter is the basic building block of a digital integrated circuit and its performance reflects the type of transistors used as switches. Complementary inverter using vertical heterostructure transistors as switches can be used similar to a CMOS inverter. Figure 11.9(a) and (b) shows the symbols of p-type graphene JTET and n-type graphene JTET, respectively. Since graphene JTET has similarity with a ballistic nanoscale MOSFET with respect to source-drain ballistic transport, such symbols are partially designed based on the conventional depletion type MOSFET symbols. However, since the channel barrier control is carried out through the vertical interlayer tunneling, we have adopted the conventional sign of tunneling between top and bottom gate electrodes. Therefore, the symbols drawn in Figure 11.9(a) and (b) combine both the concept of vertical interlayer tunneling between gates at the top and bottom graphene layers and source-drain ballistic transport. Figure 11.9(c) shows the schematic of a complementary graphene JTET vertical inverter.

The gate bias (V_G) is defined as the difference between the top (V_{GT}) and bottom gate voltages (V_{GB}) of the transistor ($V_G = V_{GT} - V_{GB}$). The bottom gate of

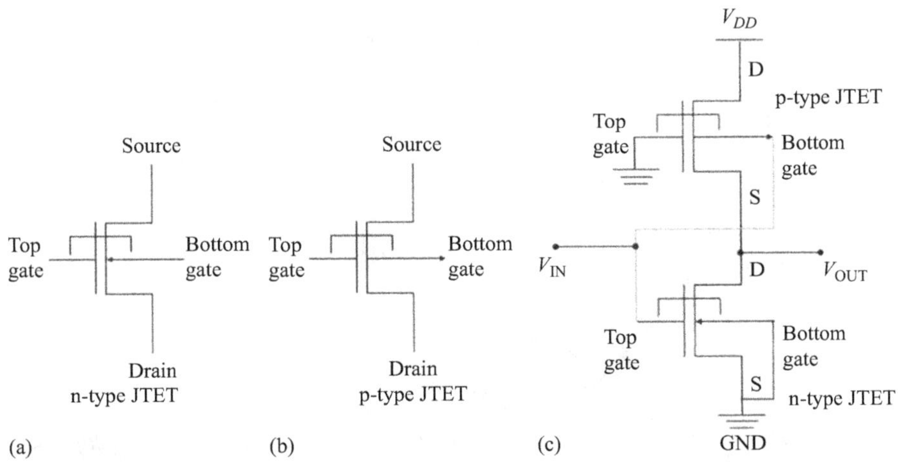

Figure 11.9 (a) Symbol for p-type JTET, (b) n-type JTET, and (c) schematic of complementary graphene JTET-based vertical logic inverter

p-type graphene JTET is connected with the top gate of the n-type graphene JTET for which it is termed as common gate contact. An input voltage (V_{IN}) applied at the common gate contact will generate two opposite type of shifts in Fermi levels in each of these transistors independently. For example, a positive bias at the common gate will generate a positive gate voltage, V_G ($V_G = V_{IN} - 0 = V_{IN}$) resulting in n-type characteristics in bottom JTET whereas a negative gate voltage, V_G ($V_G = 0 - V_{IN} = -V_{IN}$) resulting in p-type characteristics in top JTET. Drain of the n-type graphene JTET is connected to the source of p-type graphene JTET. Drain of p-type graphene JTET is connected to the supply voltage (V_{DD}) and source of the n-type graphene JTET is grounded (0 V). Being vertically connected, a single gate contact is necessary for graphene JTET vertical inverter. In this way, no additional interconnect is required to connect the two gates of the two complementary transistors. Figure 11.10(a) shows the voltage transfer characteristics (VTC) of the complementary graphene JTET inverter operating at different supply voltages. The inverter gain (A_V) of 4.35 is obtained for $V_{DD} = 0.5$ V whereas the gain in 3.15 for $V_{DD} = 0.1$ V. This reflects the capability of graphene JTET inverter to operate at reduced supply voltage with higher gain.

Compared to a conventional CMOS inverter where gain plummets as supply voltage goes below 0.5 V, graphene JTET vertical inverter can retain its gain at low supply voltages. It is also noted from the transfer characteristics that sharp transition from off- to on-state is obtained at all supply voltages. Figure 11.10(b) shows the extraction of noise margin for $V_{DD} = 0.1$ V for the graphene JTET inverter. We have calculated the low noise margin, N_{ML} as 0.021 V and high noise margin, N_{MH} as 0.022 V. Both of these values are more than 20% of the original signal which substantiates strong noise immunity.

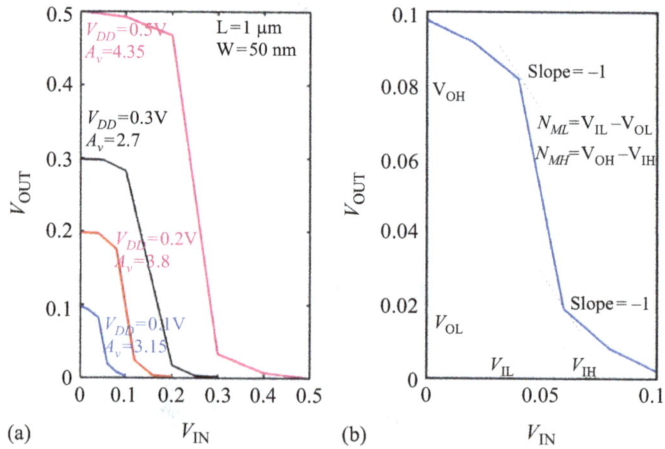

Figure 11.10 (a) Voltage transfer characteristics of a complementary graphene JTET vertical inverter for different supply voltages with corresponding inverter gain and (b) noise margin for the supply voltage of 0.1 V

11.6 Conclusion

A new type of graphene-switching transistor termed as "junctionless tunnel effect transistor (JTET)" based on graphene-hBN-graphene vertical heterostructure and interlayer tunneling is proposed and an analytical current transport model has been developed. The drain current in graphene JTET flows between the source and drain of bottom graphene layer. The current in the channel is regulated by the shift in channel Fermi level which depends on the net vertical tunneling of carriers from top graphene to bottom graphene layers through hBN. Performance of graphene JTET is evaluated for different numbers of hBN layers. A comparison between graphene JTET and ITRS projected 2020 nMOSFET is also provided apart from graphene JTET performance comparison with similar iTFETs. Current saturation is observed in graphene JTET output characteristic for both p- and n-type operations, which makes graphene JTET suitable for digital circuit design. Graphene JTET is also capable of suppressing NDR effect, and shows steep subthreshold slope with high on/off current ratio and normal operation at room temperature. A complementary vertical inverter is presented similar to a CMOS inverter and analyzed for its performance. Graphene JTET vertical inverter gives inverter gain higher than unity at the low supply voltage and both low and high noise margins. It is concluded that with an average 25 mV/decade subthreshold slope at 0.1 V supply voltage and a current ratio of $\sim 10^4$, graphene interlayer junctionless tunnel effect transistor meets the ITRS requirement of device scaling for energy-efficient circuit design.

References

[1] *International Technology Roadmap for Semiconductor*. Available at: www.itrs.net, 2013.

[2] L. Britnell, R. V. Gorbachev, R. Jalil *et al.*, "Field-effect tunneling transistor based on vertical graphene heterostructures," *Science*, vol. 335, pp. 947–50, 2012.

[3] A. K. Geim and I. V. Grigorieva, "Van der Waals heterostructures," *Nature*, vol. 499, pp. 419–25, 2013.

[4] T. Georgiou, R. Jalil, B. D. Belle *et al.*, "Vertical field-effect transistor based on graphene-WS_2 heterostructures for flexible and transparent electronics," *Nat Nano*, vol. 8, pp. 100–03, 2013.

[5] P. Zhao, R. M. Feenstra, G. Gong, and D. Jena, "SymFET: a proposed symmetric graphene tunneling field-effect transistor," *IEEE Trans. Elect. Dev.*, vol. 60, pp. 951–57, 2013.

[6] T. Roy, Z. R. Hesabi, C. A. Joiner, A. Fujimoto, and E. M. Vogel, "Barrier engineering for double layer CVD graphene tunnel FETs," *Microelec. Eng.*, vol. 109, pp. 117–19, 2013.

[7] G. Fiori, S. Bruzzone, and G. Iannaccone, "Very large current modulation in vertical heterostructure graphene/hBN transistors," *IEEE Trans. Elect. Dev.*, vol. 60, pp. 268–73, 2013.

[8] N. Ghobadi and M. Pourfath, "A comparative study of tunneling FETs based on graphene and GNR heterostructures," *IEEE Trans. Elect. Dev.*, vol. 61, pp. 186–92, 2014.

[9] A. Mishchenko, J. S. Tu, Y. Cao *et al.*, "Twist-controlled resonant tunnelling in graphene/boron nitride/graphene heterostructures," *Nat Nano*, vol. 9, pp. 808–13, 2014.

[10] B. Fallahazad, K. Lee, S. Kang *et al.*, "Gate-tunable resonant tunneling in double bilayer graphene heterostructures," *Nano Let.*, vol. 15, pp. 428–33, 2015.

[11] S. Kang, B. Fallahazad, L. Kayoung *et al.* "Bilayer graphene-hexagonal boron nitride heterostructure negative differential resistance interlayer tunnel FET," *IEEE Elect. Dev. Lett.*, vol. 36, pp. 405–07, 2015.

[12] G. Giovannetti, P. A. Khomyakov, G. Brocks, V. M. Karpan, J. van den Brink, and P. J. Kelly, "Doping graphene with metal contacts," *Phys. Rev. Lett.*, vol. 101, pp. 026803, 2008.

[13] Md S. Fahad and A. Srivastava, "A graphene switching transistor for vertical circuit design," *ECS J. Sol. Sci. and Tech.*, vol. 5, pp. M13–M21, 2016.

[14] C. R. Dean, A. F. Young, I. Meric *et al.* "Boron nitride substrates for high-quality graphene electronics," *Nat Nano*, vol. 5, pp. 722–26, 2010.

[15] G. Giovannetti, P. A. Khomyakov, G. Brocks, P. J. Kelly, and J. van den Brink, "Substrate-induced band gap in graphene on hexagonal boron nitride: Ab initio density functional calculations," *Phys. Rev. B*, vol. 76, pp. 073103, 2007.

[16] Y. W. Son, M. L. Cohen, and S. G. Louie, "Energy gaps in graphene nanoribbons," *Phys. Rev. Lett.*, vol. 97, pp. 216803, 2006.

[17] A. Sciambi, M. Pelliccione, M. P. Lilly *et al.*, "Vertical field-effect transistor based on wave-function extension," *Phys. Rev. B*, vol. 84, pp. 085301, 2011.

[18] L. Britnell, R. V. Gorbachev, R. Jalil *et al.*, "Electron tunneling through ultrathin boron nitride crystalline barriers," *Nano Let.*, vol. 12, pp. 1707–1710, 2012.

[19] N. Kharche and S. K. Nayak, "Quasiparticle band gap engineering of graphene and graphone on hexagonal boron nitride substrate," *Nano Letters*, vol. 11, pp. 5274–78, 2011.

[20] M. I. Katsnelson, K. S. Novoselov, and A. K. Geim, "Chiral tunnelling and the Klein paradox in graphene," *Nat Phys*, vol. 2, pp. 620–25, 2006.

[21] A. H. Castro Neto, F. Guinea, N. M. R. Peres, K. S. Novoselov, and A. K. Geim, "The electronic properties of graphene," *Rev. of Modern Phys.*, vol. 81, pp. 109–62, 2009.

[22] J. G. Simmons, "Generalized formula for the electric tunnel effect between similar electrodes separated by a thin insulating film," *J of App. Phys.*, vol. 34, pp. 1793–803, 1963.

[23] S. Datta, *Quantum Transport: Atom to Transistor*. United Kingdom: Cambridge University Press, 2005.

[24] M. Lundstrom and C. Jeong, *Lessons from Nanoscience: A Lecture Note Series Near-Equilibrium Transport*. New Jersey: World Scientific, 2012.
[25] M. Fahad and A. Srivastava, "Subthreshold slope of vertical graphene interlayer tunnel transistor," *Nano*, vol. 6, pp. 1750069, 2017.
[26] J. Appenzeller, Y. M. Lin, J. Knoch, and P. Avouris, "Band-to-band tunneling in carbon nanotube field-effect transistors," *Phys. Rev. Let.*, vol. 93, pp. 196805, 2004.
[27] D. Newns, B. Elmegreen, X. Hu Liu, and G. Martyna, "A low-voltage high-speed electronic switch based on piezoelectric transduction," *J. of App. Phys.*, vol. 111, p. 084509, 2012.
[28] S. Kim, I. Jo, J. Nah, Z. Yao, S. K. Banerjee, and E. Tutuc, "Coulomb drag of massless fermions in graphene," *Phys. Rev. B*, vol. 83, pp. 161401, 2011.

Chapter 12

Molybdenum disulfide–boron nitride junctionless tunnel effect transistor

Ashok Srivastava[1] *and Muhammad Shamiul Fahad*[1]

12.1 Introduction

Scaling of planar metal-oxide semiconductor field-effect transistor (MOSFET) is predicted to face its near end as the Moore's law continues, down to the technology node of 7 nm and below [1]. In addition to shrinking MOSFET channel length to sub-10 nm for high transistor density, vertical integration of MOSFETs based on the stacking of two-dimensional layered materials has recently been explored [2–16]. Novel two-dimensional material systems such as graphene and non-graphene have largely made this feasible [17]. These transistors hold the promise for vertical integration, providing an alternative approach for maintaining the lifeline of Moore's law and beyond. Compared to conventional inversion mode of operation, field effect tunneling-based current transport has been studied in these vertical FETs. Majority of these vertical FETs consider two graphene layers separated by a thin tunnel barrier, mostly hex boron nitride (*h*BN).

Considering Bose condensation of Fermions (electron–hole pairs) between two graphene layers, BiSFET proposed by Banerjee *et al.* [5] was one of the theoretical graphene-based interlayer FETs. The theoretical model of an interlayer tunneling transistor, SymFET, proposed by Zhao *et al.* [7] was another graphene/*h*BN heterostructure. With an on/off current ratio of ~100, SymFET provides a large resonant current peak. However, the model in [7] does not provide any insight on SymFET subthreshold slope. Operating frequency of SymFET was also not reported in [7]. Recently, Fiori *et al.* [9] have studied very large current modulation in graphene/*h*BN vertical heterostructure from the multiscale simulation approach. A large subthreshold slope of 385 mV/decade, with an on/off current ratio of ~15 is reported. The intrinsic cut-off frequency also falls below 1 GHz.

Ghobadi and Pourfath [10] studied a vertical heterostructure similar to [9] considering both graphene and quantum-confined graphene nanoribbon (GNR) separated by *h*BN with a focus on high-frequency operation. However, low on/off

[1]Department of Electrical and Computer Engineering, Louisiana State University, Baton Rouge, LA, USA

current ratio (~3–10) and high subthreshold slope (>1,000 mV/decade) were obtained for ~100 GHz cutoff frequency. Compared to graphene, atomically thin molybdenum disulfide (MoS_2)-based planer FET has already shown promise [18–21]. However, unlike graphene, the study of vertical FET based on interlayer tunneling between two MoS_2 layers separated by a thin tunnel barrier has remained largely unexplored. Moreover, the current transport mechanism proposed for graphene JTET requires additional understanding for the case of JTET with large band-gap material. Graphene is a zero band semiconductor. Therefore, the performance of JTET other than graphene as top and bottom electrode separated by tunneling barrier structure needs further description.

In this chapter, the operating principle of JTET discussed in Chapter 11 has been extended for the study of MoS_2 JTET considering MoS_2/*h*BN/MoS_2 for reduced subthreshold slope operation and sustainable leakage. The interlayer tunneling-based barrier control mechanism as proposed for graphene JTET in Chapter 11 and [16] is used for the current transport study of MoS_2 JTET through self-consistent simulation method [22]. Similar to graphene JTET, multilayer *h*BN is considered as the gate dielectric for MoS_2 JTET. The performances of MoS_2 JTET are compared with the earlier reported graphene-based iTFET reported in [9] and [10].

12.2 Device structure and operation

Figure 12.1 shows the schematic of MoS_2 JTET where the channel is a monolayer MoS_2 of 10 nm length and 5 nm width [21]. Compared to the graphene JTET device structure discussed in Chapter 11, MoS_2 JTET considers as single-layer MoS_2 as both the top and bottom electrodes. Following the work in [3] and [16], gate dielectric comprises of 20 layers of *h*BN (~7 nm). Monolayer *h*BN is considered as the vertical tunneling barrier between two MoS_2 layers. Compared to conventional interlayer tunneling field-effect transistor (iTFET), MoS_2 JTET considers source and drain ohmic contacts on the bottom MoS_2 layer.

Recently, it has been experimentally observed that chemical vapor deposition-based direct growth of monolayer MoS_2 on *h*BN provides smaller lattice strain, low doping level, and clean and sharp interface [23]. Moreover, monolayer MoS_2 is stable over monolayer hexagonal BN (*h*BN) substrate for an inter-planer distance of 4.89 Å [24]. Based on the density functional theory (DFT), an energy bandgap of 1.83 eV is observed between the MoS_2 and *h*BN [24]. This is little more than the energy bandgap (1.5 eV) between graphene and *h*BN valence bands. A hybridization between d_{x-y} orbital of MoS_2 and the p_z orbital of *h*BN originates such bandgap [24]. Recently, it is demonstrated that monolayer MoS_2 retains high carrier mobility free of surface scattering on *h*BN substrate. The substrate layer of *h*BN protects MoS_2 layer from Coulomb scattering from charge impurities in SiO_2 [25].

In a fully planar two-dimensional FET based on layered semiconductors, *h*BN has also been used as the top gate dielectric layer providing superior gate control

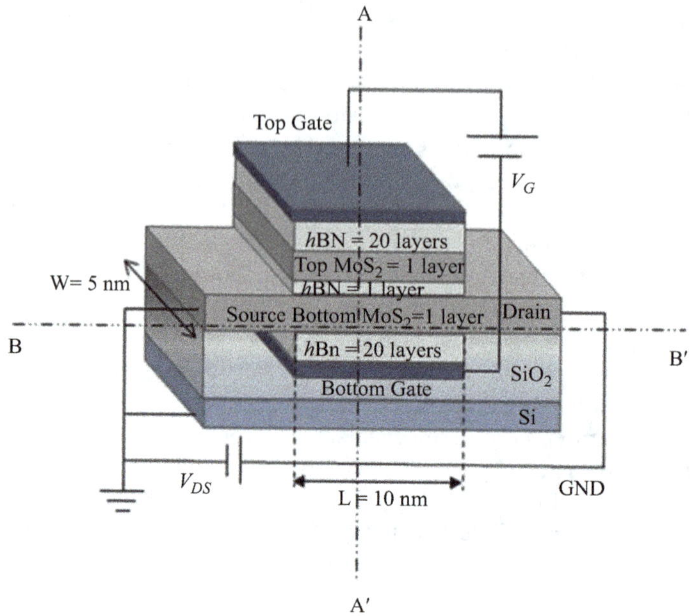

Figure 12.1 Schematic of MoS_2 JTET considering MoS_2/hBN/MoS_2. The dashed line AA′ refers to vertical direction of interlayer tunneling and BB′ refers to lateral direction of source-drain ballistic transport

over the channel [26]. Therefore, *h*BN is considered as both top and bottom gate dielectric in MoS_2 JTET. Experimentally it is found that single-layer *h*BN is a potential candidate for interlayer tunneling barrier for vertical iTFET [27,28]. Such thin tunnel barrier not only allows wave function extension between two semiconducting layers but also preserves the coherent length of tunneling [4].

Operation of MoS_2 JTET is twofold [16], i.e. (a) gate bias (V_G) between the top and bottom MoS_2 layers initiates the vertical interlayer tunneling of carriers which changes the channel Fermi level and (b) the corresponding shift in channel Fermi level controls the height of the barrier between source and drain. In Figure 12.1, dashed line A–A′ refers to the band diagram in vertical direction of interlayer tunneling and B–B′ refers to the lateral direction of source-drain ballistic transport.

Figure 12.2(a) and (b) shows the MoS_2/*h*BN vertical energy band diagram for $V_G = 0$ V and $|V_G| \neq 0$ V, respectively. For $V_G = 0$ V, Fermi levels of both top and bottom MOS_2 layers are assumed to be in equilibrium as shown in Figure 12.2(a). As bias is applied between these two layers, the tunnel barrier *h*BN screens out some electric field, however, a shift in Fermi level at the bottom (channel) MoS_2 layer is still observed. This is shown in Figure 12.2(b).

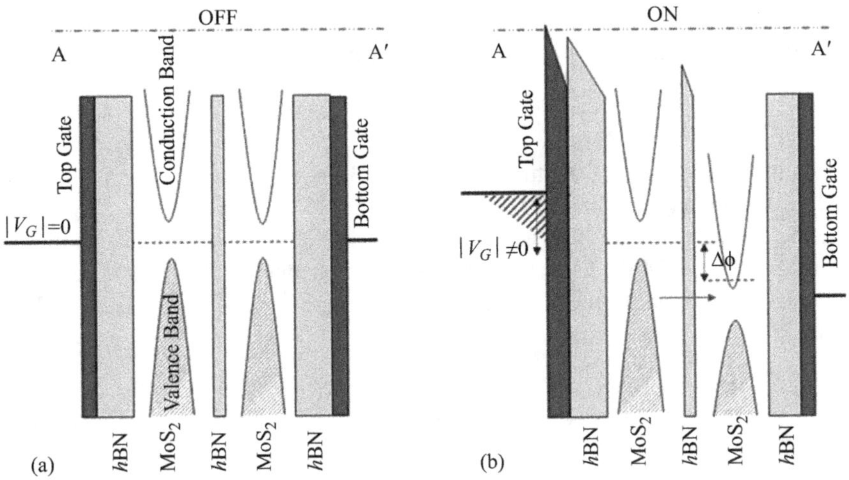

Figure 12.2 (a) Energy band diagram along vertical AA′ direction in the off-state in MoS_2 JTET and (b) in on-state. Note that $\Delta\phi$ denotes change in the Fermi level at bottom (channel) Fermi level

As the gate bias is applied, a finite amount of carrier tunnels from top MoS_2 layer to bottom MoS_2 which is estimated as follows [29]:

$$N_1 = \int_0^{\Delta\phi} \rho_{\mathrm{MoS_2}} T_{\mathrm{WKB}}(E) f_{\mathrm{T}}(E) dE \tag{12.1}$$

Similarly, tunneling of carriers from bottom MoS_2 to top MoS_2 layer is estimated from:

$$N_2 = \int_0^{\Delta\phi} \rho_{\mathrm{MoS_2}} T_{\mathrm{WKB}}(E) f_{\mathrm{B}}(E) dE \tag{12.2}$$

The net amount of tunnel carrier concentration at the bottom MoS_2 channel is described as follows:

$$N = \int_0^{\Delta\phi} \rho_{\mathrm{MoS_2}} T_{\mathrm{WKB}}(E) (f_{\mathrm{T}}(E) - f_{\mathrm{B}}(E)) dE \tag{12.3}$$

where $\rho_{\mathrm{MoS_2}} = g_s g_v m^*_{\mathrm{MoS_2}}/(2\pi\hbar^2)$ is the density of states (DOS) in MoS_2, g_s (=2) and g_v (=2) are spin and valley degeneracy, respectively, $m^*_{\mathrm{MoS_2}}$ is effective mass in MoS_2 (0.57 m_o) and $\hbar$ is the reduced Planck's constant [29]. $T_{WKB}(E)$ is the tunneling probability between two MoS_2 layers through *h*BN barrier and $f_T(E)$ and $f_B(E)$ are Fermi functions at the top and bottom MoS_2 layers (with the generic expression of $\{1/(1+\exp((E\text{-}E_F)/k_BT))\}$ and k_B is Boltzmann's constant), respectively. Interlayer tunneling probability is determined as in [6]:

$$T_{\mathrm{WKB}}(E) = \exp(-2d\sqrt{2m * \Delta}/\hbar) \tag{12.4}$$

where d is the thickness of the tunnel barrier (1.3 nm in this work), m^* is the carrier effective mass inside the barrier (=0.5 m_o inside hBN) [3] and Δ is the height of the tunneling barrier (1.83 eV between MoS_2 and hBN) [24]. Effective change in Fermi level of the bottom MoS_2 layer (which is also the channel MoS_2 layer) is expressed as $\Delta\phi$. Using proper limits of integration, net doping density (N) from (12.3) is integrated as follows:

$$N = \frac{2qV_{\mathrm{T}}m^*_{\mathrm{MoS_2}}}{\pi(\hbar)^2}T_{\mathrm{WKB}}(E)\left[\ln\left(1+\exp\left(-\frac{\Delta\phi}{V_{\mathrm{T}}}\right)\right)+\ln\left(4/\left(1+\exp\left(\frac{\Delta\phi}{V_{\mathrm{T}}}\right)\right)\right)\right] \tag{12.5}$$

where V_{T} (=k_BT/q) is the thermal voltage. Compared to a doped MoS_2 layer, we have estimated the position of Fermi level for a biased and non-doped MoS_2 channel. The objective is to study the gate induced channel degeneracy due to an applied bias in an intrinsic MoS_2 layer. For a positive bias, an n-type degeneracy in channel Fermi level is observed whereas for a negative bias, p-type degeneracy in channel Fermi level is observed. Change in the Fermi level in n-type channel is determined as follows [30]:

$$E_{\mathrm{Fn}} = E_{\mathrm{C}} + qV_{\mathrm{T}}\ln[\exp(N/(\rho_{\mathrm{MoS_2}}k_{\mathrm{B}}T))] \tag{12.6}$$

and in p-type, the expression is

$$E_{\mathrm{Fp}} = E_{\mathrm{V}} - qV_{\mathrm{T}}\ln[\exp(N/(\rho_{\mathrm{MoS_2}}k_{\mathrm{B}}T))] \tag{12.7}$$

In both types of interlayer tunneling transistors and vertical band-to-band tunneling transistors, tunneling phenomena is dependent on temperature [6,19]. Using (12.4)–(12.7), Figure 12.3(a) is plotted which shows the change in Fermi level with temperature at different interlayer gate biases. Figure 12.3(b) shows the induced carrier concentration from interlayer tunneling. It is found that Fermi level curve for an intrinsic MoS_2 channel biased at 0.74 V matches with that of an unbiased MoS_2 channel doped at $10^{17}/\mathrm{cm}^2$. Considering the bandgap of 1.8 eV of single-layer MoS_2, the conduction or valence band lies at $\pm E_G/2$. However, using interlayer tunneling technique, the Fermi level of an intrinsic MoS_2 can shift above the conduction band or below the valence band for positive or negative gate bias, respectively.

Temperature effect on carrier concentration is also studied in Figure 12.3(b). The zero gate bias carrier concentration increases as the temperature increases and gets saturated at higher gate bias. At high temperature, more carriers gain higher energy resulting in interlayer tunneling between the two MoS_2 layers which raises the zero bias carrier concentration. Furthermore, impurity scattering and electron–hole interaction at higher gate bias cause the carrier concentration to saturate.

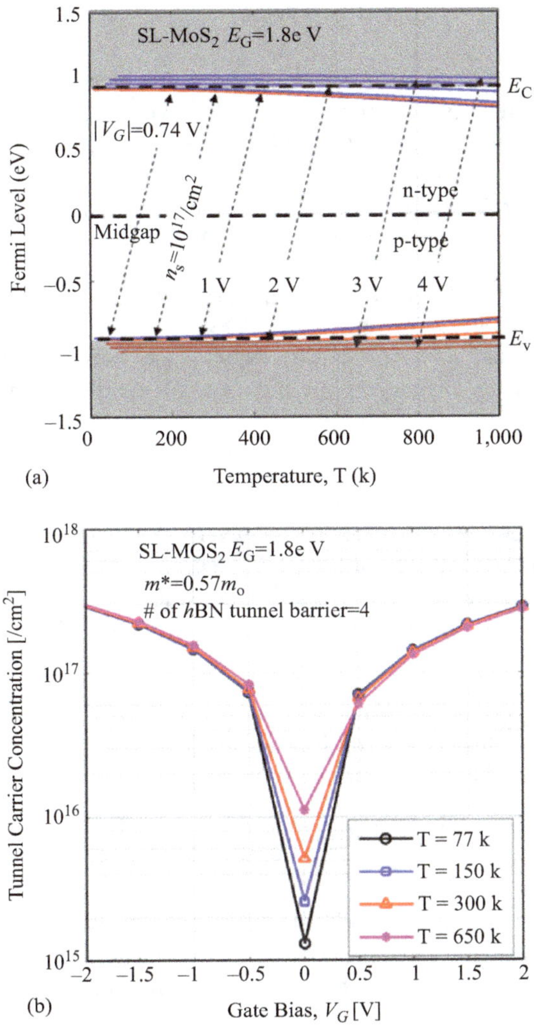

Figure 12.3 (a) Change in Fermi level in n-type (above 0 eV) and p-type (below 0 eV) for a single-layer (SL) MoS_2 channel with change in temperature (T) for different gate bias (V_G). The Fermi level for a doped SL-MoS_2 of $n_s = 1 \times 10^{17}/cm^2$ at zero gate bias matches with non-doped SL MoS_2 JTET operating at $|V_G| = 0.74$ V. (b) Induced interlayer tunnel carrier concentration (N) with change in gate bias (V_G) for different temperatures (T)

12.3 Estimation of drain current

The effective change in channel Fermi level not only depends on the gate bias but also on the associated voltage drops between the two gate contacts [30]. In order to model and calculate drain current of iTFET, these voltage drops are necessary to calculate as follows in this section.

The voltage drop in the channel (V_{ch}) due to the interlayer tunneling-based doping density (N), is determined as follows [30]:

$$V_{ch} = V_o \pm V_T \ln[\exp((N/(\rho_{MoS_2}k_B T)) - 1)] \tag{12.8}$$

where $V_0 = E_0/q$ and $E_0 = E_G/2$ [28]. Note, V_0 is termed as intrinsic mid-gap bias [30]. We refer the channel charge induced voltage drop along A–A′ as in [30] as follows:

$$V_V = qN/C_V \tag{12.9}$$

where C_V is the net vertical capacitance between the top and bottom gate electrodes. Having similarity with MOSFET, iTFET is also assumed to suffer the effect of drain induced barrier lowering (DIBL). We consider DIBL as

$$\lambda_{DIBL} = \alpha V_{DS} \tag{12.10}$$

where α is the fractional coefficient of DIBL and lies between 0 and 1, where 0 stands for no drain bias effect and 1 stands for full-drain bias effect [31]. Now, the effective change in channel Fermi level $\Delta\phi$ becomes:

$$\Delta\phi = V_G - V_{ch} - V_V - \lambda_{DIBL} \tag{12.11}$$

Equation (12.11) is dependent on (12.5) and is a transcendental equation which needs to be solved both numerically and self-consistently. Considering transverse mode along the channel for an energy window between 0 and $\Delta\phi$, using Landauer's expression, lateral drain current between source, and drain of MoS_2 JTET can be written as follows [31]:

$$I = \int dE[(G(E)(f_S(E) - f_D(E)))] \tag{12.12}$$

Here, $G(E)$ is the channel conductance and expressed as

$$G(E) = (2q^2/\hbar)T_B(E)M(E) \tag{12.13}$$

where $f_S(E)$ and $f_D(E)$ are the source and drain Fermi levels, respectively. $T_B(E)$ is the ballistic transmission coefficient in the channel and is taken 1 for the ballistic transport. $M(E)$ is the number of modes in the channel and written as follows [32]:

$$M(E) = g_v W \sqrt{2m^*_{MoS_2}(E - E_C)}/\pi\hbar \tag{12.14}$$

where W is the width of the channel and E_C is the position of the channel conduction band. Combining (12.12)–(12.14), drain current becomes:

$$I_D = \int dE \left[\left(\frac{2q^2}{\pi\hbar^2} g_v W \sqrt{2m^*_{MoS_2}(E - E_C)} (f_S(E) - f_D(E)) \right) \right] \tag{12.15}$$

The Fermi functions in the source and drain are described as follows:

$$f_S(E) = \frac{1}{1 + e^{(E - E_F^S)/k_B T}} \tag{12.16}$$

and

$$f_D(E) = \frac{1}{1 + e^{(E - E_F^D)/k_B T}} \tag{12.17}$$

Equation (12.15) becomes:

$$I_D = \frac{2q^2}{\pi\hbar^2} g_v W \sqrt{2m^*_{MoS_2}} \int \left(\frac{\sqrt{(E - E_C)}}{1 + e^{(E - E_F^S)/k_B T}} - \frac{\sqrt{(E - E_C)}}{1 + e^{(E - E_F^D)/k_B T}} \right) dE \tag{12.18}$$

Now considering:

$$\xi = (E - E_C)/k_B T \tag{12.19}$$

$$\eta_{FS} = (E_F^S - E_C)/k_B T \tag{12.20}$$

$$\eta_{FD} = (E_F^D - E_C)/k_B T \tag{12.21}$$

Drain current in (12.18) can be written as follows:

$$I_D = \frac{q^2}{\sqrt{\pi}\hbar^2} g_v W \sqrt{2m^*_{MoS_2} q V_T} [\Im_{1/2}(\eta_{FS}) - \Im_{1/2}(\eta_{FD})] \tag{12.22}$$

where:

$$\Im_{1/2}(\eta_{FS}) = \frac{2}{\sqrt{\pi}} \int_0^{\Delta\phi} \frac{\xi^{1/2}}{1 + e^{(\xi - \eta_{FS})}} d\xi \tag{12.23}$$

and:

$$\Im_{1/2}(\eta_{FD}) = \frac{2}{\sqrt{\pi}} \int_0^{\Delta\phi} \frac{\xi^{1/2}}{1 + e^{(\xi - \eta_{FD})}} d\xi \tag{12.24}$$

Both (12.20) and (12.21) are the expressions of Fermi–Dirac integral of order ½, which needs to be solved numerically. Solving (12.19) for ξ from 0 to $\Delta\phi$, drain current can be written as follows:

$$I = \frac{2q^2}{\sqrt{\pi}\hbar^2} W \sqrt{2m^*_{MoS_2} q V_T} [\sqrt{\chi_1} - \sqrt{\chi_2}] \tag{12.25}$$

$$\chi_1 = (\Delta\phi - E_C)[\ln(1 + \exp(\Delta\phi - E_F^D)) - \ln(1 + \exp(\Delta\phi - E_F^S))] \quad (12.26)$$

$$\chi_2 = (-E_C)[\ln(1 + \exp(-E_F^D)) - \ln(1 + \exp(-E_F^S))] \quad (12.27)$$

From (12.25), the drain current depends on both (12.5) and (12.11) for which it needs to be solved self-consistently in order to account for both interlayer tunneling induced charge density and source-drain ballistic transport.

12.4 Results and discussion

Using (12.5), (12.11) and (12.25), transfer characteristics of iTFET are plotted in Figure 12.4. A small negative differential resistance (NDR) region is observed at different drain bias at room temperature as shown in Figure 12.4(a). For $V_{DS} = 1.2$ V, an on/off current ratio of 17 with a subthreshold slope of 57 mV/decade is obtained for $V_G > 0$ which is 70 mV/decade for $V_G < 0$ with an on/off current ratio of 18. The off-state leakage current of MoS_2 JTET is calculated as 25.2 μA for $V_{DS} = 1.2$ V. Subthreshold slope is calculated from SS = $\log(10)[I_D/(dI_D/dV_G)]$, where I_D is the drain current and V_G is the gate bias. Compared to a conventional MOSFET, a reduced subthreshold slope at low on/off current ratio in MoS_2 JTET is observed and explained through Figure 12.5(a)–(c).

The intrinsic MoS_2 channel in Figure 12.5(a) considers the source (E_{FS}), channel (E_{FC}) and drain (E_{FD}) Fermi levels in equilibrium. As the negative gate

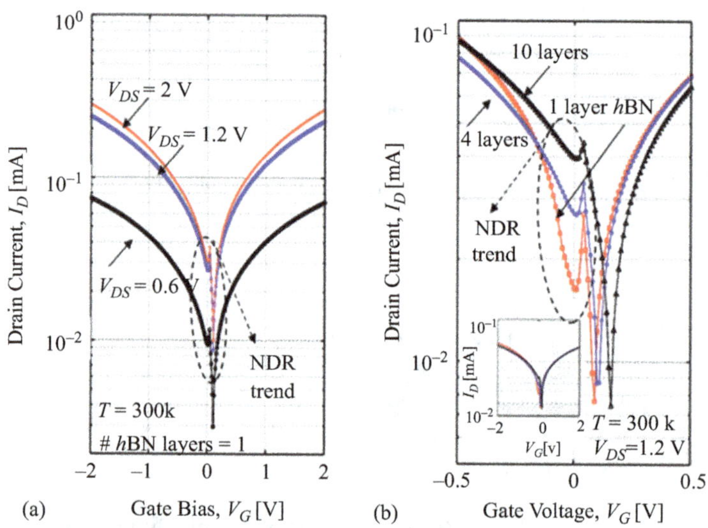

Figure 12.4 Transfer characteristics of MoS_2 JTET. (a) I_D–V_G curve for different drain biases (V_{DS}) and (b) I_D–V_G curve for different number of hBN layers as tunnel barrier between top and bottom MoS_2 layers. Inset in (b) shows drain current for complete bias operation where the effect of number of hBN layers on drain current is non-differentiable

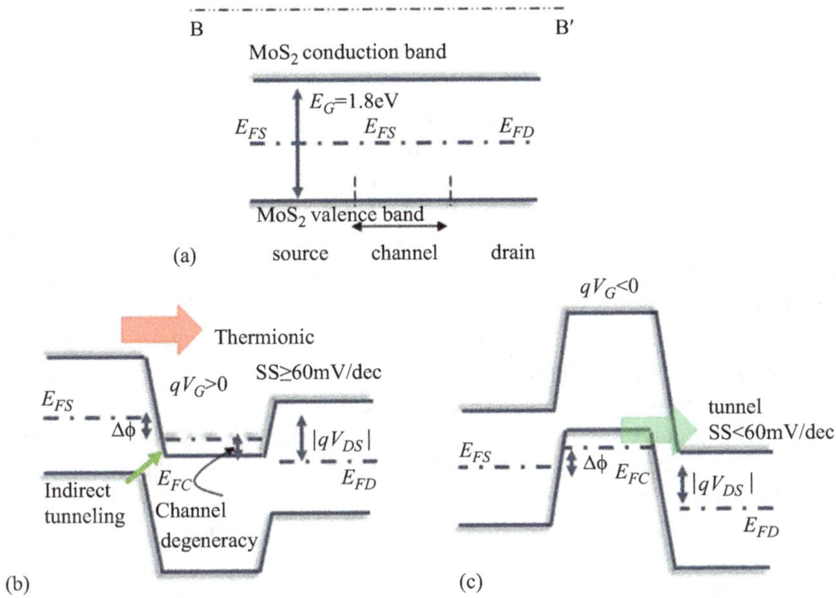

Figure 12.5 (a) Energy band diagram of the bottom (channel) MoS_2 layer in equilibrium at off-state of MoS_2 JTET, (b) energy band diagram at on-state for $qV_G > 0$, and (c) energy band diagram at on-state for $qV_G < 0$. Red arrow points to thermionic transport and green arrow to band-to-band tunneling transport. BB′ refers to the lateral direction of ballistic transport between source and drain

bias ($V_G < 0$ giving $qV_G > 0$) is applied, the degenerately doped (from the interlayer tunneling) n-type channel Fermi level (E_{FC}) moves down which is shown in Figure 12.5(b). The $|qV_{DS}|$ is the amount of shift between E_{FS} and E_{FD} due to drain-source bias. Similar to a MOSFET, thermionic transport (red arrow) dominates the source-drain ballistic transport. For this reason, a subthreshold slope more than the thermionic limit of 60 mV/decade is observed. A small amount of phonon-assisted indirect band-to-band tunneling (BTBT) is assumed which occurs between the source and channel and is shown by a single green arrow in Figure 12.5(b). Note that similar BTBT contributes toward the NDR trend which is also found in ATLAS TFET for a p + Ge source and n-MoS_2 channel [19]. As the positive gate bias ($V_G > 0$ giving $qV_G < 0$) is applied, the degenerately doped (from interlayer tunneling) p-type Fermi level (E_{FC}) of the channel moves below the channel valence band. Hence, the channel valence band comes opposite to the drain conduction band and channel-drain BTBT has occurred. A subthreshold slope of 57 mV/decade is observed due to this BTBT dominated drain current which is shown by the green arrow in Figure 12.5(c).

Number of *h*BN layers as tunnel barriers also affects the MoS_2 JTET transfer characteristics which are shown clearly in Figure 12.4(b). As the number of *h*BN layers as tunnel barrier increases, the tunneling probability exponentially decreases which results in less charge density. Therefore, with a shallow degeneracy, less NDR is observed at higher number of *h*BN layers. The output characteristics of MoS_2 JTET are plotted in Figure 12.6(a) and (b) considering change of gate bias and change in number of *h*BN layers, respectively. Since the operation of MoS_2 JTET is more controlled by the gate bias than drain bias, insignificant effect is observed in output characteristics as the number of *h*BN layers varies in Figure 12.6(b).

Compared to the benchmarked performance of monolayer MoS_2 transistor [20,21], MoS_2 JTET provides low on/off current ratio. This can be understood from the field effect mobility (μ_{FE}) diagram in Figure 12.7. Field effect mobility is estimated from $\mu_{FE} = \mathrm{d}I_D/\mathrm{d}V_G(L/W)(1/C_G)$, considering both quantum and geometric capacitances [30]. As V_G increases, μ_{FE} drops. Based on the semi-classical Drude formula, conductivity $\sigma(=\mu_{FE}Nq)$ is linearly dependent on μ_{FE}. Therefore, as the channel MoS_2 becomes degenerately doped, conductivity drops as μ_{FE} decreases.

Moreover, a further study of metal-insulator transition in the channel MoS_2 layer of MoS_2 JTET can be understood by Ioffe–Regel criterion [33,34]. According to this criterion, MoS_2 is metallic for $k_Fl_e \gg 1$ and is insulating for $k_Fl_e \ll 1$. Here $k_F = \sqrt{(2\pi N)}$ is the Fermi wave vector and $l_e = \hbar k_F\sigma/Nq^2$ is the mean free path [33]. Two points are selected to check the criteria ($V_G = 0.2$ V and 0.5 V) between which the mobility drops. Using Figure 12.7, $V_G = 0.2$ V, $k_Fl_e \sim 294$ ($\gg$1); and at $V_G = 0.5$ V, $k_Fl_e \sim 5.16 \times 10^{-4}$ ($\ll$1) are found, providing a metal-insulator transition in the channel MoS_2 layer at high gate bias. Therefore, a low on-state drive current is obtained resulting in low on/off current ratio in MoS_2 JTET. The low subthreshold slope of MoS_2 JTET is comparable with the standard MOSFET

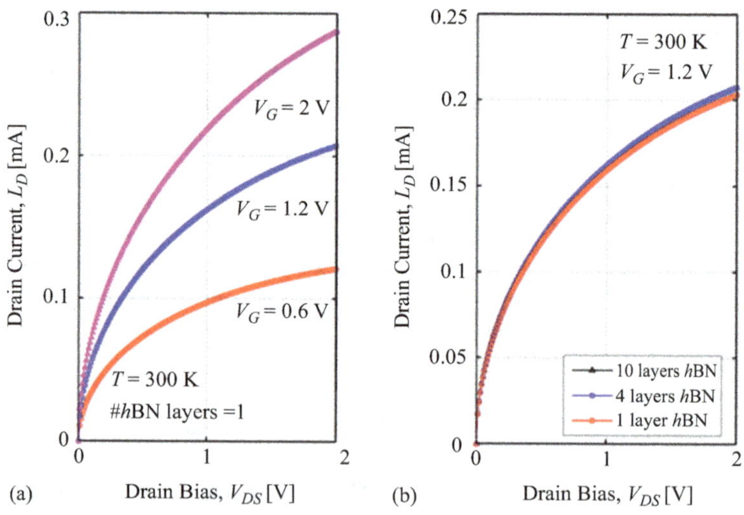

Figure 12.6 Output characteristics of MoS_2 JTET: (a) I_D–V_{DS} curve for different gate biases (V_G) and (b) I_D–V_{DS} curve for different number of hBN layers as tunnel barrier between top and bottom MoS_2 layers

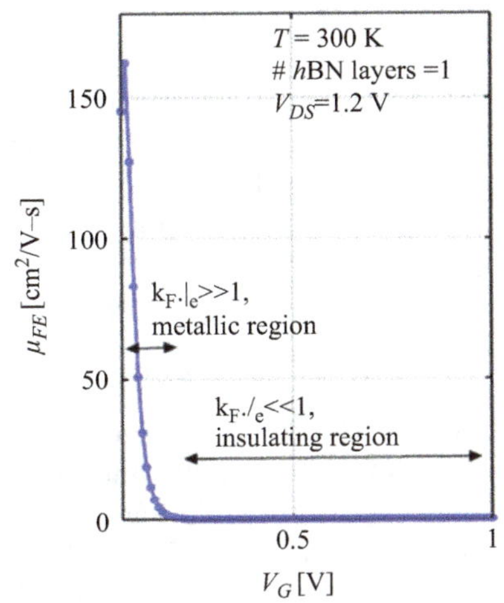

Figure 12.7 Gate bias-dependent field effect mobility in MoS_2 JTET

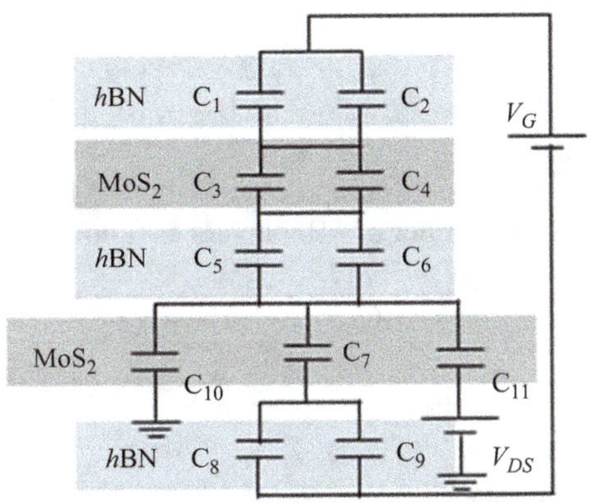

Figure 12.8 Capacitive network of MoS_2 JTET

subthreshold slope of 60 mV/decade. However, metal insulator transition and mixed mode of thermionic and BTBT current transport limits achieving high on/off current ratio in MoS_2 JTET.

The capacitance network for the MoS_2 JTET is shown in Figure 12.8. The total gate capacitance is estimated as follows: $1/C_G = 1/C_V + 1/Cq_{ch}$, where $1/C_V = 1/(C_1+C_2) + 1/(C_3+C_4) + 1/(C_5+C_6) + 1/(C_8+C_9)$ considering series–parallel network of all the vertical capacitances. The geometric and quantum capacitances of top and bottom gate *h*BN layers and top MoS_2 layer are expressed as $C_1 = C_8 = \varepsilon_0\varepsilon_{hBN}/(Z_g t_{hBN})$, $C_2 = C_9 = C_{q,hBN}/Z_g$, $C_3 = \varepsilon_0\varepsilon_{MoS2}/t_{MoS2}$ and $C_4 = C_{q,MoS2} = q^2\rho_{MoS2}$, respectively. $C_{q,hBN} = q^2\rho_{hBN}$ (where $\rho_{hBN} = g_s g_v m^*_{hBN}/(2\pi\hbar^2)$) is the quantum capacitance of single-layer *h*BN [35]. Z_g is the number of *h*BN layers ($Z_g = 20$) in gate dielectric in AA′ direction. Similarly, the geometric and quantum capacitances of the single-layer hBN as tunnel barrier are expressed as $C_5 = \varepsilon_0\varepsilon_{hBN}/(Z_t t_{hBN})$ and $C_6 = C_{q,hBN}/Z_t$ (where Z_t is the number of hBN layers ($Z_t = 1$) in tunnel barrier), respectively. C_{10} and C_{11} are source and drain quantum capacitance of MoS_2. Note that t_{MoS2} (=0.65 nm) and ε_{MoS2}(=2.8) [36], t_{hBN}(=0.325 nm) and ε_{hBN}(=4) are the thickness and dielectric permittivity of MoS_2 and hBN, respectively. Based on the work of Ma and Jena [30], the gate-dependent channel quantum capacitance (Cq_{ch}) is estimated as follows:

$$Cq_{\mathrm{ch}} = C_7 = q^2\rho_{\mathrm{MoS_2}}\left[1 + \frac{\exp(E_{\mathrm{G}}/2k_{\mathrm{B}}T)}{2\cosh(q\Delta\phi/k_{\mathrm{B}}T)}\right]^{-1} \tag{12.28}$$

Using (12.5) and (12.11), (12.28) is solved and is plotted in Figure 12.9. For $V_G = 1.2$ V, C_G is estimated to be 0.0952 F/m^2. Intrinsic cut-off frequency ($f_T = g_m/2\pi C_G$) dependence on gate bias is shown in Figure 12.10 for transconductance, $g_m = dI_D/dV_G$. For a supply voltage of 1.2 V, $f_T = 19.73$ THz has been calculated which increases as V_G reduces. This value is higher than the reported f_T in [9] and [10]. Intrinsic frequency of MoS_2 transistors is independent of on/off current ratio [37,38] and is related to gate capacitance. From Figure 12.9, the gate capacitance (C_G) is nearly two orders less than the channel quantum capacitance (Cq_{ch}) for which MoS_2 JTET achieves very low gate capacitance providing high intrinsic cutoff frequency. Using $\tau = C_G V_{DD}/I_{ON}$, intrinsic gate delay is plotted in Figure 12.11(a) from which the power delay product ($PDP = \tau V_{DD} I_{ON}$) is plotted in Figure 12.11(b). From Figures 12.9 and 12.10, beyond THz operation of MoS_2 JTET can be observed.

Performance of MoS_2 JTET is compared in Table 12.1 with the results reported in [9] and [10] for an equal number of hBN layers as tunnel barrier and gate bias. In

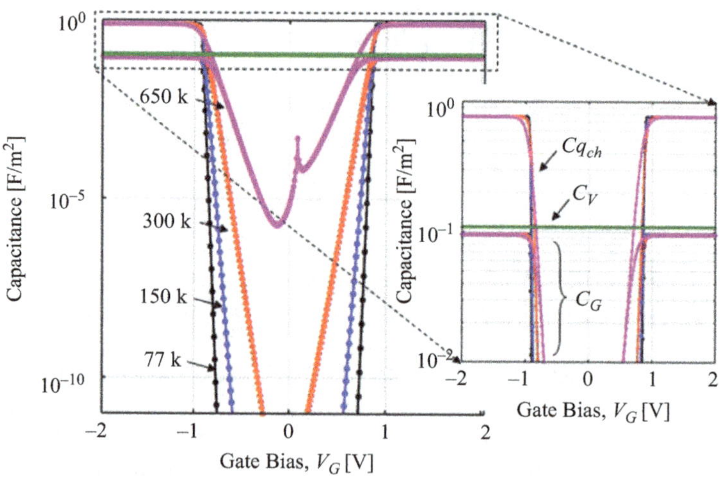

Figure 12.9 Change in channel quantum capacitance (Cq_{ch}) and total gate capacitance (C_G) with gate bias (V_G) for different temperatures of MoS_2 JTET. Note: Non-channel fixed vertical capacitance (C_V) is shown in green line

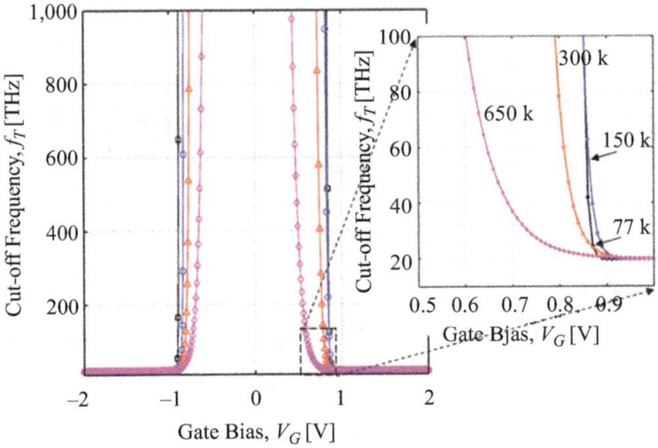

Figure 12.10 Intrinsic cut-off frequency (f_T) variation of MoS_2 JTET with change in gate bias (V_G)

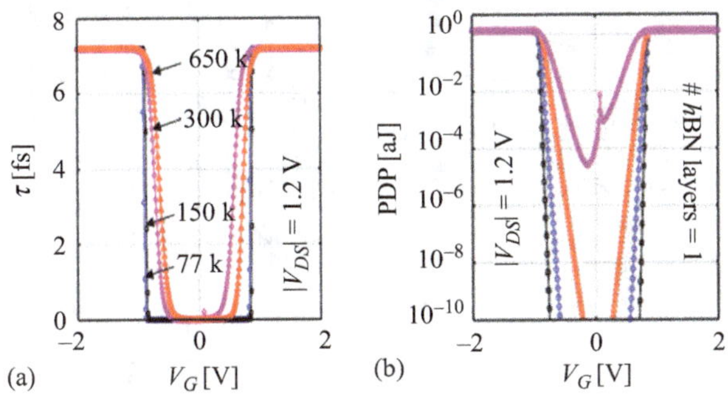

Figure 12.11 (a) Intrinsic gate delay (τ) versus the gate bias (V_G) and (b) corresponding power delay product (PDP) for different temperatures of MoS_2 JTET

Table 12.1 Comparison of MoS_2 JTET performance with earlier similar models

Parameters	**Fiori *et al.* [9]**	**VTGFET [10]**	**VTGNRFET [10]**	**MoS_2 JTET**
Gate voltage	1.2 V	1.2 V	1.2 V	1.2 V
#*h*BN layers	3	3	3	3
Subthreshold slope	386 mV/dec	1,535 mV/dec	1,297 mV/dec	57 mV/dec
I_{on}/I_{off}	~15	3	4	17
Cutoff frequency	0.5 GHz	58 GHz	97 GHz	19.73 THz

Table 12.2 *Comparison of MoS_2 JTET with existing two-dimensional high-frequency devices*

Ref. [Year]	Device transport type	Material system/channel	Channel length/tunneling barrier thickness	Bias voltage	I_{On}/I_{Off}	f_T
[9] [2013]	iTFET	Graphene-*h*BN-graphene	1.03 nm (tunneling barrier thickness)	1.2 V	15	0.5 GHz
[10] [2014]	iTFET	GNR-*h*BN-GNR	1.03 nm (tunneling barrier thickness)	1.2 V	4	97 GHz
[37] [2016]	FET	CVD MoS_2 on flexible substrate	1 μm (channel)	2 V	105	5.6 GHz
[39] [2009]	FET	Graphene	500 nm (channel)	1.6 V	~2	4 GHz
[40] [2014]	FET	MoS_2	240 nm (channel)	2 V	~300	8.2 GHz
[41] [2012]	FET	Bilayer graphene	40 nm (channel)	1 V	~800	1.5 THz
[42] [2010]	FET	Graphene	140 nm (channel)	1 V	~3	300 GHz
[43] [2011]	FET	Graphene	40 nm (channel)	1.5 V	~800	155 GHz
[44] [2013]	FET	Epitaxial graphene from SiC	100 nm (channel)	0.8 V	~2	110 GHz
[45] [2013]	BJT type	Graphene-based heterojunction	2–5 nm (SiO_2 tunneling barrier thickness)	1 V	104	1 THz
[46] [2013]	Hot electron transistor	Graphene base	2 nm (Al_2O_3 tunneling barrier thickness)	1.5 V	>105	Unspecified
[47] [2015]	TFET	Graphene nanoribbon (GNR)	20 nm (channel)	0.1 V	122	~1 THz
[48] [2012]	TFET	Graphene-hBCN	7 nm (channel)	0.6 V	104	~2 THz
[49] [2016]	Interlayer excitonic generation	MoS_2-*h*BN- MoS_2	5 nm (tunneling barrier thickness)	-	-	-
[50] [2014]	FET	Black phosphorus	300 nm (channel)	2 V	2 × 103	12 GHz
[51] [2012]	iTFET-plasma resonance-based	Graphene-barrier-graphene	10 nm (tunneling barrier thickness) 500 nm (channel)	0.5 V	Unspecified	1.42 THz
[52] [2014]	FET	Bilayer graphene	2.5 μm (channel)	0.001	Unspecified	0.29–0.38 THz
[53] [2014]	FET	Exfoliated MoS_2 on SiO_2	68 nm (channel)	5 V	104	42 GHz
[54] [2015]	FET	CVD MoS_2 on SiO_2	250 nm (channel)	3.5 V	200	6.7 GHz
[22] [2016] This work	iTFET-interlayer tunneling-based barrier control	MoS_2-*h*BN- MoS_2	10 nm (channel) 1.03 nm (tunneling barrier thickness)	1.2 V	17	19.73 THz

terms of subthreshold slope, MoS_2 JTET provides ~7 and ~27 times less than that of the reported in [9] and [10], respectively, for graphene vertical FETs. Due to a small bandgap at 5 nm width, subthreshold slope of MoS_2 JTET is 23 times less than that of the vertical GNR iTFET reported in [10]. Compared to both [9] and [10], MoS_2 JTET provides THz operation due to very low gate capacitance. The on/off current ratio is nearly the same as reported in [9] and [10]. Furthermore, high-frequency performance of this MoS_2 JTET is also compared with the existing two-dimensional materials (both graphene and non-graphene)-based high-frequency devices and is summarized in Table 12.2.

Based on the data in Table 12.2, MoS_2 JTET outperforms other devices at a comparable supply voltage, on/off current ratio and channel length. The only similar device structure like MoS_2 JTET is found in the work of Calman *et al.* [49] which studies controlled excitonic generation in similar van der Waals heterostructure. However, the work in [49] does not account for any high-frequency performance estimation and transistor-type electronic behavior and hence becomes unsuitable for comparison. The high-frequency performance of MoS_2 JTET originates from interlayer tunneling-based barrier control mechanism and the use of two-dimensional layered materials (in this work *h*BN) as the gate dielectric providing low gate-capacitance.

12.5 Conclusion

Current transport MoS_2 JTET is studied in this chapter which is controlled by the gate induced interlayer tunneling-dependent charge density unlike inversion mode operation in MOSFETs. The current transport between source and drain is ballistic. Compared to recently reported device structures in [9] and [10], the present device structure gives subthreshold slope close to 60 mV/decade and demonstrates upper GHz operation with relatively comparable on/off current ratio. Low bandgap insulator or wide bandgap-layered semiconductor materials can be used as an interlayer tunneling barrier to improve the on/off current ratio and making MoS_2 JTET suitable for digital applications. A comparison of the performance of MoS_2 JTET with other types of device structures exhibits superior performance and high-frequency THz operation.

References

[1] M. Waldrop, "The chips are down for Moore's law," *Nature*, vol. 530, pp. 144–7, 2016.

[2] A. B. Sachid, M. Tosun, S. B. Desai, *et al.* "Monolithic 3D CMOS using layered semiconductors," *Advanced Materials*, vol. 28, pp. 2547–54, 2016.

[3] L. Britnell, R. V. Gorbachev, R. Jalil, *et al.*, "Field-effect tunneling transistor based on vertical graphene heterostructures," *Science*, vol. 335, pp. 947–50, 2012.

[4] A. Sciambi, M. Pelliccione, M. P. Lilly, *et al.*, "Vertical field-effect transistor based on wave-function extension," *Physical Review B*, vol. 84, p. 085301, 2011.

[5] S. K. Banerjee, L. F. Register, E. Tutuc, D. Reddy, and A. H. MacDonald, "Bilayer pseudospin field-effect transistor (BiSFET): a proposed new logic device," *IEEE Elect. Dev. Lett.*, vol. 30, pp. 158–60, 2009.

[6] T. Georgiou, R. Jalil, B. D. Belle, *et al.*, "Vertical field-effect transistor based on graphene-WS2 heterostructures for flexible and transparent electronics," *Nat Nano*, vol. 8, pp. 100–03, 2013.

[7] P. Zhao, R. M. Feenstra, G. Gong, and D. Jena, "Symfet: a proposed symmetric graphene tunneling field-effect transistor," *IEEE Transactions on Electron Devices*, vol. 60, pp. 951–57, 2013.

[8] S. C. de la Barrera, Q. Gao, and R. M. Feenstra, "Theory of graphene–insulator–graphene tunnel junctions," *J. of Vac. Sci. & Tech. B*, vol. 32, p. 04E101, 2014.

[9] G. Fiori, S. Bruzzone, and G. Iannaccone, "Very large current modulation in vertical heterostructure graphene/hBN transistors," *IEEE Transactions on Electron Devices*, vol. 60, pp. 268–73, 2013.

[10] N. Ghobadi and M. Pourfath, "A comparative study of tunneling FETs based on graphene and GNR heterostructures," *IEEE Transactions on Electron Devices*, vol. 61, pp. 186–92, 2014.

[11] A. Mishchenko, J. S. Tu, Y. Cao, R. V. Gorbachev, *et al.* "Twist-controlled resonant tunnelling in graphene/boron nitride/graphene heterostructures," *Nat Nano*, vol. 9, pp. 808–13, 2014.

[12] B. Fallahazad, K. Lee, S. Kang, *et al.* "Gate-tunable resonant tunneling in double bilayer graphene heterostructures," *Nano Letters*, vol. 15, pp. 428–33, 2015.

[13] K. Sangwoo, B. Fallahazad, L. Kayoung, *et al.* "Bilayer graphene-hexagonal boron nitride heterostructure negative differential resistance interlayer tunnel FET," *IEEE Elect. Dev. Lett.*, vol. 36, pp. 405–07, 2015.

[14] G. Iannaccone, Q. Zhang, S. Bruzzone, and G. Fiori, "Insights on the physics and application of off-plane quantum transport through graphene and 2D materials," *Solid-State Electronics*, vol. 115, Part B, pp. 213–18, 2016.

[15] S. Bruzzone, D. Logoteta, G. Fiori, and G. Iannaccone, "Vertical transport in graphene-hexagonal boron nitride heterostructure devices," *Scientific Reports*, vol. 5, p. 14519, 2015.

[16] M. Fahad and A. Srivastava, "A Graphene Switching Transistor for Vertical Circuit Design," *ECS Journal of Solid State Science and Technology*, vol. 5, pp. M13–M21, 2016.

[17] G. Fiori, F. Bonaccorso, G. Iannaccone, *et al.* "Electronics based on two-dimensional materials," *Nat Nano*, vol. 9, pp. 768–79, 2014.

[18] B. Radisavljevic, A. Radenovic, J. Brivio, V. Giacometti, and A. Kis, "Single-layer MoS_2 transistors," *Nat Nano*, vol. 6, pp. 147–50, 2011.

[19] D. Sarkar, X. Xie, W. Liu, *et al.* "A subthermionic tunnel field-effect transistor with an atomically thin channel," *Nature*, vol. 526, pp. 91–95, 2015.

[20] K. Alam and R. K. Lake, "Monolayer MoS_2 transistors beyond the technology road map," *IEEE Transactions on Electron Devices*, vol. 59, pp. 3250–54, 2012.

[21] Y. Yoon, K. Ganapathi, and S. Salahuddin, "How good can monolayer MoS_2 transistors be?," *Nano Letters*, vol. 11, pp. 3768–73, 2011.

[22] A. Srivastava and Md Fahad, "Vertical MoS_2/hBN/MoS_2 interlayer tunneling field effect transistor," *Solid State Electronics*, vol. 126, pp. 96–103, 2016.

[23] S. Wang, X. Wang, and J. H. Warner, "All chemical vapor deposition growth of MoS_2:h-BN vertical van der Waals heterostructures," *ACS Nano*, vol. 9, pp. 5246–54, 2015.

[24] Jaarsma J. Monolayer MoS_2 on a substrate of hexagonal boron nitride, B.S. Thesis. Computational Materials Science University of Twente, 2014.

[25] G.-H. Lee, Y.-J. Yu, X. Cui, *et al.* "Flexible and transparent MoS_2 field-effect transistors on hexagonal boron nitride-graphene heterostructures," *ACS Nano*, vol. 7, pp. 7931–36, 2013.

[26] T. Roy, M. Tosun, J. S. Kang, *et al.* "Field-effect transistors built from all two-dimensional material components," *ACS Nano*, vol. 8, pp. 6259–64, 2014.

[27] L. Britnell, R. V. Gorbachev, R. Jalil, *et al.* "Electron tunneling through ultrathin boron nitride crystalline barriers," *Nano Letters*, vol. 12, pp. 1707–10, 2012.

[28] C. M. Orofeo, S. Suzuki, and H. Hibino, "Ultrathin chemical vapor deposition (CVD)-grown hexagonal boron nitride as a high-quality dielectric for tunneling devices on rigid and flexible substrates," *The Journal of Physical Chemistry C*, vol. 118, pp. 3340–46, 2014.

[29] J. G. Simmons, "Generalized formula for the electric tunnel effect between similar electrodes separated by a thin insulating film," *Journal of Applied Physics*, vol. 34, pp. 1793–803, 1963.

[30] N. Ma and D. Jena, "Carrier statistics and quantum capacitance effects on mobility extraction in two-dimensional crystal semiconductor field-effect transistors," *2D Materials*, vol. 2, p. 015003, 2015.

[31] S. Datta, Quantum transport: Atom to transistor. United Kingdom: Cambridge University Press, 2005.

[32] M. Lundstrom and C. Jeong, Lessons from nanoscience: a lecture note series near-equilibrium transport, New Jersey: World Scientific. 2012: 20–31.

[33] B. Radisavljevic and A. Kis, "Mobility engineering and a metal–insulator transition in monolayer MoS_2," *Nat Mater*, vol. 12, pp. 815–20, 2013.

[34] M. Gurvitch, "Ioffe–Regel criterion and resistivity of metals," *Physical Review B*, vol. 24, pp. 7404–07, 1981.

[35] S. Luryi, "Quantum capacitance devices," *Applied Physics Letters*, vol. 52, pp. 501–03, 1988.

[36] F. A. Rasmussen and K. S. Thygesen, "Computational 2d materials database: electronic structure of transition-metal dichalcogenides and oxides," *The Journal of Physical Chemistry C*, vol. 119, pp. 13169–83, 2015.

[37] H. Y. Chang, M. N. Yogeesh, R. Ghosh, *et al.*, "Large-area monolayer MoS_2 for flexible low-power RF nanoelectronics in the GHz regime," *Advanced Materials*, vol. 28, pp. 1818–23, 2016.

[38] L. Jongho, C. Hsiao-Yu, H. Tae-Jun, *et al.*, "High-performance flexible nanoelectronics: 2D atomic channel materials for low-power digital and high-frequency analog devices," *IEDM Tech. Digest*, pp. 19.2.1–19.2.4, 2013.

[39] Y.-M. Lin, K. A. Jenkins, A. Valdes-Garcia, J. P. Small, D. B. Farmer, and P. Avouris, "Operation of graphene transistors at gigahertz frequencies," *Nano Letters*, vol. 9, pp. 422–26, 2009.

[40] D. Krasnozhon, D. Lembke, C. Nyffeler, Y. Leblebici, and A. Kis, "MoS_2 transistors operating at gigahertz frequencies," *Nano Letters*, vol. 14, pp. 5905–11, 2014.

[41] G. Fiori and G. Iannaccone, "Insights on radio frequency bilayer graphene FETs," *IEDM Tech. Digest*, pp. 17.3.1–7.3.4, 2012.

[42] L. Liao, Y.-C. Lin, M. Bao, *et al.*, "High-speed graphene transistors with a self-aligned nanowire gate," *Nature*, vol. 467, pp. 305–08, 2010.

[43] Y. Wu, Y. M. Lin, A. A. Bol, *et al.*, "High-frequency scaled graphene transistors on diamond-like carbon," *Nature*, vol. 472, pp. 74–78, 2011.

[44] Z. Guo, R. Dong, P. S. Chakraborty, *et al.* Heer, "Record maximum oscillation frequency in c-face epitaxial graphene transistors," *Nano Letters*, vol. 13, pp. 942–47, 2013.

[45] V. D. Lecce, R. Grassi, A. Gnudi, *et al.* "Graphene-based heterojunction transistor: an attractive device for terahertz operation," *IEEE Electron Device Letters*, vol. 60, pp. 4263–68, 2013

[46] C. Zeng, E. B. Song, M. Wang, *et al.* "Vertical graphene-base hot-electron transistor," *Nano Letters*, vol. 13, pp. 2370–75, 2013.

[47] M. S. Fahad, A. Srivastava, A. K. Sharma, C. Mayberry, "Analytical current transport modeling of graphene nanoribbon tunnel field-effect transistors for digital circuit design," *IEEE Transactions on Nanotechnology*, vol. 15, pp. 39–50, 2016.

[48] G. Fiori, A. Betti, S. Bruzzone, and G. Iannaccone, "Lateral graphene–hBCN heterostructures as a platform for fully two-dimensional transistors," *ACS Nano*, vol. 6, pp. 2642–48, 2012.

[49] E. V. Calman, C. J. Dorow, M. M. Fogler, *et al.*, "Control of excitons in multi-layer van der Waals heterostructures," *Applied Physics Letters*, vol. 108, p. 101901, 2016.

[50] H. Wang, X. Wang, F. Xia, *et al.*, "Black phosphorus radio-frequency transistors," *Nano Letters*, vol. 14, pp. 6424–29, 2014.

[51] V. Ryzhii, T. Otsuji, M. Ryzhii, and M. S. Shur, "Double graphene-layer plasma resonances terahertz detector," *Journal of Physics D: Applied Physics*, vol. 45, p. 302001, 2012.

[52] D. Spirito, D. Coquillat, S. L. De Bonis, *et al.*, "High performance bilayer-graphene terahertz detectors," *Applied Physics Letters*, vol. 104, p. 061111, 2014.

[53] R. Cheng, S. Jiang, Y. Chen, *et al.*, "Few-layer molybdenum disulfide transistors and circuits for high-speed flexible electronics," *Nature Communications*, vol. 5, pp. 1–9, 2014.

[54] A. Sanne, R. Ghosh, A. Rai, *et al.*, "Radio frequency transistors and circuits based on CVD MoS_2," *Nano Letters*, vol. 15, pp. 5039–45, 2015.

Index

www.ingramcontent.com/pod-product-compliance
Lightning Source LLC
LaVergne TN
LVHW011848200926
848484LV00002B/418